Practical Time-Frequency Analysis

Gabor and Wavelet Transforms with an Implementation in S

Wavelet Analysis and Its Applications

The subject of wavelet analysis has recently drawn a great deal of attention from mathematical scientists in various disciplines. It is creating a common link between mathematicians, physicists, and electrical engineers. This book series will consist of both monographs and edited volumes on the theory and applications of this rapidly developing subject. Its objective is to meet the needs of academic, industrial, and governmental researchers, as well as to provide instructional material for teaching at both the undergraduate and graduate levels.

Wavelets have evolved to become a popular tool for the electrical engineers, both in the academia and industry, for processing and analyzing non-stationary signals. Although there is already a fair amount of published work in wavelet literature, including monographs and editor volumes, this attractive and comprehensive writing by three experts in the field is the very first book devoted to the important aspects of time-frequency analysis with an accompanying library of S functions and computer programs. The book is a very valuable addition to this series, and will prove to be useful for both beginners and experts who can benefit from a good understanding of time-frequency/time-scale analysis, their applications to solve practical problems, as well as computer programming. The series editor would like to thank the authors for this outstanding contribution to the wavelets literature.

This is a volume in
WAVELET ANALYSIS AND ITS APPLICATIONS

Charles K. Chui, Series Editor

A list of titles in this series appears at the end of this volume.

Practical Time-Frequency Analysis

Gabor and Wavelet Transforms with an Implementation in S

René Carmona

Princeton University

Princeton NJ

Wen-Liang Hwang

Academia Sinica

Taiwan

Bruno Torrésani

CNRS-Luminy

France

ACADEMIC PRESS

San Diego London Boston
New York Sydney Tokyo Toronto

This book is printed on acid-free paper. 

ACADEMIC PRESS
525 B Street, Suite 1900, San Diego, CA 92101-4495, USA
http://www.apnet.com

24–28 Oval Road, London NW1 7DX, UK
http://www.hbuk.co.uk/ap/

Library of Congress Cataloging-in-Publication Data
Carmona, R. (René)
Practical time-frequency analysis : Gabor and wavelet transforms with an implementation in S / René Carmona, Wen-Liang Hwang, Bruno Torrésani.
p. cm. — (Wavelet analysis and its applications ; v. 9)
Includes bibliographical references and index.
ISBN 0-12-160170-6 (alk. paper)
1. Signal processing—Mathematics. 2. Time-series analysis. 3. Wavelets (Mathematics) 4. Frequency spectra. I. Hwang, Wen-Liang. II. Torrésani, Bruno. III. Title. IV. Series.
TK5102.9.C34 1998
621.382'2—dc21 98-22862
CIP

Printed in the United States of America
98 99 00 01 02 IP 9 8 7 6 5 4 3 2 1

Preface

Wavelets have become known as a powerful tool with which to manipulate signals that have a complex structure; the wavelet transform, like other time-frequency transforms, makes it possible to disentangle different components in such signals. The most widespread applications of the wavelet transform use orthonormal or biorthogonal wavelet bases, and these are indeed the tool of choice for many applications involving straightforward compression. When one wants to recognize in or extract features from a signal, a redundant wavelet transform or another redundant time-frequency transform, such as the Gabor transform, is often more useful, however.

This book illustrates applications of both redundant and nonredundant time-frequency transforms; it focuses mostly on the redundant case. It gives many different types of applications of these time-frequency transforms, implemented with S-tools that are made readily available in the free companion toolbox Swave. All the applications are illustrated with a wealth of examples. Many examples are revisited several times, using different tools that highlight different features.

This book is useful both for those, like myself, who already know something about time-frequency transforms but who are not fluent in S, and for the many practitioners of S who would like to learn about time-frequency analysis (the theory of which is carefully explained here as well), and at the same time plunge into applications. Don't forget to try out many different tools on the same application – it is a particularly instructive practice that I recommend highly to all readers!

Ingrid Daubechies
Princeton

Reader's Guide

The purpose of this book (and of the accompanying library of S functions) is to give a self-contained presentation of the techniques of time-frequency/time-scale analysis of 1-D signals and to provide a set of useful tools (in the form of computer programs) to perform the analyses. Such a package should be especially attractive to that part of the scientific community interested in mathematical and practical issues, especially if they involve random or noisy signals with possibly nonstationary features.

Our use of the S language is a reflection of our initial intent to reach the statistical community which, despite the traditional interest in the spectral analysis of time series (and some attempts at the understanding of nonstationary processes) and the pervasive use of orthonormal wavelet bases, has seen very few attempts to understand the benefits of the continuous transforms. For quite a long time the electrical engineers have used the continuous Gabor and wavelet transforms to statistical ends. They used them to detect, to denoise, and to reconstruct signals, and most importantly to perform spectral analyses of nonstationary signals. We believe that statisticians did not get a fair share in the benefits.

The first part of the book is intended to be a hands-on crash course on some of the major components of the time-frequency analysis of signals. A special emphasis is put on the analyses of noisy signals, and great care is taken to address the stationarity issue and to describe the statistical significance of the spectral analyses and the denoising procedures.

The second part of the book should be used as a reference manual for the library of S functions which we wrote to perform all the computations relative to the examples described in the first part of the monograph.

We now give a quick guided tour of the various chapters of this book.

Part I

The first two chapters are intended to set the stage for the main course served in Part II. The classical spectral analysis of deterministic and stationary random processes is reviewed with a special emphasis on the issues which will be crucial in the remaining of the book: sampling of continuous signals, stationarity versus nonstationarity, time/frequency representations, ... The material is classical in nature, but we decided to present it anyway for the sake of completeness and with the hope of easing the way through the maze of notation and terminology which we use throughout the book.

Part II

The first two chapters of this second part give a crash course on two of the most important time-frequency signal representations, namely the (continuous) Gabor and wavelet transforms. Few theorems are proved, but all the results are stated with great care and precise references are given. One of our goals is to bring these transforms to the attention of the part of the scientific community which has overlooked their potential. As explained earlier, we believe that they provide a very versatile toolbox for the spectral analysis of nonstationary signals. Unfortunately, the corresponding statistical theory is not fully developed yet. We revisit the classical elements of the spectral theory of stationary random processes in the light of these new tools. We then illustrate our presentation with examples of time/frequency analyses of real-life nonstationary signals. These examples are used to illustrate the main features of these transforms and to introduce the use of the Swave library of S functions which we wrote and which we make available free of charge on the Internet. We also devote a chapter to material already existing in book form. For the sake of completeness we discuss frames and orthonormal bases, wavelet packets and pursuit techniques. We make an effort to address the computational issues associated with the implementation of these theoretical concepts.

Part III

The final three chapters are devoted to signal analysis applications, and the bias of the authors will presumably show in the choice of the methods and illustrations. In particular, a detailed discussion of some of the recent works of the authors on ridge detection and statistical reconstructions of noisy signals is included. We review the little that is known (at least to the authors) on the statistical theory of nonstationary stochastic processes from the point of view of time-frequency analysis, and we devote a chapter to frequency modulated signals. The latter are of crucial importance in many medical and military application, but our emphasis is on the problems of speech analysis.

Part IV

The last part of the book contains the library of S programs which we wrote to perform all the computations relative to the examples described in the first part of the monograph. We call this library Swave.

StatSci (the developer of the only commercially available version of S) has recently added a Wavelet toolkit to the latest version of Splus. It is called S+wavelet. There is a definite overlap with our package Swave in the sense that both packages deal with time-frequency analysis of signals and both packages contain tool for denoising, smoothing, and so forth, of noisy signals. But there are important differences. First (and perhaps above all), Swave is free. Also, while the Stat Sci library is mostly concerned with wavelet decompositions and discrete transforms (whether or not they are subsampled) our emphasis is more on the time-frequency time-scale applications using the continuous wavelet transform and the Gabor transform. As of today, these transforms are not available in the commercial package offered by StatSci.

The short Chapter 9 contains the explanations needed to download the C-code, the S-code, and the Makefile needed to install the Swave package on your system (detailed explanations are given in a README file contained in the package). They should come handy when trying to create the module you will want to use with your Splus implementation of S. The archive also contains the data files used throughout the monograph as well as the help files. Chapter 10 can be viewed as a hard copy of the on-line help of Swave. It contains all the descriptions of all the functions included in the package. If the installation has been successful, these files are available on line via the help command of Splus. Chapter 11 contains documentation on additional S functions which are either simple utilities or called as subroutines by the functions described in Chapter 10.

Bibliographies and Indexes

The references are organized in three distinct bibliographical lists. The first one gives all the references quoted in the text as long as they are of a general nature. It includes neither wavelet-related books nor S-related books. A special bibliography devoted exclusively to wavelets. It gives a list of books, monographs, conference proceedings, and special issues of journals which discuss time-frequency and/or time-scale analysis of signals in some detail. We also prepared a list of books and monographs which present statistical applications based on the use of the S language.

In a similar way, the index has been divided in four subindexes: a notation index, an index of all the Swave functions and utilities, an author index, and a subject index.

Acknowledgments

The present book grew out of a collaboration which started at the University of California and continued at Princeton University. Part of the work of René Carmona was funded by the Office of Naval Research. Bruno Torrésani and Wen Liang Hwang would like to thank the University of California at Irvine and Princeton University for their warm hospitality. They also acknowledge the partial support of the National Science Foundation and the Office of Naval Research.

The authors are indebted to L. Hudgins (Northrop Corp.), V. Chen (NRL), A. Cakmak (Princeton University), D. Lake (ONR), J. Ax and S. Strauss (Axcom Inc), Medical Diagnostics Inc, P. Flandrin (CNRS), C. Noel (Semantic), G. Hewer (China Lake NAWCWPNS), and W. Willinger (Bellcore) for providing us with valuable data sets which we used to illustrate the concepts presented in the book.

Before and during the completion of this work, we benefited from many enlightening discussions with colleagues and friends. It would be difficult to list them all here. At the risk of forgetting some of the most significant contributions, we would like to extend special thanks to A. Antoniadis, G. Beylkin, C. Noel, I. Daubechies, A. Grossmann, M. Holschneider, J.M. Innocent, S. Jaffard, S. Mallat, Y. Meyer, S. Schwartz, Ph. Tchamitchian, M.V. Wickerhauser, and S. Zhong.

Contents

List of Figures

Part I

Background Material

Chapter 1

Time, Frequency, and Time-Frequency

This first chapter is devoted to the presentation of a set of background concepts and results in the Fourier theory of deterministic signals. We introduce the notation and the definitions which we use to describe and analyze one-dimensional signals. Even though the signal analyst and the statistician are generally faced with data in the form of finite discrete sets of samples, we also consider signals as functions of an unbounded continuous variable and we introduce the Fourier transform in this generality. We make this choice for mathematical convenience: we want to be able to use the full force of sophisticated tools from functional analysis. But since most of the applications we present are dealing with signals with finitely many samples, we also discuss Fourier series and the discrete finite Fourier transform and we address with extreme care the issues of sampling and aliasing. We also discuss the notion of autocovariance and of periodogram of a finite signal in the context of Wiener's deterministic spectral theory. We end this chapter by introducing the concept of time-frequency representation of a signal: after all, time-frequency is our motto!

1.1 First Integral Transforms and Function Spaces

The basic function space for the mathematical analysis of signals will be the space of complex valued square-integrable functions on the real line $\mathbb{R}$. It will be denoted by $L^2(\mathbb{R})$ in general and $L^2(\mathbb{R}, dx)$ when we want to emphasize the variable. It is equipped with a natural inner product which turns it into a Hilbert space. Our convention for the inner product is as follows. Given two functions f and g in $L^2(\mathbb{R})$, their inner product is given by

$$\langle f, g \rangle = \int f(x)\overline{g(x)}\, dx \ , \tag{1.1}$$

where the bar stands for complex conjugation. We often use the term "signal of finite energy" for a generic element of $L^2(\mathbb{R})$. More generally, we shall often use the word "energy" to denote the square of the L^2 norm of such a function.

Finally, despite the fact that the variable has the interpretation of time in most applications, we shall nevertheless use the letter x. This is our way to conform with the practice of functional analysis. Indeed, the latter is the right framework for the discussion of this book. In addition, we shall describe in the last two chapters a set of iterative methods for post-processing time-frequency representations, and we shall reserve the letter t for the iteration counter in these methods.

1.1.1 Fourier Transforms

Our convention for the Fourier transform is the following: if $f \in L^2(\mathbb{R})$, its Fourier transform $\hat{f}$ is defined as

$$\hat{f}(\xi) = \int_{-\infty}^{\infty} f(x) e^{-i\xi x}\, dx, \qquad \xi \in \mathbb{R}. \tag{1.2}$$

Note that this formula implies that $\hat{f}(\xi) = \overline{\hat{f}(-\xi)}$ when the function f is real valued. Technically speaking, the right-hand side of (1.2) defines a function $\hat{f}$ only when the original signal f is absolutely integrable, but this notation is so suggestive that it is used even when this right-hand side has to be defined through a limiting argument using convergence in the mean quadratic sense (i.e., in the Hilbert space $L^2(\mathbb{R})$.) Let us give a more precise definition for the the Fourier transform of a signal with finite energy. If f is merely square integrable, since the defining integral (1.2) does not always exist, one first defines for each $n > 0$ the function

$$\hat{f}_n(\xi) = \int_{-n}^{n} f(x) e^{-i\xi x}\, dx,$$

which makes perfectly good sense, and one shows that the sequence $\{\hat{f}_n\}_n$ of functions so-defined has a limit and we use this limit as the definition of the Fourier transform of the function f. So in the general case, the integral formula (1.2) is only a notation, though a suggestive one. Unfortunately, there is no guarantee that for a given ξ the numerical sequence $\{\hat{f}_n(\xi)\}_n$ converges and the only convergence which can be proved is in the sense of L^2, namely,

$$\lim_{n\to\infty} \|\hat{f} - \hat{f}_n\|^2 = \lim_{n\to\infty} \int_{-\infty}^{\infty} |\hat{f}(\xi) - \hat{f}_n(\xi)|^2 d\xi = 0,$$

and this is the rigorous mathematical definition of the Fourier transform of a general signal with finite energy.

One of the first (and most important) result proved for the Fourier transform is its unitarity:

$$\int |f(x)|^2\, dx = \frac{1}{2\pi} \int |\hat{f}(\xi)|^2\, d\xi. \tag{1.3}$$

The quantity $|\hat{f}(\xi)|^2$ has the interpretation of the energy or power of the signal at frequency ξ, and the function $\xi \hookrightarrow |f(\xi)|^2$ is sometimes called the power spectrum. This formula is easy to prove when f is a smooth function decaying rapidly at infinity and the limiting argument alluded to earlier can then be used to extend this formula to the whole class of signals of finite energy. Relation (1.3) (the so-called *Parseval identity*) plays a fundamental role. Indeed, it shows that the Fourier transform can be extended to the whole space $L^2(\mathbb{R})$ in an isometry and more precisely an unitary equivalence between the complex Hilbert spaces $L^2(\mathbb{R}, dx)$ and $L^2(\mathbb{R}, d\xi/2\pi)$. If f and g are two signals of finite energy, the following identity is easily obtained from Parseval identity (1.3) by polarization (i.e., by using the identity $4\langle f, g\rangle = \|f+g\|^2 - \|f-g\|^2 + i\|f+ig\|^2 - i\|f-ig\|^2$)

$$\int f(x)\overline{g(x)}\, dx = \frac{1}{2\pi} \int \hat{f}(\xi)\overline{\hat{g}(\xi)}\, d\xi,$$

and the latter can be rewritten in the form

$$\langle f, g\rangle_{L^2(dx)} = \langle f, g\rangle = \frac{1}{2\pi}\langle \hat{f}, \hat{g}\rangle = \langle \hat{f}, \hat{g}\rangle_{L^2(d\xi/2\pi)}, \tag{1.4}$$

which emphasizes the fact that the Fourier transform preserves the Hilbert spaces inner products. Pushing this fact further, we see that

$$\begin{aligned}
\int f(x)\overline{g(x)}\, dx &= \frac{1}{2\pi} \int \left(\int f(x) e^{-ix\xi} dx \right) \overline{\hat{g}(\xi)}\, d\xi \\
&= \int \frac{1}{2\pi} \int \hat{f}(\xi) \overline{\left(\int g(x) e^{-ix\xi} dx \right)}\, d\xi \\
&= \int \left(\frac{1}{2\pi} \int \hat{f}(\xi) e^{ix\xi} d\xi \right) \overline{g(x)}\, dx
\end{aligned}$$

if we can interchange the two integral signs. This interchange can be justified by first considering integrable signals and then using a limiting argument (i.e. Fatou's lemma) to obtain this formula for general signals of finite energy. Since the preceding equality has to be true for all the signals g with finite energy, this implies that

$$f(x) = \frac{1}{2\pi} \int \hat{f}(\xi) e^{i\xi x} \, d\xi. \tag{1.5}$$

This shows not only that the Fourier transform is invertible, but that its inverse transform is given by

$$\check{f}(x) = \frac{1}{2\pi} \int f(\xi) e^{i\xi x} \, d\xi. \tag{1.6}$$

Parseval identity is important because of its numerous functional analytic consequences. But for us, its main feature is that it states a *conservation of energy* between the *time* and the *frequency* domains.

Remark 1.1 *Aside for probabilists and statisticians.* One is usually exposed to Fourier theory very early in the first (graduate) probability courses because of the possible characterizations of probability distributions and their convergence properties in terms of their Fourier transforms (so-called characteristic functions in the probabilistic jargon). We shall not need or use these features of the Fourier theory. The second instance in which probabilists and statisticians need Fourier analysis is the spectral analysis of time series and stationary processes. This theory is reviewed in Chapter 2. It is very much in the spirit in which we shall use the Fourier transform and in fact, all of the transforms which we consider in this monograph. The spectral analysis familiar to the statistician is restricted to stationary series and processes. Most of the efforts leading to the notions and methods presented in this book have been motivated by the desire to free the spectral analysis from the stationarity assumption and to make it possible to analyze in the same spirit nonstationary signals.

Localization: Heisenberg's Inequality

The main notion we shall deal with throughout this text is that of *signal localization.* We shall say that the (finite energy) signal f is well localized in the time domain if it satisfies some decay properties away from a fixed value. For example, we say that f is polynomially localized near $x = x_0$ if

$$|f(x)| \leq K \frac{1}{(1 + (x - x_0)^2)^{k/2}}, \qquad x \in \mathbb{R},$$

for some positive constants K and k. Similarly, we say that it is exponentially localized if

$$|f(x)| \leq K e^{-\alpha|x-x_0|}$$

for some positive constants K and α. Equivalently, we shall talk of polynomial or exponential frequency localization if similar bounds hold in the Fourier domain.

It is well known that it is not possible to achieve optimal localization simultaneously in the time and the frequency domains. For instance, contracting a function in the time domain in order to improve its time localization cannot be done without dilating it in the frequency domain, i.e., weakening its frequency localization. This fact is a consequence of the so-called *Heisenberg uncertainty principle.* Let us consider $f \in L^2(\mathbb{R})$ and let us assume that its derivative f' and its Fourier transform are also in $L^2(\mathbb{R})$. Then we introduce the notation

$$\overline{x} = \frac{1}{||f||^2} \int x|f(x)|^2 dx, \qquad \overline{\xi} = \frac{1}{||\hat{f}||^2} \int \xi|\hat{f}(\xi)|^2 d\xi$$

for the so-called time and frequency averages and

$$\Delta_x = \frac{1}{||f||} \sqrt{\int (x-\overline{x})^2 |f(x)|^2 dx}, \qquad \Delta_\xi = \frac{1}{||\hat{f}||} \sqrt{\int (\xi-\overline{\xi})^2 |\hat{f}(\xi)|^2 d\xi} \tag{1.7}$$

for the corresponding time and frequency variances. The Heisenberg uncertainty principle states that

$$\Delta_x \Delta_\xi \geq \frac{1}{2}\,. \tag{1.8}$$

Thus, an improvement of the time localization (i.e., a decrease of Δ_x) is likely to be accompanied by a deterioration in the frequency localization (i.e., an increase of Δ_ξ.) Note that the inequality which occurs in (1.8) becomes an equality in the case of Gaussian (or modulated Gaussian) functions.

The proof of the Heisenberg uncertainty principle is essentially a consequence of Cauchy-Schwarz inequality, i.e.,

$$\left|\Re \int_{\mathbb{R}} x f(x) f'(x) dx\right|^2 \leq \left\{\int_{\mathbb{R}} |xf(x)|^2 dx\right\} \left\{\int_{\mathbb{R}} |f'(x)|^2 dx\right\}\,.$$

A complete proof can be found, for example, in Theorem 3.5 of [Chui92a], where a detailed discussion of its consequences in terms of time-frequency localization can also be found. Another proof can be obtained from Problem 78, p. 132, of [255], where the quantum-mechanical interpretation of this principle is stressed.

Fourier Series

We now consider the case of signals given by doubly infinite sequences $f = \{f_j\}_{j=-\infty}^{j=\infty}$. Our motivation is the following (see also Section 1.2). Let us assume that $f_j = f(x_0 + j\Delta x)$ is obtained by sampling a continuous signal at regularly spaced values $x_j = x_0 + j\Delta x$. For the sake of notation we shall assume that $x_0 = 0$. A formal computation gives

$$\begin{aligned}
\hat{f}(\xi) &= \int_{-\infty}^{+\infty} f(x)e^{-ix\xi}\,dx \\
&= \sum_{j=-\infty}^{\infty} \int_{x_j-\Delta x/2}^{x_j+\Delta x/2} f(x)e^{-ix\xi}\,dx \\
&\approx \sum_{j=-\infty}^{\infty} \left(\int_{x_j-\Delta x/2}^{x_j+\Delta x/2} f(x)dx \right) e^{-ix_j\xi} \\
&\approx \Delta x \sum_{j=-\infty}^{\infty} f(x_j)e^{-ix_j\xi}
\end{aligned}$$

if one is willing (a) to approximate $e^{-ix\xi}$ by $e^{-ix_j\xi}$ in the interval $[x_j - \Delta x/2, x_j + \Delta x/2]$ and (b) to approximate the average of $f(x)$ over the interval $[x_j - \Delta x/2, x_j + \Delta x/2]$ by $(\Delta x)f(x_j)$. This motivates the following definition for the Fourier transform of the (doubly infinite) sequence $f = \{f_j\}_{j=-\infty}^{j=\infty}$:

$$\hat{f}(\xi) = \Delta x \sum_{j=-\infty}^{\infty} f_j e^{-ij\Delta x\xi}, \qquad \xi \in \mathbb{R}. \tag{1.9}$$

Note that this function is periodic with period $2\pi/\Delta x$. As a consequence, its values on the entire line $\mathbb{R}$ are completely determined by its values over an interval of length equal to $2\pi/\Delta x$. To take advantage of possible symmetries in the transform (for example, when the original signal is real), we choose most often to consider the values over an interval symmetric around the origin and restrict the definition to

$$\hat{f}(\xi) = \Delta x \sum_{j=-\infty}^{\infty} f_j e^{-ij\Delta x\xi}, \qquad \xi \in (-\frac{\pi}{\Delta x}, \frac{\pi}{\Delta x}). \tag{1.10}$$

In particular, one needs only to consider the interval $[0, \frac{\pi}{\Delta x})$ when the signal is real. In the present situation the finite-energy condition becomes

$$\sum_{j=-\infty}^{\infty} |f_j|^2 < \infty.$$

In other words, the space of (discrete) signals of finite energy is the Hilbert space $\ell^2(\mathbb{Z})$ of (complex) square summable (doubly infinite) sequences. Note that this space does not depend upon the sampling interval Δx. In fact, the value of Δx is of crucial importance only when we consider sequences occurring as the result of the sampling of a real-life signal (see, for example, Section 1.2), and this is the main reason for including its value in the definition of the Fourier transform. However, it is irrelevant when it comes to the mathematical properties of the Fourier transform of a sequence of finite energy. As a consequence, mathematicians will want to set $\Delta x = 1$ and define the Fourier transform of such a sequence by the formula

$$\hat{f}(\xi) = \sum_{j=-\infty}^{\infty} f_j e^{-ij\xi}, \qquad \xi \in [-\pi, \pi). \tag{1.11}$$

We say that the function $\hat{f}(\xi)$ is defined as the sum of the *Fourier series* of the square summable sequence $f = \{f_j\}_j$, and it can be viewed either as an element of the Hilbert space $L^2([0, 2\pi))$ or as a locally square integrable 2π periodic function. For any such function $\hat{f}(\xi)$, it is customary (at least in the mathematical literature) to define its Fourier coefficients c_j's by

$$c_j = \frac{1}{2\pi} \int_0^{2\pi} \hat{f}(\xi) e^{ij\xi} d\xi \ . \tag{1.12}$$

Using the same arguments as in the case of the functions defined on the whole axis $\mathbb{R}$, one easily shows that this formula gives the inverse of the transform defined by formula (1.11). In other words, $f_j = c_j$ and the original sequence $f = \{f_j\}_j$ can be recovered from the sum of its Fourier series. As before, this reconstruction formula gives rise to an "energy conservation" formula (also called Parseval's formula):

$$\sum_{j \in \mathbb{Z}} |c_j|^2 = \frac{1}{2\pi} \int_{-\pi}^{\pi} |\hat{f}(\xi)|^2 d\xi \ . \tag{1.13}$$

The study of the convergence of Fourier series is a subject in itself. See, for example, [302, 189]. As defined earlier, it establishes an isometry between the Hilbert space $\ell^2(\mathbb{Z})$ of square summable doubly infinite sequences and the Hilbert space $L^2([-\pi, \pi))$ (which can be identified to the space of locally square integrable 2π periodic functions on the real line). Notice that, by switching the roles of the direct and the inverse transforms, it also provides a theory for the Fourier transform for locally square integrable periodic functions (or equivalently, for square integrable functions on a bounded interval.) We conclude this short review of Fourier series by rewriting the inversion formula with the sampling interval length Δx because this formula will be needed in our discussion of the sampling theorem. The inversion formula reads

$$f_j = \frac{1}{2\pi} \int_{-\pi/\Delta x}^{\pi/\Delta x} \hat{f}(\xi) e^{ij\Delta x \xi} d\xi \ . \tag{1.14}$$

Finite Fourier Transform

There is still another Fourier transform to be considered. Indeed, if *doubly infinite* sequences are useful because they give a good foundation for the phenomenon of sampling, they are not practical, since real-life applications produce only *finite* sequences. This explains why we now turn to the case of finite sequences. The effect of the truncation of a doubly infinite sequence on the spectral characteristics of the signal will be considered in the next chapter. For the time being, let us consider the case of unit interval sampling (i.e., let us assume that $\Delta x = 1$) and let us consider a finite sequence $\{f_0, f_1, \ldots, f_{N-1}\}$. Fourier transform is a linear transform and, as such, cannot produce more than N linearly independent values from the sequence $\{f_0, f_1, \ldots, f_{N-1}\}$. Therefore, the (finite) Fourier transform of $\{f_0, f_1, \ldots, f_{N-1}\}$ is defined as the sequence $\{\hat{f}_0, \hat{f}_1, \ldots, \hat{f}_{N-1}\}$ whose entries are given by the formula

$$\hat{f}_k = \sum_{j=0}^{N-1} f_j e^{-2i\pi jk/N} \ . \tag{1.15}$$

Note that we merely rewrote the definition of the Fourier series assuming that all the coefficients f_j with $j < 0$ or $j \geq N$ were equal to zero, and we computed the sum only at the discrete values $\xi_k = 2\pi k/N$ of the frequency variable ξ. In other words, in the case of a finite sequence of length N we merely sample the frequency interval $[0, 2\pi)$ with N equally spaced frequencies which we will call the natural frequencies. Notice that we work with the interval $[0, 2\pi)$ instead of $[-\pi, +\pi)$. The inversion of such a transform can be done as before. But in the present situation it can be done in a *pedestrian way* using basic trigonometry. In any case the result reads

$$f_j = \frac{1}{N} \sum_{k=0}^{N-1} \hat{f}_k e^{2i\pi jk/N} \ . \tag{1.16}$$

Obviously, the length of the sampling interval Δx can be included in the formulae (1.15) and (1.16). It is a remarkable property that such a finite Fourier transform and its inverse can be computed fast. Indeed, the simple application of Equations (1.15) or (1.16) requires $O(N^2)$ complex multiplications. A good organization of the computations (see [247] for example) may reduce this to $O(N \log N)$ computations (at least for some values of N). We shall see applications of these algorithms and especially of the so-called *FFT* (which stands for Fast Fourier Transform) algorithms in Section 5.6.

1.1.2 Hilbert Transform, Analytic Signal

Later in this chapter we discuss abstract time-frequency transforms which have an energy conservation characteristic. This allows them to provide energy representations of signals in a domain which is neither the time domain nor the frequency domain, but instead a mixed domain in which information concerning the properties of the signals in both domains can be read off. In the meantime, we introduce the notion of Hilbert transform and of analytic signal. These notions are not standard in the probabilistic and statistical literatures. They are standard in signal analysis because of their usefulness when it comes to finding intrinsic and rigorous definitions for time-varying quantities which have very natural and intuitive interpretations but which are difficult to capture mathematically.

In addition to $L^2(\mathbb{R})$, we shall often make use of the *complex Hardy space*,

$$H^2(\mathbb{R}) = \left\{ f(x) \in L^2(\mathbb{R}); \hat{f}(\xi) = 0,\ \xi \leq 0 \right\}, \tag{1.17}$$

sometimes called the *space of analytic signals*. In other words, this space is made of the signals which do not have negative frequency components. The space $H^2(\mathbb{R})$ is intimately related to the *Hilbert transform* H, defined by

$$H \cdot f(x) = \frac{1}{\pi} P. \int f(x+y) \frac{dy}{y}, \qquad x \in \mathbb{R}, \tag{1.18}$$

where P denotes the principal value integral. The Hilbert transform is conveniently expressed in the Fourier domain as

$$\widehat{H \cdot f}(\xi) = -i \operatorname{sgn}(\xi) \hat{f}(\xi), \qquad \xi \in \mathbb{R}. \tag{1.19}$$

Notice that the Hilbert transform maps sine waves into cosine waves, and vice versa. Given a real-valued signal of finite energy f, the associated *analytic signal* is defined (up to a factor 2) as its orthogonal projection Z_f onto the subspace $H^2(\mathbb{R})$ of $L^2(\mathbb{R})$. It is given by the formula

$$Z_f(x) = [I + iH] f(x), \tag{1.20}$$

where I denotes the identity operator. Equivalently, its Fourier transform is given by

$$\widehat{Z_f}(\xi) = 2\Theta(\xi) \hat{f}(\xi), \tag{1.21}$$

where $\Theta(\xi)$ denotes the Heaviside step function which is equal to 1 when $\xi \geq 0$ and to 0 otherwise. The analytic signal representation has been proven to be useful in many applications. Indeed, physical phenomena are causal and measurements (accelerations, acoustic pressure, temperature, photon counts, ...) are real, and because of the symmetry of the Fourier transform of real signals, all the information is contained in the positive frequencies. Using analytic signals can be viewed as a way of taking advantage of the versatility of complex function spaces without having to include too much extra information. Moreover, the notion of analytic signal provides a way to make sense of the notion of time-dependent frequency, or *instantaneous frequency*, or to the "dual" quantity, the *group delay* $\tau(\xi)$. We shall come back to these concepts later in this chapter and again in Chapter 7.

1.2 Sampling and Aliasing

The problem of the discretization of continuous time signals (i.e., the sampling of analog signals) is the subject of this section. The basic theoretical result is the so-called *sampling theorem*. It may be attributed to several authors (among them Shannon, Nyquist, Whittaker, and Kotel'nikov). This theorem (which is essentially a consequence of Poisson's summation formula) describes the sampling of band limited (also called Paley-Wiener) functions.

Let us consider a signal $f \in L^2(\mathbb{R})$ of finite energy defined on the whole real line and let us also consider the (doubly infinite) sequence $f^{(\Delta x)} = \{f_j\}_{j=-\infty}^{j=+\infty}$ obtained by sampling the continuous time signal f at regular time intervals of lengths Δx

$$f_j = f(x_0 + j\Delta x), \qquad j = -\infty, \cdots, +\infty.$$

Since both the continuous time signal f and the sequence of samples have a Fourier transform (recall the definitions of the first section of the chapter), it is natural to investigate the possible relations between these two transforms. Because of the inversion formula (1.5) for functions defined on the whole real axis $\mathbb{R}$, it holds that (as before we assume $x_0 = 0$ for the sake of simplicity)

$$\begin{aligned} f_j = f(j\Delta x) &= \frac{1}{2\pi}\int_{-\infty}^{\infty}\hat{f}(\xi)e^{i\xi j\Delta x}d\xi \\ &= \frac{1}{2\pi}\sum_{k=-\infty}^{\infty}\int_{(2k-1)\pi/\Delta x}^{(2k+1)\pi/\Delta x}\hat{f}(\xi)e^{i\xi j\Delta x}d\xi \\ &= \frac{1}{2\pi}\int_{-\pi/\Delta x}^{\pi/\Delta x}\sum_{k=-\infty}^{\infty}\hat{f}(\xi-\frac{2k\pi}{\Delta x})e^{i\xi j\Delta x}d\xi. \end{aligned} \tag{1.22}$$

Restating the inversion formula (1.14) for Fourier series with the notation $\hat{f}^{(d)}(\xi)$ for the sum of the Fourier series of the sequence $f^{(\Delta x)}$, we get

$$f_j = \frac{1}{2\pi}\int_{-\pi/\Delta x}^{\pi/\Delta x}\hat{f}^{(d)}(\xi)e^{i\xi j\Delta x}d\xi\ . \tag{1.23}$$

Identifying (1.22) and (1.23), one gets

$$\hat{f}^{(d)}(\xi) = \sum_{k=-\infty}^{\infty}\hat{f}(\xi+\frac{2k\pi}{\Delta x}). \tag{1.24}$$

This formula tells us that $\hat{f}^{(d)}(\xi)$, the Fourier transform at frequency ξ of the sampled sequence $f^{(\Delta)} = \{f_j\}_j$, is the sum of the contributions from the Fourier transform $\hat{f}$ of the continuous signal f, not only at ξ but also at the countably infinite set of frequencies $\xi + 2k\pi/\Delta x$ for $k = \pm1, \pm2, \cdots$. These frequencies are called *aliases* of the frequency ξ and this folding of the Fourier transform of f to produce the Fourier transform of the sampled sequence $f^{(\Delta)} = \{f_j\}_j$ is called *aliasing.* The frequency

$$\xi_{(N)} = \frac{\pi}{\Delta x} \tag{1.25}$$

plays a special role. It is called the *Nyquist frequency* or *folding frequency.*

Formula (1.24) shows that it may not always be possible to recover the function $\hat{f}$ on the whole real line $\mathbb{R}$ from the values of the function $\hat{f}^{(d)}$ on a period interval of length 2π. Since the Fourier transforms realize one-to-one correspondences between functions and/or sequences and their respective transforms, this implies that it may not always be possible to recover the whole function f from its sample f_j's. Figure 1.1 gives a typical example of such a situation.[1] It shows the graphs of the five functions

$$f^{(-2)}(x) = -\sin(14\pi x) \qquad f^{(-1)}(x) = -\sin(6\pi x)$$
$$f^{(0)}(x) = \sin(2\pi x)$$
$$f^{(1)}(x) = \sin(10\pi x) \qquad f^{(2)}(x) = \sin(18\pi x).$$

[1] Nevertheless, notice that the functions in Figure 1.1 are not in $L^2(\mathbb{R})$.

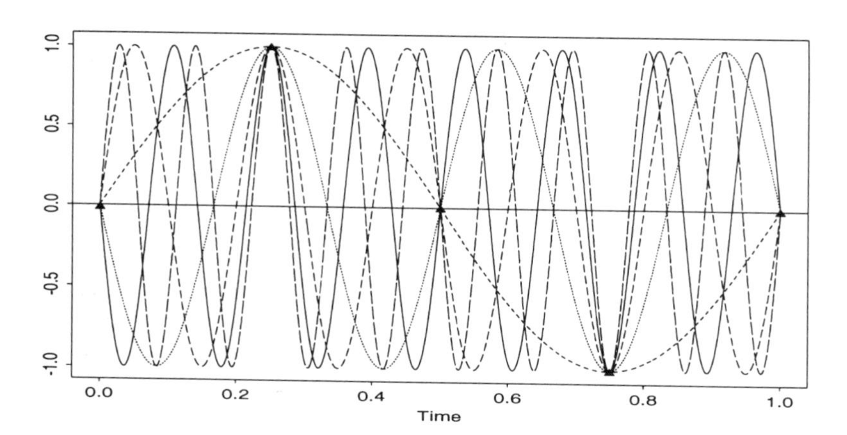

Figure 1.1. Graphs over the unit interval $[0,1]$ of the five sinusoids described in the text. They all have the same values at the sampling points marked by black triangles.

The frequencies of these five sinusoids are $-7,\ -3,\ 1,\ 5,\ 9$ Hz, which we can write in the form

$$\xi^{(m)} = 1 + 4m, \qquad m = -2, -1, 0, 1, 2,$$

and we can write the sinusoids with the single formula

$$f^{(m)}(x) = \sin(2\pi(1 + 4m)x), \qquad m = -2, -1, 0, 1, 2.$$

sampling these signals at the rate 4 Hz (i.e., restricting ourselves to x's of the form $j/4$ seconds), we get

$$f_j^{(m)} = f^{(m)}(\frac{j}{4}) = \sin(2\pi(1 + 4m)j/4) = \sin(j\frac{\pi}{2}),$$

which are the same independently of the value of m. We now address the fundamental question:

> When can we recover the all the values of the function f from the knowledge of the samples $f_j = f(j\Delta x)$?

Definition 1.1 *We say that a function $f \in L^2(\mathbb{R})$ is band limited with limit frequency ξ_f if its Fourier transform $\hat{f}$ vanishes outside the interval $[-\xi_f, \xi_f]$.*

Band-limited functions are analytic; in fact, they are analytic functions of exponential type. See, for example, [255]. We are now in a position to give a generic form of the sampling theorem.

Theorem 1.1 *Let f be a band limited function and let us consider the sampled sequence $f^{(\Delta x)} = \{f_j\}_j$ obtained as earlier with $f_j = f(j\Delta x)$. Then the whole function f can be recovered from its samples f_j's whenever the* sampling frequency $\xi_s = 1/\Delta x$ *is not smaller than the limit frequency ξ_f, i.e., whenever $\xi_s \geq \xi_f$.*

Notice that, under the condition of the theorem, there is no aliasing since when $|\xi| < \pi/\Delta x$ only one aliased frequency is in the interval $[-\xi_f, \xi_f]$ outside which $\hat{f}$ is identically zero. We shall not give a detailed proof of the sampling theorem. Instead, we give an intuitive argument which contains all the ingredients of the proof (see, e.g., [Meyer89a] for a complete proof.) If we assume that the function f is band-limited and if we denote by ξ_f its frequency limit, then, if $\xi_s > \xi_f$, formula (1.24) is the key for understanding why $\hat{f}(\xi)$ may be simply recovered from $\hat{f}^{(d)}(\xi)$. Indeed, let us, for example, multiply $\hat{f}^{(d)}(\xi)$ by a compactly supported function, say $\hat{\varphi}(\xi)$, such that $\hat{\varphi}(\xi) = 1$ if $\xi \in [-\xi_f, \xi_f]$ and $\hat{\varphi}(\xi) = 0$ whenever $\xi \notin [-\xi_s, \xi_s]$. Then one argues that $\hat{f}(\xi)$ and $\hat{\varphi}(\xi)\hat{f}^{(d)}(\xi)$ are analytic functions which coincide on a non-empty open interval so they coincide everywhere, and one recovers the whole function f by computing the inverse Fourier transform of $\hat{\varphi}(\xi)\hat{f}^{(d)}(\xi)$. This argument also yields an interpolation formula for the values of $f(x)$ from the values of the samples:

$$f(x) = \sum_n f(n\Delta x)\varphi(x - n\Delta x) .$$

Now, if $\xi_s = \xi_f$, the same result holds, with the restriction that the only possible choice for $\hat{\varphi}(\xi)$ is the characteristic function of the interval $[-\nu, \nu]$, i.e., $\varphi(x)$ is the cardinal sine $\sin(\nu x)/(\nu x)$. Finally, if $\xi_s < \xi_f$, then the function f cannot be recovered without additional assumptions, because the folding mixes the interval contributions to $\hat{f}(\xi)$ as we saw in our discussion of aliasing.

A Practical Application of the Sampling Theorem

We use the data from a sound file of dolphin clicks to illustrate one of the applications of the sampling theorem. This data set was obtained from Doug Lake's Web page. The data comprise $N = 2499$ samples collected in a T-second recording. Hence, the sampling frequency was $\xi_s = N/T$ Hz.

Figure 1.2 shows the original time series of the 2499 samples of the original dolphin click data and the modulus of the (discrete) Fourier transform of this series. The values of the Fourier transform $\hat{f}^{(\Delta x)} = \{\hat{f}_k;\ k = 0, \cdots, N-1\}$ turn out to be essentially zero for $k > 600$ and $k < 1898$. For all practical purposes we can assume that this Fourier transform vanishes for $K < k < N - 1 - K$ with $K = 600$. Notice that in order to recast

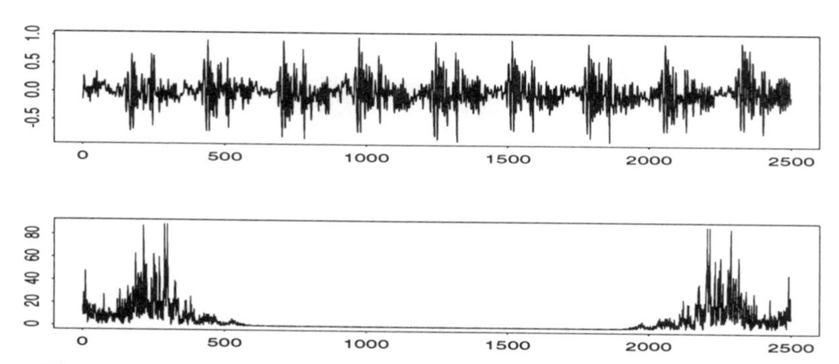

Figure 1.2. Dolphin click data and modulus of its Fourier transform

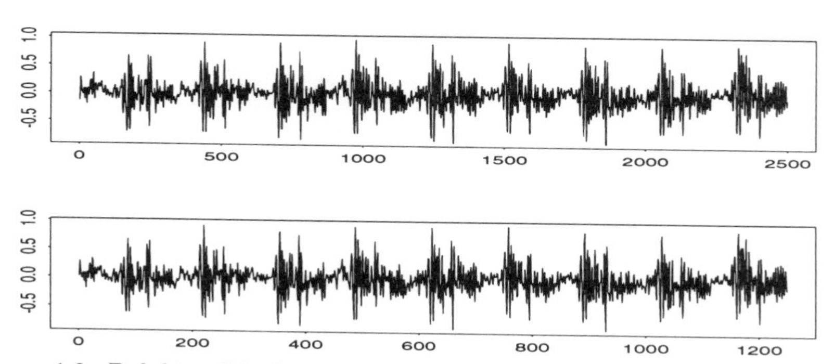

Figure 1.3. Dolphin click data: original dolphin click data (top) and the result of the subsampling (bottom).

this practical computation in the theoretical framework presented earlier, we need to shift the values $\hat{f}_k$ for $k = 1300, 1301, \cdots, 2499$ to the left to make sure that our Fourier transform is computed on an interval symmetric with respect to the origin. After we do so, the Fourier transform looks like the plot on top of Figure 1.3. The function $f = \{f(x); x \in \mathbb{R}\}$ from which we are seeing a sample is clearly band limited with a limit frequency $\xi_f = K/T$ Hz. The sampling theorem tells us that one could reduce the sampling frequency ξ_s all the way down to ξ_f without losing any information. If we just do that, the new signal will have $2K$ samples; in other words, this would amount to subsampling the dolphin click data set at the subsampling rate of $2K/N = 1200/2499 = 0.48$. So we shall subsample the signal by keeping only every other sample from the original data set. The result is a shorter signal (which we give at the bottom of Figure 1.3) containing the same information when it comes to inferring properties of

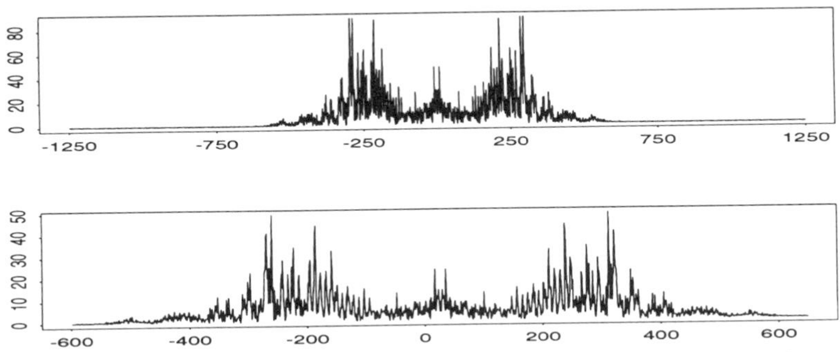

Figure 1.4. Dolphin click data: the top plot gives the modulus of the (discrete) Fourier transform of the dolphin click data in the form needed to use the sampling theorem while the bottom plot gives the same thing after subsampling.

the function $f = \{f(x), x \in \mathbb{R}\}$.

1.3 Wiener's Deterministic Spectral Theory

Wiener's spectral theory of deterministic signals is a beautiful set of mathematical ideas which we now review for the sake of completeness (see, e.g., [195] for more details). It is best understood in the framework of continuous time signals. Let us consider a real-valued function f defined on the real line and let us assume that the limit

$$C_f(\tau) = \lim_{T\to\infty} \frac{1}{2T} \int_{-T}^{+T} f(x+\tau) f(x) dx \tag{1.26}$$

exists and is finite for every $\tau \in \mathbb{R}$. The function $C_f(\tau)$ so defined is called the *autocovariance function* of f and its value at the origin,

$$C_f(0) = \lim_{T\to\infty} \frac{1}{2T} \int_{-T}^{+T} f(x)^2 dx, \tag{1.27}$$

is called the *power* of the signal f. It is finite by hypothesis. The autocovariance function of a real valued signal is even since $C_f(\tau) = C_f(-\tau)$. If we restrict ourselves to the class of (possibly complex valued) functions f with finite power, i.e., for which the quantity $C_f(0)$ defined by

$$C_f(0) = \lim_{T\to\infty} \frac{1}{2T} \int_{-T}^{+T} |f(x)|^2 dx$$

exists, then the formula

$$\langle f, g \rangle = \lim_{T \to \infty} \frac{1}{2T} \int_{-T}^{+T} f(x)\overline{g(x)}dx \tag{1.28}$$

defines an inner product in this space of functions and Schwarz's inequality gives

$$|C_f(\tau)| \leq C_f(0), \qquad \tau \in \mathbb{R}.$$

The autocovariance function C_f is nonnegative definite in the sense that

$$\sum_{j,k=1}^{n} z_j \overline{z_k} C_f(x_j - x_k) \geq 0 \tag{1.29}$$

for all possible choices of the complex numbers z_j's and the n elements x_j of the real line. Because of Bochner's theorem (see, for example, [141]), this implies the existence of a nonnegative finite measure ν_f on the real line satisfying

$$C_f(\tau) = \frac{1}{2\pi} \int_{-\infty}^{+\infty} e^{i\tau\omega} \nu_f(d\omega), \qquad \tau \in \mathbb{R} . \tag{1.30}$$

This measure ν_f is called the *spectral measure* of the signal f. Throughout the book we shall mean *nonnegative finite measure* when we say measure. The spectral analysis of the signal f consists in finding the properties of this measure. A first natural question concerns the Lebes-gue decomposition of this measure (see, e.g., [255]). It reads

$$\nu_f = \nu_{f,pp} + \nu_{f,sc} + \nu_{f,ac} , \tag{1.31}$$

where $\nu_{f,pp}$ is a pure point measure (weighted sum of Dirac point masses), $\nu_{f,sc}$ is a singular (i.e., concentrated on a set of Lebesgue measure zero), continuous, i.e., $\nu_{f,sc}(\{\omega\}) = 0$ for any singleton ω measure and $\nu_{f,ac}$ is absolutely continuous, i.e., given by a density with respect to the Lebesgue measure in the sense that $d\nu_{f,ac}(\omega) = \nu_{f,ac}(\omega)d\omega$ for some nonnegative integrable function ν_f. The Lebesgue decomposition of the spectral measure gives a corresponding decomposition of the autocovariance function

$$C_f = C_{f,pp} + C_{f,sc} + C_{f,ac} ,$$

with

$$C_{f,\cdot\cdot}(\tau) = \int_{-\infty}^{+\infty} e^{i\tau\omega} \nu_{f,\cdot\cdot}(d\tau) .$$

The meaning of this decomposition is the following: The original signal can be understood as the sum of three orthogonal (i.e., uncorrelated) signals with *pure* spectra given by the components of the original spectral components in the Lebesgue decomposition. The function $\rho_f(\omega) = \nu_f(\{\omega\})$ is called the ***spectral mass function*** while the density $\nu_{f,ac}(\omega)$ is called the ***spectral density function***. It is extremely difficult to give a reasonable interpretation to the singular continuous component $C_{f,sc}$ of the covariance. Fortunately, this singular continuous component is rarely present and $\nu_{f,sc} = 0$ in most practical applications. Since $\nu_{f,pp}$ is of the form

$$\nu_{f,pp} = \sum_k \rho_k \delta_{\omega_k} ,$$

for some nonnegative weights $\rho_k = \rho_f(\omega_k)$ and (possibly infinitely many) unit point masses δ_{ω_k}, the pure point part of the autocovariance function has the form

$$C_{f,pp}(\tau) = \frac{1}{2\pi} \sum_k \rho_k e^{i\omega_k \tau}$$

of a (possibly complex) trigonometric polynomial when the number of frequencies ω_k is finite and of an almost periodic function in general. Notice that

$$\frac{1}{2\pi} \sum_k \rho_k = C_{f,pp}(0) < +\infty.$$

Example: Spectrum of an Almost Periodic Function

Let us consider a general almost periodic function of the form

$$f(x) = \sum_{j=-\infty}^{+\infty} c_j e^{i\lambda_j x} , \tag{1.32}$$

where the λ_j are distinct real numbers and where the coefficients c_j are (possibly complex) numbers satisfying

$$\sum_{j=-\infty}^{+\infty} |c_j|^2 < \infty.$$

A direct calculation shows that

$$C_f(\tau) = \sum_{j=-\infty}^{+\infty} |c_j|^2 e^{i\lambda_j \tau} ,$$

which shows that the autocovariance function is pure point (in the sense that $C_{f,sc}(\tau) = C_{f,ac}(\tau) = 0$) and that the spectrum is concentrated on the set of λ_j's. More precisely, the spectral mass function $\rho_f(\omega)$ is given by

$$\rho_f(\omega) = \begin{cases} |c_j|^2 & \text{if} \qquad \omega = \lambda_j \\ 0 & \text{otherwise.} \end{cases}$$

It is important to notice the loss of the phase information. Indeed, the autocovariance function $C_f(\tau)$ is only a function of the modulus of the coefficients c_j and the arguments of these complex numbers cannot be recovered from the knowledge of the spectrum. Indeed, these phase values measure the displacements of the harmonic components relative to a fixed time origin, and they cannot be retained by the spectral analysis because the latter is based on a definition of the autocovariance function which wipes out any natural notion of time origin.

It is worth pointing out that Wiener's theory extends to random signals as long as the limit (1.26) defining the autocovariance function exists almost surely and the limit is nonrandom. We shall see that this is indeed the case for stationary and ergodic processes. But the spectral theory of these processes has a definite advantage over Wiener's spectral theory of deterministic signals which we reviewed earlier. Indeed, it allows for a representation of the original signal as a (random) superposition of complex exponentials (see Section 2.2) and this representation is extremely useful in the spectral analysis of these random signals. Unfortunately, there is no convenient decomposition of this type in the Wiener theory of deterministic signals.

1.4 Deterministic Spectral Theory for Time Series

Let us first present a concrete example of a time series which we use for illustration purposes throughout this section.

Example 1.1 Figure 1.5 gives the sequential plot of the values of the logarithms of the price of a futures contract of Japanese yen. Note that these data have been preprocessed to remove the effect of the proximity of the expiration date of a given contract. Approaching expiration of a contract may not have a significant effect on financial contracts, but it is important for other types of commodities because contract holders do not want to bear the risk of having to take delivery of the goods specified by the contract (speculators are avoiding delivery like the plague). Roughly speaking, when a contract comes within a month of its expiration date the price is replaced by the price of a contract which is identical except for the delivery date, which is pushed forward in time. The data contains 500 positive

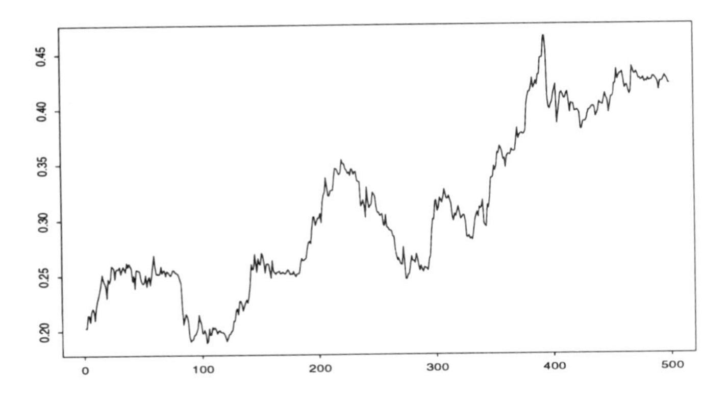

Figure 1.5. Logarithms of the prices of a contract of Japanese yen from 1/2/89 to 12/31/90 (corresponding to 500 working days).

numbers for the log-prices of this Japanese yen contract from the beginning of year 1988 to the end of 1990 (250 working days per year).

We postpone the discussion of the mathematical tools needed to handle the randomness and we concentrate here on the manipulations which do not attempt to model the sources of randomness.

1.4.1 Sample Autocovariance and Autocorrelation Functions

Throughout this section we work with a finite time series,

$$f = \{f_0, f_1, \cdots, f_{N-1}\},$$

of real numbers and we discuss various ways in which the statistical correlations between the successive values of the f_j's can be quantified.

The measure of dependence most commonly used is the *sample autocovariance function,* defined by the formula

$$c_f(j) = \frac{1}{N} \sum_{k=0}^{N-1-|j|} (f_k - \overline{f})(f_{k+|j|} - \overline{f}), \qquad |j| < N, \tag{1.33}$$

where the notation $\overline{f}$ is used for the *sample mean,* defined in the usual way by:

$$\overline{f} = \frac{1}{N}\sum_{j=0}^{N-1} f_j. \tag{1.34}$$

We are aware of the conflict in notation created by the use of a bar for the sample mean, but this notation is so standard in the statistical literature that we did not want to replace it by a home-grown ersatz. We hope that the context will help and that it will not be confused with the notation for the complex conjugate. We shall explain later in this chapter why the definition (1.33) was preferred to the more natural one:

$$\tilde{c}_f(j) = \frac{1}{N-|j|}\sum_{k=0}^{N-1-|j|}(f_k - \overline{f})(f_{k+|j|} - \overline{f}) = \frac{N}{N-|j|}c_f(j). \tag{1.35}$$

The (signed) integer j appearing in the definitions of the autocovariance functions is usually called the *lag*. It is often restricted to nonnegative values and the autocovariance is often rescaled to remove overwhelming effects of size on dependence measurements. This leads to the notion of *sample autocorrelation function* of the series f. It will be denoted ρ_f. It is defined by the formula

$$\rho_f(j) = \frac{c_f(j)}{c_f(0)} = \frac{\sum_{j'=0}^{N-1-j}(f_{j'} - \overline{f})(f_{j+j'} - \overline{f})}{\sum_{j'=0}^{N-1-j}(f_{j'} - \overline{f})^2}, \qquad j = 0, 1, \cdots, N-1. \tag{1.36}$$

The plot of the autocorrelation $\rho_f(j)$ versus j is sometimes called the *correlogram*. Note that $\rho_f(0) = 1$ and that the maximum lag j for which $\rho_f(j)$ makes sense is $N-1$, even though we use much smaller values in practice to make sure that there are enough terms in the sums used in the definition (1.36). We say that the series has a long (resp. short) memory when the correlogram decays slowly (resp. fast.)

Example 1.1 Revisited. The daily prices of a Japanese yen futures contract are strongly correlated and, as one can see from Figure 1.6 for example, the correlogram shows that the series has long memory. The tools presented in this section (and the theory presented in the next) are more appropriate for the series of the daily differences of the log-prices (in other words the logarithms of the ratios of the prices on two consecutive days). Figure 1.7 shows the sequential plot of this series together with the plot of the values of the autocorrelation function for the first values of the lag. As one can see, the correlogram shows very short memory: the autocorrelation function seems to be nonzero only for the zeroth lag. We

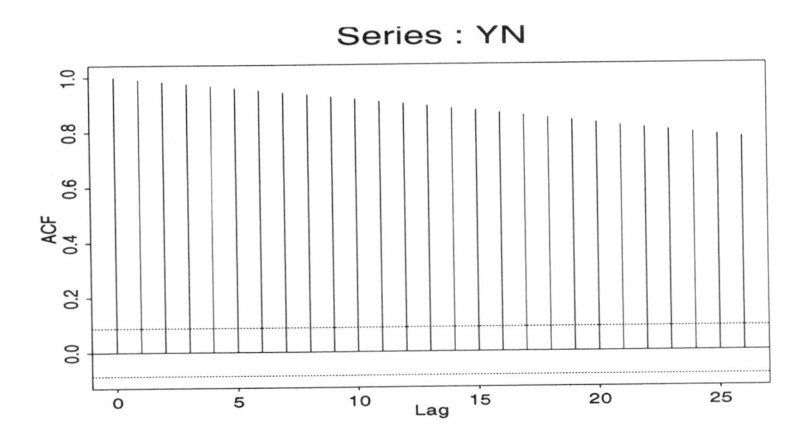

Figure 1.6. Plot of the first 27 values of the autocorrelation function of the daily log-prices of the Japanese yen futures contract shown earlier.

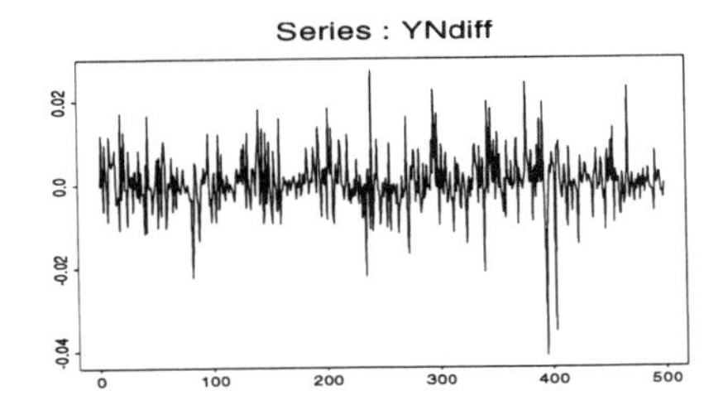

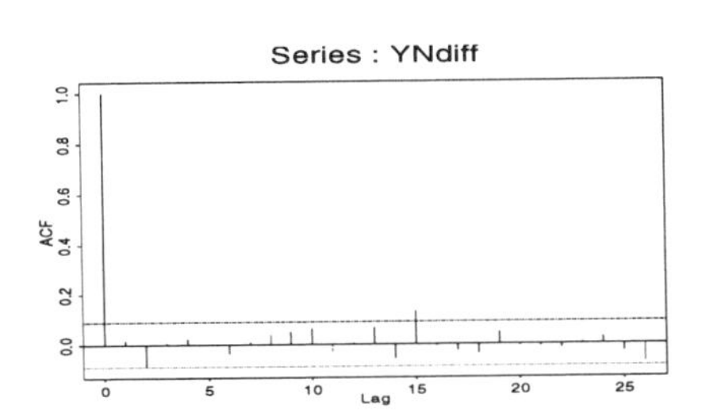

Figure 1.7. Japanese yen future contracts. Left: Daily differences of the logarithms of the prices of the Japanese yen futures contract considered earlier. Right: Plot of the first values of the autocorrelation function of the daily differences of the log-prices of the Japanese yen futures contract. The dotted lines limit a confidence interval: only the values outside this interval should be regarded as (significantly) different from 0.

shall give an interpretation of that fact in terms of the white noise series introduced in the next chapter. Compare the plot of this correlogram with the corresponding plot given in the middle of Figure 2.1 for the case of a white noise.

1.4.2 Spectral Representation and Periodogram

The main goal of spectral theory is to represent each signal as the sum of trigonometric functions with specific phases and amplitudes. The set of phases involved in the representation is called the *spectrum* of the series and the relative importance of the amplitudes is summarized in what is usually called the *periodogram*. For pedagogical reasons we believe that it is important to give the details of this representation. Let us recall the definition of the *discrete Fourier transform* (DFT for short) given in (1.15),

$$\hat{f}_k = \sum_{j=0}^{N-1} f_j e^{-2i\pi j\omega_k}, \qquad k = 0, 1, \cdots, N-1, \tag{1.37}$$

where we now use the notation ω_k for the *natural frequency* $\omega_k = k/N$. For notation purposes one should think of the frequency variable ω as being linked to the Fourier variable ξ by the scaling relation $\xi = 2\pi\omega$. Recall also that the original sequence f can be recovered from its transform by the *inverse Fourier transform*

$$f_j = \frac{1}{N} \sum_{k=0}^{N-1} \hat{f}_k e^{2i\pi j\omega_k}, \qquad k = 0, 1, \cdots, N-1. \tag{1.38}$$

This inversion formula shows that any finite series f can be written as the linear combination of complex exponentials. We now work on this decomposition, especially when the original series is real. We set:

$$a_k = \frac{1}{N}\Re\hat{f}_k \qquad \text{and} \qquad b_k = \frac{1}{N}\Im\hat{f}_k, \qquad k = 0, 1, \cdots, N-1.$$

Notice that $a_1 = \overline{f}$ and $b_1 = 0$. Moreover, we have

$$f_j - \overline{f} = \sum_{k=2}^{N-1} e_j(k) ,$$

where the trigonometric function e_j are defined by

$$e_j(k) = r_k \cos(2\pi j\omega_k - \varphi_k) ,$$

with $r_k = \sqrt{a_k^2 + b_k^2}$ and $\varphi_k = \tan^{-1}(b_k/a_k)$. It is easy to see that for $k = 1, \cdots, [N/2]$ the following relationships hold:

$$a_k = a_{N-k} \qquad \text{and} \qquad b_k = -b_{N-k} ,$$

and consequently,

$$r_k^2 = r_{N-k}^2 \qquad \text{and} \qquad e_j(k) = e_j(N-k) .$$

This implies the trigonometric representation

$$f_j - \overline{f} = \begin{cases} 2\sum_{k=1}^{[N/2]} r_k \cos(2\pi j\omega_k) & \text{if } N \text{ is odd,} \\ 2\sum_{k=1}^{[N/2]-1} r_k \cos(2\pi j\omega_k) + r_{[N/2]} \cos(2\pi j\omega_{[N/2]}) & \text{if } N \text{ is even.} \end{cases} \tag{1.39}$$

The reason for the derivation of the representation (1.39) is motivated by the desire to untangle the dependence among the complex exponentials appearing in the inverse Fourier transform (1.38). This dependence is no longer present in (1.39) because (as it is easy to see) the trigonometric functions $e_j(k) = r_k \cos(2\pi j\omega_k)$ appearing in the representation (1.39) are mutually orthogonal! As announced earlier, any series f can be written as the sum of cosine functions of frequencies $\omega_1 = 1/N$, $\omega_2 = 2/N$, $\cdots$, $\omega_{[N/2]} = [N/2]/N$. The plot of Nr_k^2 versus ω_k (recall that

$$r_k^2 = \frac{1}{N^2}\left|\sum_{j=0}^{N-1} f_j e^{-2i\pi j\omega_k}\right|^2, \qquad k = 0, 1, \cdots, [N/2] \tag{1.40}$$

and $\omega_k = k/N$) for $k = 0, 1, \cdots, [N/2]$ is called the *periodogram* of the series f, and the function $\nu_f(\omega)$ defined for $\omega \in [0, 1]$ by:

$$\nu_f(\omega) = \frac{1}{N}\left|\sum_{j=0}^{N-1} f_j e^{-2i\pi j\omega}\right|^2 \tag{1.41}$$

for $\omega \in [0, 1/2]$ and by $\nu_f(\omega) = \nu_f(1-\omega)$ for $\omega \in [1/2, 1]$ is called the *sample spectral density* or *sample spectral function* of f. Notice that, because $\nu_f(\omega_k) = Nr_k^2$, the function $\nu_f(\omega)$ is a natural way to extend the periodogram into a function of the continuous variable.

It happens very often that a few values of the r_k are very large compared to the rest. This makes it very difficult to grasp the information contained in the periodogram. Plotting the (natural) logarithm of the values of r_k^2 instead of the values themselves helps in seeing more of the features of the frequencies. Moreover, an easy computation shows that

$$\frac{1}{N}\sum_{j=1}^{N-1} \frac{Nr_k^2}{\sigma^2} = 1$$

if σ^2 denotes as usual the *sample variance*,

$$\sigma^2 = \frac{1}{N} \sum_{j=0}^{N-1} (f_j - \overline{f})^2, \tag{1.42}$$

and because of this last formula, in practice one plots

$$\log\left(\frac{Nr_k^2}{\sigma^2}\right) \qquad \text{against} \qquad \omega_k$$

to have a more informative graphical display of the frequency content of the series. In Chapter 6 we shall also plot the values of the logarithm of the periodogram against the logarithm of the frequencies. The following remarks are useful when it comes to the interpretation of a periodogram.

- It is smooth when the amplitudes of the cosine functions of low frequency are large compared to the other frequencies.
- It is erratic and wiggly when the amplitudes of the cosine functions of high frequency are large compared to the other frequencies.
- It has a peak at the frequency ω when the signal is a cosine or a sine function of period $1/\omega$.
- It has a peak at the frequency ω and peaks at the frequencies multiples of ω (harmonics) when the signal is a periodic function of period $1/\omega$ without being a cosine or a sine function.

Example 1.2 Heart Rate Variability. As the examples discussed in this book will show, the spectral analysis tools presented herein have been used extensively in many fields. In medical applications they were traditionally the tools of choice in ECG (ElectroCardioGram) and EEG (ElectroEncephaloGram) analyses. For the sake of illustration we mention the analysis of *heart rate variability*. See, for example, [237] or [300] for details. Figure 1.8 shows the time series of the lengths of the time intervals in between two successive heart beats of a typical male subject, together with the corresponding periodogram. This periodogram has been slightly smoothed for the purpose of the figure (for mathematical justifications of the smoothing, see also the discussion of the practical spectral estimation procedures for random stationary signals in the next chapter). Also, as explained earlier, the vertical axis is on a logarithmic scale. In fact, the periodogram $\nu_f(\omega)$ is plotted in *decibels* in the sense that the quantity $10\log_{10}(\nu_f(\omega))$ appears on the vertical axis.

The data in Figure 1.8 are from a normal patient. The detection of heart rate variability has proven of crucial importance in many applications such as fetal monitoring. In this case diminished beat-to-beat variation signifies

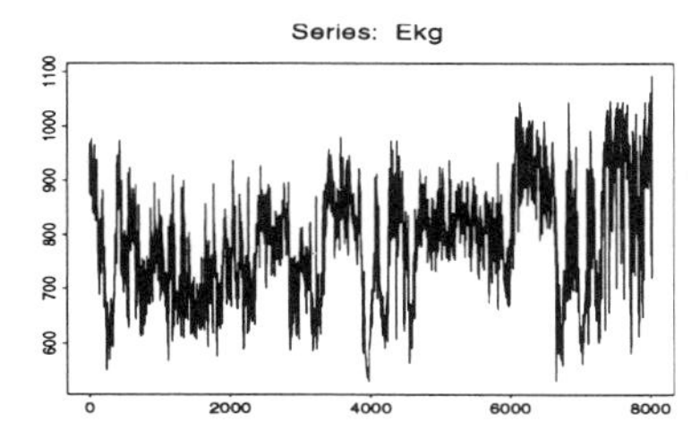

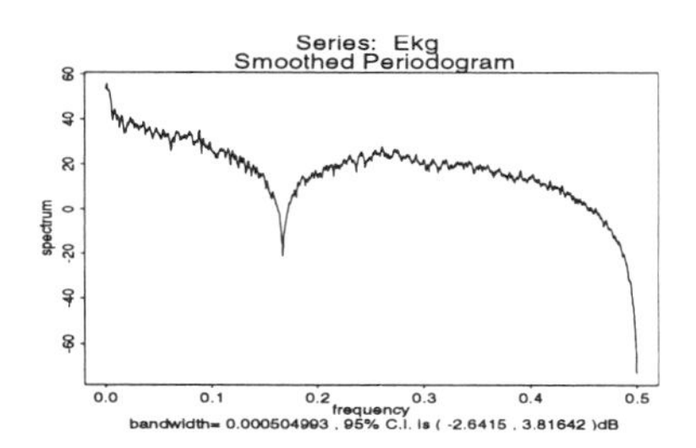

Figure 1.8. Successive beat-to-beat intervals for a normal patient. Smoothed periodogram. The separation between high frequency (0.2 – 0.5 Hz) mid-frequency (0.07 – 0.15 Hz) appears clearly and the peak at low frequency (0.01 – 0.05 Hz) is also obvious.

fetal distress and the need for rapid delivery. Usually, the power in the spectrum is distributed in three separate regions: high frequency (0.2 – 0.5 Hz), mid-frequency (0.07 – 0.15 Hz), and low frequency (0.01 – 0.05 Hz), and any departure from this distribution is regarded as anomalous and subject to further investigation and action. Obviously, the availability of reliable spectral estimators is of crucial importance for this type of medical application. See [237], [300] and the references therein for details.

1.4.3 The Correlogram-Periodogram Duality

We now come back to the discussion of the two possible definitions of the autocovariance function of a (finite) time series. Notice that there is no significant difference between the two candidates when the size N of the data is large. We nevertheless justify our choice on the basis of mathematical consistency. The second definition $\tilde{c}_f$ in (1.35) is the natural candidate from the point of view of the statistical analysis discussed in the next section when we deal with samples f_j's of random variables forming a stationary stochastic process. Indeed, this second definition provides an unbiased estimate of the true autocovariance function. Despite all that, we choose the first definition and we use c_f for the sample autocovariance function. The reason for this choice is the following: except for the pathological case for which all the f_j's are equal to each other, the sequence $c_f = \{c_f(j)\}_j$ defined by (1.33) when $|j| < N$ and 0 otherwise is *nonnegative definite.* Consequently, Bochner's theorem implies the existence of a nonnegative (finite) measure ν_f on $[0,1]$ whose Fourier transform is the sequence c_f. This measure is obviously absolutely continuous (recall that $c_f(j)$ is nonzero for at most finitely many j's.) We denote its density by

$\nu_f(\omega)$.

$$\nu_f(\omega) = \sum_{j=-(N-1)}^{N-1} c_f(j) e^{2i\pi j\omega}, \qquad \omega \in [0, 1]. \tag{1.43}$$

This density $\nu_f(\omega)$ is called the *sample spectral density* because it is equal to the quantity defined earlier with the same name. Indeed, it is possible to use simple trigonometric manipulations to show that the function $\nu_f(\omega)$ defined by formula (1.43) satisfies

$$\nu_f(\omega) = \frac{1}{N} \left| \sum_{j=0}^{N-1} f_j e^{-2i\pi j\omega} \right|^2 .$$

The content of this statement is that

the sample autocovariance function c_f *and the sample spectral density* $\nu_f(\omega)$ *form a Fourier pair*

in the sense that they are Fourier transforms of each other. This is very important both from the conceptual and the practical (i.e., computational) point of views.

1.5 Time-Frequency Representations

Our goal is now to introduce the concept of a time-frequency representation. Its introduction may be traced back to the mid-1940s. It originated in the work of Wigner [293] in quantum mechanics. It can also be found in the implementation of these ideas in the context of signal analysis by J. Ville [288] and throughout the fundamental work of Gabor [139].

1.5.1 The Concept of Time-Frequency

Specifying a time domain representation and/or a frequency domain representation of a signal is not necessarily the best first step of the analysis. Indeed, the limitations of these representations can greatly hinder our understanding of the properties of the signal. Let us consider, for example, a musical signal. The human ear interprets it as a series of notes, i.e., a series of "atoms of sound" appearing at given times, each having a finite duration and a given height (the frequency of the tone). If the signal contains a given note once, say A, a Fourier representation of the signal will exhibit a peak at the corresponding frequency, without any indication of location or duration. As stressed by J. Ville, "the representation is mathematically

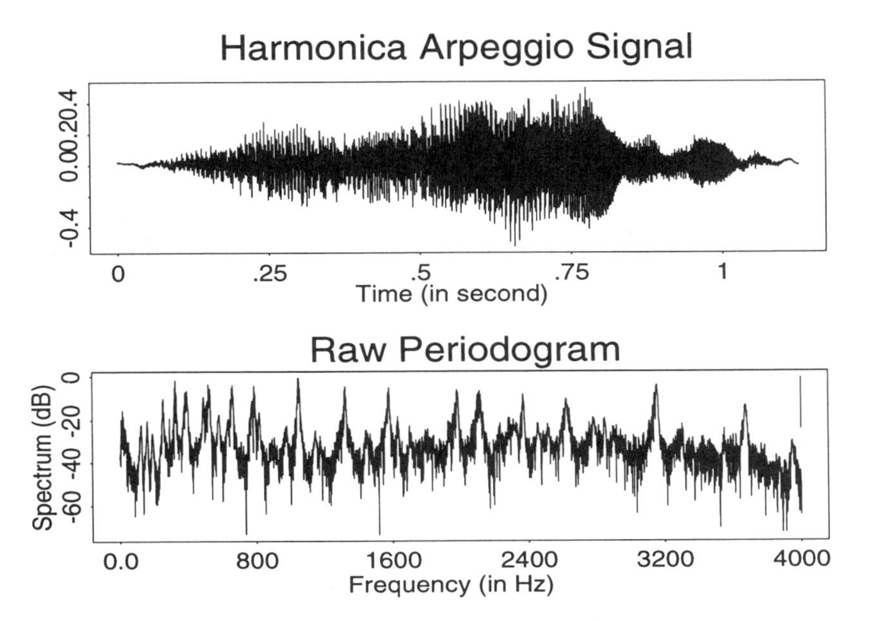

Figure 1.9. Harmonica arpeggio signal and its power spectrum.

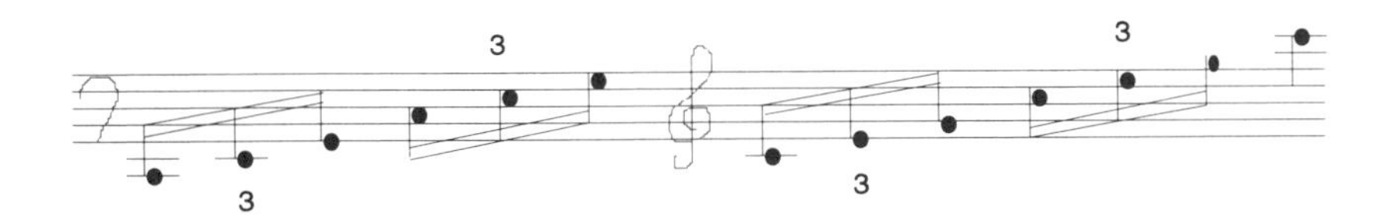

Figure 1.10. Score for the arpeggio signal.

correct because the phases of the tones close to A have managed to suppress it by interference phenomena before it is heard, and to enforce it, again by interferences, when it is heard .... However there is here a deformation of reality: when the A is not heard, it is simply because it is not emitted ...". Music is generally represented on a score. We (and many others) claim that time-frequency is precisely the information which is coded in the musical notation (see Figure 1.10, where an arpeggio is displayed) and this is why one may then say that the musical notation is the prototype of time-frequency representations of signals.

When the signal displayed in Figure 1.10 is played with an instrument (say with a harmonica), it is difficult to envision its mathematical structure. The latter is not always easy to detect. Indeed, the plots of the signal and

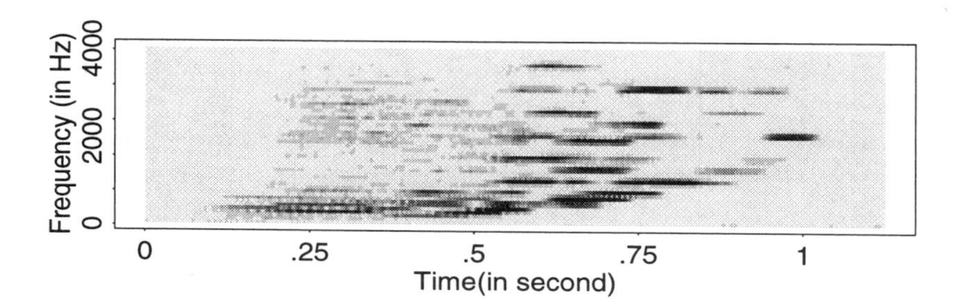

Figure 1.11. Periodogram of the harmonica arpeggio signal.

its Fourier spectrum (which we will define later in the book) which are displayed in Figure 1.9 are not capable of revealing this fine structure. On the contrary, the fact that it is an arpeggio is much clearer in a time-frequency representation such as a periodogram displayed in Figure 1.11, in which the time and the frequency variables are displayed simultaneously. We can clearly see that the time-frequency representation of the signal under consideration is localized around a certain number of points, and it is tempting to associate such points with the individual notes (or their harmonics.) We can also follow the evolution of the fundamental frequency of the signal. The latter seems to follow a curve that may be put in correspondence with the score in Figure 1.10. Of course, the time frequency representation given in Figure 1.11 is not as simple as the score given in Figure 1.10. But extracting salient features of signals from such time-frequency representations is precisely what time-frequency analysis is about! The purpose of this book is to give an overview of some of the techniques developed in this context, and to emphasize the statistical nature of most of the processing involved.

Throughout this text, we shall be concerned with mathematical idealizations of such time-frequency representations. In other words, we shall describe methods for representing functions or signals simultaneously in terms of a time variable and a frequency (or scale) variable. Let us immediately point out that, because of the Heisenberg uncertainty principle already discussed, obtaining an intrinsic and infinitely precise description of the "time-frequency content" of a signal is out of the question. Some arbitrariness is necessarily introduced into the analysis. Let us emphasize some of the main features of time-frequency representations:

- First, there is no uniqueness of a time-frequency representation: there are many different ways of describing the "time-frequency content" of a signal. We shall see several illustrations of this point with a particular emphasis on speech examples.

- Second, for a given time-frequency representation, it is impossible to achieve perfect time-frequency localization, because of the Heisenberg uncertainty principle. This means that we shall always have to look for a compromise between time localization and frequency localization. Nevertheless, we shall describe in Chapters 6 and 7 some attempts to "fool" Heisenberg's uncertainty restrictions.

Throughout the text, we describe several approaches to time-frequency representation, as well as some applications to specific problems. But before going into more specifics we review some practical situations in which time-frequency methods are clearly relevant.

1.5.2 Instantaneous Frequency and Group Delay

Let us start with some deterministic examples. Signals with time-varying frequencies are the best examples to show the limitations of standard Fourier analysis. The notion of frequency has an obvious physical interpretation, and it is a major success of Fourier analysis to be able to account for this notion. However, Fourier analysis has a hard time accounting for frequencies varying with time. From a purely intuitive point of view, the latter are as relevant as constant frequencies (think about a violin "glissando"!). Since Ville's original paper [288], the (desperate) quest for the *instantaneous frequency* (and its dual characteristic the *group delay*) has been a major goal of the signal analysis community. Associating an instantaneous frequency to a signal amounts to expressing it in the form

$$f(x) = a(x) \cos \varphi(x) \tag{1.44}$$

and defining the frequency as the derivative of the phase function φ (up to a factor of 2π). Unfortunately, it is obvious that such a representation is far from unique. For example, if we set $A = \sup |f(x)|$, then one can find $\phi(x)$ such that $f(x) = A \cos \phi(x)$, which is another representation of the type (1.44). As a consequence, one has to identify a canonical representation among all the possible representations. This may be done by using the so-called *analytic signal* Z_f of the signal f defined in (1.20), and writing

$$Z_f(x) = A(x) e^{i\phi(x)} \ . \tag{1.45}$$

The instantaneous frequency is then defined by

$$\omega(x) = \frac{1}{2\pi} \frac{d\phi(x)}{dx} \ . \tag{1.46}$$

We refer to [53, 54, 242, 243] for a careful analysis of the notion of instantaneous frequency. The group delay is defined in a similar way,

$$\tau(\xi) = -\frac{1}{2\pi}\frac{d \arg \widehat{Z_f}(\xi)}{d\xi}, \tag{1.47}$$

and may be intuitively interpreted as the time of appearance of the frequency ξ. It is sometimes called the "dual quantity" since it provides a time as a function of frequency, whereas the instantaneous frequency gives a frequency as a function of time; however, except in specific cases, they are *not* inverse of each other. Such quantities are of crucial importance in physical applications. For instance, in radar and sonar applications, the group delay corresponds to the true propagation time of a signal, the speed of propagation being the speed of the group.

1.5.3 Nonstationary and Locally Stationary Processes

For the purpose of the present discussion we use freely (i.e., without defining them) terms and concepts which we will introduce only in Chapter 2 when we consider in detail the spectral analysis of stationary random processes. When a stochastic process is stationary, its autocovariance defines a convolution operator which is diagonalized by Fourier expansions. This is the starting point of all nonparametric spectral estimation methods.

In a number of practical situations, the autocovariance function does not give rise to a convolution operator and the problem of spectral estimation cannot be solved by standard Fourier analysis. Stationarity may be broken in several different ways. Let us mention two cases where time-frequency (or time-scale) representations are of some help.

The first one is the case of *locally stationary processes*, i.e., processes to which a local spectrum slowly varying in time may be associated. Such processes have been studied in [213, 248, 249], and it has been shown that time-frequency representations provide powerful tools for the corresponding *local spectral estimation* problems (see also [Flandrin93].)

The second one is the case of *self-similar processes* such as fractional Brownian motion or fractional Brownian noise [217], discussed for example in [108, 34, 35]. These processes have stationary increments and we shall see that after wavelet transform, the nonstationarity essentially disappears.

1.5.4 Wigner-Ville and Related Representations

One of the most popular time-frequency representations is provided by the so-called *Wigner-Ville transform*. Given a signal f, its Wigner-Ville transform is defined by

$$W_f(b,\omega) = \int f\left(b+\frac{x}{2}\right)\overline{f\left(b-\frac{x}{2}\right)}e^{-i\omega x}dx\ . \tag{1.48}$$

The Wigner-Ville transform $W_f(b,\omega)$ essentially amounts to considering inner products of copies $f(x+b)e^{-i\omega x}$ of the original signal shifted in time-frequency with the corresponding reversed copies $f(-x+b)e^{i\omega x}$. Simple geometrical considerations show that such a procedure provides insights on the time-frequency content of the signal.

Ville proposed this choice to make sense of the notion of instantaneous spectrum of a signal. Indeed, if we assume momentarily that the signal is random, taking the expectation in both sides of (1.48), we get

$$\mathbb{E}\{W_f(b,\omega)\} = \int C_f(b+x/2, b-x/2)\exp\{-i\omega x\}dx$$

if we use the notation C_f for the autocovariance function of the process (see Chapter 2 for the definition). In the stationary case, this quantity is independent of b and it is equal to the spectral density. See formula (2.6) in Chapter 2. We shall also come back to that in Chapter 6. Ville's original suggestion was to consider the Wigner-Ville transform of a signal as a probability measure on the time-frequency plane. This proposal proved to be inadequate, because of the non positivity of the representation. However the Wigner-Ville transform has a number of interesting properties, which we list here for the sake of completeness (the terms "Dirac signal" and "chirp" will be defined in Chapter 3 when we give the first examples of Gabor transform computations). Among these, the "localization properties" are considered to be of great importance. They express that the transform is well adapted to some specific classes of signals; for example, one may say that the Fourier representation is optimal for sines and cosines, in the sense that all the energy of the Fourier transform of such functions is localized in a single point. Optimality in this sense makes easier further tasks such as denoising or detection for example. Achieving such an "optimal localization" for different classes of signals is one of the goals of time-frequency analysis.

1. The Wigner-Ville transform is optimally localized in the time domain for "Dirac signals."
2. The Wigner-Ville transform is optimally localized in the frequency domain for pure monochromatic waves, and for "linear chirps." In particular, if $f(x) = \exp\{i\lambda(x+\alpha x^2/2)\}$, then we have

$$W_f(b,\omega) = 2\pi\delta(\omega-\lambda(1+\alpha b))\ .$$

3. The time marginal reproduces the power spectrum:

$$\int W_f(b,\omega)db = |\hat{f}(\omega)|^2 \ .$$

4. The frequency marginal reproduces the square modulus of the signal:

$$\int W_f(b,\omega)d\omega = 2\pi|f(b)|^2 \ .$$

The Wigner-Ville representation was generalized to a large class of time-frequency representations called the *Cohen's class* representations,

$$W_f(w;b,\omega) = \int e^{i\xi(x-b)} w(t,x,\xi) f\left(b+\frac{x}{2}\right) \overline{f\left(b-\frac{x}{2}\right)} e^{-i\omega t} dx\, dt\, d\xi \ , \tag{1.49}$$

which depend upon the choice of a weight function $w(t, x, \xi)$. Different choices of the weight function yield representations with different properties. Other possible generalizations may be introduced as well. Generally, the goal is to propose a representation which would be optimal for given classes of signals (say, signals with a certain type of frequency modulation). We will not elaborate on those generalizations. However, we have to stress the fact that the field of bilinear time-frequency representations is a very active field of research and that a significant part of the publications in time-frequency methods are devoted to the construction of representations with certain optimal properties (concerning, for example, invariance properties of the transform, properties of the marginals, time-frequency localization for certain types of signals, ...). For more details on such aspects, we refer to [Flandrin93] and [83, 293, 288, 53, 54, 164] and references therein.

1.5.5 The Ambiguity Function

The ambiguity function of a signal of finite energy is naturally associated to its Wigner-Ville transform. It is defined by the formula

$$A_f(\xi, x) = \int f(b+\frac{x}{2})\overline{f(b-\frac{x}{2})}e^{-i\xi b}\, db. \tag{1.50}$$

The ambiguity function essentially amounts to taking inner products of the signal with time-frequency-shifted copies of itself. As such it gives an indication on the spreading of the signal over the time-frequency plane. Moreover, it is related to the Wigner representation by a simplectic Fourier transform:

$$A_f(\xi, x) = \frac{1}{2\pi}\int W_f(b,\omega)e^{i(\omega x-\xi b)}\, db\, d\omega. \tag{1.51}$$

The importance of the ambiguity function comes from its application to radar technology.[2] We present an oversimplified description of such an application in the hope of illustrating the importance of the role of this function. We shall assume for the purpose of this discussion that the purpose is to detect a single object with range r and velocity v, so for us $r = r_0 + tv$. Let us consider a function of time $s(t)$ as a pulse transmitted by a radar system, let us denote by $e(t)$ the amplitude of the return (i.e., the echo received), and as before we denote by Z_f the analytic signal associated to a signal of finite energy f. Notice that we use the notation t for the variable because of the obvious interpretation as time. The quantities

$$t_0 = \int_{-\infty}^{+\infty} t|Z_s(t)|^2\, dt \qquad \text{and} \qquad \xi_0 = \int_{-\infty}^{+\infty} \xi|\hat{Z}_s(\xi)|^2\, d\xi \tag{1.52}$$

are called the epoch and the carrier frequency of the pulse s. The return is assumed to be of the form

$$e(t) = Kf\left(t - \frac{2r}{c+v}\right) \tag{1.53}$$

for some constant $K > 0$, where c denotes the speed of electromagnetic waves and the echo is easily seen to be given by the formula

$$e(t) = \Re[Z_e(t)] = \Re\left[\sqrt{a}e^{-i\xi_0 x_0} u(at - t_0 - x_0) e^{-i\eta_0 t} e^{i\xi_0 t}\right], \tag{1.54}$$

provided we set $a = (c - v)/(c + v)$, and

$$x_0 = \frac{2r_0}{c+v} \qquad \text{and} \qquad \eta_0 = \frac{2\xi_0 v}{c+v}.$$

The function u is defined by

$$u(t) = Z_e(t + t_0)e^{-i\xi_0(t+t_0)}.$$

The quantities x_0 and η_0 are the time delay in the echo due to the range of the object and the Doppler or frequency shift in the echo due to the velocity of the object. In the regime of v much smaller than c, a can be approximated by 1 and the echo e can be realistically replaced by

$$e(t) \approx \Re\left[e^{-i\xi_0 x_0} u(t - t_0 - x_0) e^{-i\eta_0 t} e^{i\xi_0 t}\right].$$

Obviously, x_0 and η_0 determine the quantities r_0 and v uniquely, and finding the position and the velocity of the object can be achieved by estimating

[2] Our definition is not exactly the one used in radar theory. Our choice is motivated by the definition of the Wigner-Ville transform and the desire of expressing their relationship in a simple way.

x_0 and η_0. This is done by computing the inner product of the return $e(t)$ with all the possible echoes $e_{x,\eta}$ defined by

$$e_{x,\eta}(t) = \Re[Z_{e_{x,\eta}}(t)] = \Re\left[e^{-i\xi_0 x}u(t-t_0-x)e^{-i\eta t}e^{i\xi_0 t}\right]. \tag{1.55}$$

This is done by computing

$$I(x,\eta) = \left|\int_{-\infty}^{+\infty} Z_e(t)\overline{Z_{e_{x,\eta}}(t)}\,dt\right|^2,$$

and since

$$I(x_0,\eta_0) = 1 \qquad \text{and} \qquad I(x,\eta) \le 1,$$

one expects that an image with the gray level of a pixel (x,η) given by the value $I(x,\eta)$ will *image* the object that the radar is trying to detect. In fact, a simple calculation shows that

$$I(x,\eta) = |A_u(x_0-x,\eta_0-\eta)|^2,$$

and consequently a radar image can be viewed as a surface plot of an ambiguity function.

Remark 1.2 The optimization of the waveform so as to facilitate the interpretation of the radar image is a classical problem in radar theory. Clearly, the more localized the ambiguity function, the easier the detection. Therefore, one often prefers to use pulses $s(t)$ which have a strongly localized ambiguity function. We shall come back to this point in a different context when discussing the time-frequency analysis of nonstationary processes.

1.5.6 Linear Time-Frequency Representations

We have seen in the last subsection some of the interesting properties of the Wigner-Ville transform. Such transforms, however, suffer from a certain number of drawbacks. The latter are all consequences of the bilinearity of the transform. In particular, *the Wigner-Ville transform of the sum of two signals is* not *the sum of the Wigner-Ville transforms.* This means, for example, that if f_1 and f_2 are two signals whose Wigner-Ville transforms have optimal localization, the Wigner-Ville transform $W_{f_1+f_2}(b,\omega)$ contains an additional "ghost" term,

$$W_{f_1,f_2}(b,\omega) = 2\Re\int f_1\left(b-\frac{x}{2}\right)\overline{f_2\left(b+\frac{x}{2}\right)}e^{-i\omega x}dx.$$

In fact, if f is a real-valued signal, it contains positive and negative frequencies, and its Wigner-Ville transform will present a DC term, a consequence of the interferences between positive and negative frequencies.[3] Such an interference term contains information, but this information is quite difficult to understand, except in simple cases (see, for instance, [165, 186, 134] for a detailed analysis of the geometry of ghost terms). A huge number of publications have been devoted to the main drawbacks of the transform, trying either to provide an interpretation of the ghost terms, or to remove them by an appropriate smoothing (unfortunately decreasing the time-frequency resolution). For these reasons, we shall not pursue any further the discussion of the bilinear representations and turn instead to an alternative that presents a certain number of advantages (and also a few drawbacks, as we shall see later). We believe that this choice is much more convenient for a large number of applications (but this belief might simply be a consequence of our religious convictions ...).

We shall focus on the so-called *linear time-frequency representations.* They are called linear because they depend linearly upon the signal, i.e., are obtained via the action of a linear operator, say

$$\mathcal{L} : L^2(\mathbb{R}) \hookrightarrow \mathcal{H} ,$$

where $\mathcal{H}$ is the target space of $\mathcal{L}$, in general a space of functions of the time and frequency variables which are square-integrable with respect to a given measure. $\mathcal{L}$ is generally constructed as follows (this general construction is known as coherent states construction in theoretical physics): Let $\{\psi_\lambda \in L^2(\mathbb{R});\ \lambda \in \Lambda\}$ be a set of functions labeled by an index λ, in general a two-dimensional time-frequency parameter. Then if $f \in L^2(\mathbb{R})$, $\mathcal{L}$ is given by

$$[\mathcal{L}f](\lambda) := \langle f, \psi_\lambda \rangle , \qquad \lambda \in \Lambda.$$

This function of the time-frequency variables λ is viewed as a time-frequency transform of f. Time-frequency transforms may actually be constructed in many different ways. There are several natural requirements on which to insist in order that such a transform be useful.[4] Among these requirements, the most important ones are

- *Completeness*: The function $\mathcal{L}f$ must contain all the information about f. Equivalently, there must exist a way of "reconstructing" f

[3]This is why most of the time, the Wigner-Ville transform of the analytic signal is preferred to that of the signal itself.

[4]As before we use the terminology which is commonly in use among those who developed the theory. In particular, the term covariance used later does not have the meaning it has in probability and in statistics. It refers to the action of groups of transformations at the level of the signals and at the level of the transforms of these signals. This use of the terminology is standard in many parts of mathematics, including geometry and group theory.

from $\mathcal{L}f$. The inversion formula in general takes the form

$$f(x) = \sum_{\lambda} \mathcal{L}f(\lambda)\ \chi_\lambda(x)\ ,$$

where $\{\chi_\lambda;\ \lambda \in \Lambda\}$ is another set of functions labeled by the time-frequency parameter λ (possibly identical to the ψ_λ's), and the sum may be a discrete sum or an integral. The existence of such simple inversion formulas is a severe limitation for the existence of the transforms.

- *Covariance*: To be easily interpretable, the time-frequency transform must enjoy some covariance properties. The most common one involves time translations: the transform of a shifted copy of a given signal is the corresponding shifted copy of the transform of the original signal. This is a fairly natural assumption. In particular, there must be a way of defining the action of shifts on the λ variable. Covariance requirements are strong assumptions, which limit the possible choices of the transforms. In general, one insists on shift covariance, frequency shift, and/or scaling covariances. Such choices lead to the continuous Gabor and wavelet transforms, respectively.

Since Λ is usually chosen to be a group, the implementation of these ideas leads naturally to a group theoretical framework. We shall emphasize this feature from time to time. Throughout this text, we shall consider several examples of time-frequency representations for which these requirements are fully or partly satisfied. Most of our arguments will be reserved to wavelet or Gabor transforms. However, we shall make a few detours to visit other transforms.

1.5.7 Representing a Time-Frequency Transform

A very important aspect of time-frequency analysis is in the interpretation of the transforms. Most of the time this preliminary step precedes any serious work. Also, contrary to the case of bilinear representations which are intrinsic to the signal, linear representations depend on the choice of the analysis tools, namely the functions ψ_λ. For this reason some experience is needed in order to understand which features of the transform are intrinsic to the signal and which ones actually come from the tool. These aspects have been analyzed in [150].

This book contains a large number of illustrations, which intend to provide insights for interpreting time-frequency transforms. We shall make use of graphical representations of the time-frequency transforms. We represent the values $\mathcal{L}f(\lambda)$ of the continuous time-frequency distributions with gray-level images. Since $\mathcal{L}f(\lambda)$ is typically a complex number, we shall find it convenient to use two separate gray-level images. One represents the modulus $|\mathcal{L}f(\lambda)|$ and the other the phase, i.e., the argument arg $\mathcal{L}f(\lambda)$ of the time-frequency transform. Throughout this book, we shall essentially focus on the analysis and the interpretation of the "time-frequency localization" properties of functions and signals. One aspect will be the interpretation of a given time-frequency representation. Another aspect is the choice of the representation which makes easier the analysis of a given signal. For that reason, we focus on wavelet and Gabor transforms, which are the simplest time-frequency representations.

Remark 1.3 At this point, let us make it clear that interference (or "ghost") terms are *not* completely absent from the representations we termed "linear representations." After all, both $|\mathcal{L}f(\lambda)|$ and arg$\mathcal{L}f(\lambda)$ are nonlinear. In some cases, $|\mathcal{L}f(\lambda)|^2$ may appear as some particular smoothing of the Wigner-Ville function $W_f(\lambda)$; see, e.g., Flandrin's book [Flandrin93]. The main difference is that if the functions ψ_λ are localized in the time-frequency plane in a sense to be specified later, the ghost terms will be more localized and easier to detect and interpret. Let us consider as before the example of the sum $f = f_1 + f_2$ of two signals. Then W_f contains the cross term W_{f_1,f_2}, independently of the localization of f_1 and f_2 in the time-frequency space. In this case, $|\mathcal{L}f|^2$ also contains a cross term $2\,\Re\mathcal{L}f_1\overline{\mathcal{L}f_2}$, but this term is small whenever $\mathcal{L}f_1$ and $\mathcal{L}f_2$ are localized in different regions in time-frequency.

1.6 Examples and S-Commands

We give the details of the S-commands used to produce the plots and the results presented above for the dolphin clicks data.

The data was read from the disk and plotted with the commands:

Example 1.3 Dolphin click
Read and plot the dolphin click data:

```
> CLICK <- scan(''signals/click.asc'')
> FCLICK <- fft(CLICK)
> par(mfrow = c(2,1))
> tsplot(CLICK)
> tsplot(Mod(FCLICK))
```

The subsampling of the data and the plots of Figure 1.3 were obtained in the following way:
Subsampling of the data:

```
> K <- 1:1249
> SCLICK <- CLICK[2*K]
```

```
> tsplot(CLICK)
> tsplot(SCLICK)
```

Finally, the computation of the (discrete) Fourier transform of the subsampled signal, the rearrangement of the outputs of the function `fft` and the plots of Figure 1.4 were produced by the commands:

Fourier transforms of the original and click data:

```
> FSCLICK <- fft(SCLICK)
> NFCLICK <- c(FCLICK[1250:2499],FCLICK[1:1249])
> NFSCLICK <- c(FSCLICK[625:1249],FSCLICK[1:624])
> tsplot(Mod(NFCLICK),xaxt="n")
> axis(1,at=c(0,500,1000,1500,2000,2500),
+            labels=c("-1250","-750","-250","250","750","1250"))
> tsplot(Mod(NFSCLICK),xaxt="n")
> axis(1,at=c(0,200,400,600,800,1000,1200),
+            labels=c("-600","-400","-200","0","200","400","600"))
>par(mfrow=c(1,1))
```

1.7 Notes and Complements

The elements of Fourier theory can be found in many textbooks. We shall refer to some of the classics such as the book of Riesz and Nagy [257] or Zygmund's text [302]. A more modern presentation can be found in the second volume of the Reed and Simon series [255] or Dym and McKean [119] for a presentation more in the spirit of the spectral theory of Chapter 2 of this book.

The important problems of sampling and aliasing are discussed in detail in many textbooks. The statistically inclined reader will certainly enjoy the presentation of [239], whereas we would recommend [236] to an electrical engineer student.

A detailed account of Wiener's deterministic spectral theory is given in [195]. The spectral theory of deterministic time series can be found in a some textbooks on the statistical theory of time series (usually in the first few chapters of the book). Among them [235] and [239] are our favorites. It can also be found in most of the discrete-time signal processing textbooks on the electrical engineering shelves of any library, and we selected [236] for reference.

The study of time-frequency representations started in the mid-1940s with the two pioneer papers by D. Gabor [139] (1946) and J. Ville [288] (1948) (even though, according to [Flandrin93], a notion of instantaneous spectrum could already be found in Sommerfeld's thesis (1890)). These two approaches eventually led to the two classes of time-frequency representations which are most popular now, namely the linear and bilinear transforms. The number of contributions to the so-called "time-frequency problem" is very large and instead of trying to list them exhaustively we prefer to refer to Section 2.1 of [Flandrin93] for a lucid historical overview. In addition, the interested reader may consult the books by L. Cohen [L.Cohen95] and S. Qian and D. Chen [Qian96], or the articles [82, 83, 164] for more details on the time-frequency problem. See also the contribution of Priestley [248, 249] for related approaches in a probabilistic context.

Our discussion of the ambiguity function was modeled after the 1960 technical report by Wilcox [294] as published in the IMA volume [50].

Chapter 2

Spectral Theory of Stationary Random Processes: A Primer

This chapter is devoted to a traditional topic of the statistical culture: time series analysis. The latter is concerned with repeated measurements on the same phenomenon at different times. The goal is to take advantage of the dependence between the measurements at different times to predict, forecast, $\cdots$. Obviously, the presence of correlations in the measurements makes the statistical analysis of the time series more difficult and more challenging. We shall revisit the deterministic spectral theory of time series as presented in Chapter 1; all the descriptive statistics introduced earlier will now appear as statistical estimators computed for a particular realization of a random signal, and we shall present a systematic study of their statistical properties.

2.1 Stationary Processes

As we already remarked, in most real-life applications a signal comes in the form of a finite sequence $f = \{f_0, f_1, \ldots, f_{N-1}\}$ of real numbers. A signal is random if the numbers f_j's can be considered as outcomes of a (finite) sequence of random variables. One of the simplest examples of random signal is given by observations of a deterministic signal in the presence of an additive noise. In this case the observations are of the form

$$f_j = f_j^{(0)} + \epsilon_j, \qquad j = 0, 1, \cdots, N-1, \tag{2.1}$$

where $f^{(0)} = \{f_0^{(0)}, f_1^{(0)}, \ldots, f_{N-1}^{(0)}\}$ is the deterministic source and $\epsilon = \{\epsilon_0, \epsilon_1, \cdots, \epsilon_{N-1}\}$ is the noise perturbation. We shall devote the last chapter of the book to the analysis of this class of noisy signals, analysis meaning retrieval of the signal $f^{(0)}$ from the noisy observations f_j's. But for the time being we concentrate on the statistical analysis of the random component of these *signal + noise models.*

The goal of any mathematical theory of random time series is to develop inference tools capable of predicting, from the mere observation of one time series data set, what kind of data could have been observed. Standard approaches introduce models assumed to be driving the mechanisms generating the observations. Most of these models are based on the notion of time homogeneity of the signals. This translates mathematically into the notion of stationarity. We present the basic elements of the classical theory of stationary processes. But, because we are mostly interested in nonstationary signals for which the classical theory fails, we shall not spend much time on the inferential and the predictive parts of the stationary theory.

We develop mathematical tools and we provide computer programs to analyze the *finite* signals, but according to the discussion of Section 1.2, it is most convenient to consider the set of N samples f_j's as a finite part of a doubly infinite sequence $\cdots .f_{-1}, f_0, f_1, \cdots$, which can in turn be regarded as the set of regular samples from a continuous time signal observed at discrete times. In fact, we assume that there exists a function f of the continuous variable $x \in \mathbb{R}$ such that

$$f_j = f(x_j), \qquad j = 0, 1, \ldots, N-1, \tag{2.2}$$

where the x_j are the times of the measurements. We shall only consider regular samplings provided by observations at times $x_j = x_0 + j\Delta x$ for some fixed sampling interval Δx. A random signal $f = \{f(x); x \in \mathbb{R}\}$ is said to be *stationary in the strict sense* (stationary for short) if for every choices x and $x_1 < \cdots < x_k$ of real numbers the random vectors

$$(f(x_1), \cdots, f(x_k)) \qquad \text{and} \qquad (f(x_1 + x), \cdots, f(x_k + x))$$

have the same distributions in $\mathbb{R}^k$ (or in $\mathbb{C}^k$ if we are dealing with a complex signal). All the statistics of such a random process are thus invariant under shift. In particular the first moment $m(x) = \mathbb{E}\{f(x)\}$ (whenever it does exist) is constant since it must be independent of the variable x.

In the case of a finite signal the notion of stationarity can be defined intrinsically without having to appeal to functions of a continuous variable. It takes the following form. The random signal $f = \{f_0, f_1, \cdots, f_{N-1}\}$ is said to be stationary if for every choices j and $j_1 < \cdots < j_k$ of integers in $\{0, 1, \cdots, N-1\}$ the random vectors

$$(f_{j_1}, \cdots, f_{j_k}) \qquad \text{and} \qquad (f_{j_1+j}, \cdots, f_{j_k+j}) \tag{2.3}$$

have the same distribution in $\mathbb{R}^k$ or in $\mathbb{C}^k$. Here the integer addition is understood *mod N.* In other words, finite signals of length N are regarded as parameterized by the finite (quotient) group $\mathbb{Z}_N = \mathbb{Z}/N\mathbb{Z}$ of integers modulo N, and the stationarity of finite random signals is defined as a property of invariance of the distribution with respect of the parameter shifts of this group. If we consider the case of the *signal + noise model* given in (2.1), the stationarity assumption is obviously satisfied for the noise term $\epsilon = \{\epsilon_0, \epsilon_1, \ldots, \epsilon_{N-1}\}$, but it fails for the full signal $f = \{f_0, f_1, \ldots, f_{N-1}\}$ if the mean signal $f^{(0)} = \{f_0^{(0)}, f_1^{(0)}, \cdots, f_{N-1}^{(0)}\}$ is not constant.

The notion of stationarity of a doubly infinite random sequence $f = \{f_j;\ j \in \mathbb{Z}\}$ can be defined similarly by requiring that the random vectors in (2.3) have the same distribution, the addition of the indices being now the regular integer sum.

There is still a weaker notion of stationarity. We present it now in the case of processes indexed by the continuous variable x even though the definition obviously extends to the two other types of processes we have been dealing with. A random process $f = \{f(x); x \in \mathbb{R}\}$ of order 2 (i.e., such that all the random variables $f(x)$ are square integrable) is said to be *stationary in the wide sense* if its mean,

$$m_f(x) = \mathbb{E}\{f(x)\},$$

is independent of x, i.e., $m_f(x) = m$ for a constant m, and if its autocovariance function

$$C_f(y, x) = \mathbb{E}\{(f(y) - m)\overline{(f(x) - m)}\} \tag{2.4}$$

is a function of the difference $y - x$, i.e., if

$$C_f(y, x) = C_f(y - x) \tag{2.5}$$

for a function of one variable which we still denote by C_f. This function is nonnegative definite in the sense of (1.29), and using the same argument based on Bochner's theorem one can conclude the existence of a nonnegative finite measure ν_f on $\mathbb{R}$ satisfying (1.30). This measure ν_f is called the spectral measure of the stationary process $f = \{f(x);\ x \in \mathbb{R}\}$. Formula (1.30) states that the covariance function C_f is the Fourier transform of the spectral measure ν_f.

In most practical applications the covariance function C_f is integrable over $\mathbb{R}$ and the introduction of the spectral measure can be done in a direct manner without appealing to Bochner's theorem. Indeed, under this integrability condition, the Fourier transform

$$\mathcal{E}_f(\xi) = \int C_f(x) e^{-ix\xi}\, d\xi \tag{2.6}$$

is well defined as a function on the real line $\mathbb{R}$. Notice that, because the autocovariance function C_f is nonnegative definite, the function ν_f is nonnegative. It is called the spectral function or spectral density of the process. It is in fact the density of the spectral measure whose existence is given by Bochner's theorem in the sense that $\nu_f(d\xi) = \nu_f(\xi)d\xi$ and this justifies our use of the same notation for the spectral measure and the spectral function. We shall pursue this discussion in Chapter 6. The autocorrelation function ρ_f defined by the formula

$$\rho_f(y,x) = \frac{C_f(y,x)}{\sqrt{C_f(y,y)}\sqrt{C_f(x,x)}} \tag{2.7}$$

also becomes a function of the difference $y - x$ in the case of stationary processes and in these cases one has $\rho_f(y,x) = \rho_f(y-x)$ with:

$$\rho_f(y-x) = \frac{C_f(y-x)}{C_f(0)}.$$

The definition of stationarity in the wide sense for finite random signals or doubly infinite random sequences can be given in the same spirit. We shall always assume that the mean of a wide sense stationary signal is 0. Indeed, the mean of such a process is necessarily constant and this constant can be subtracted from the process to make it mean zero. One of the reasons of our interest in stationary processes is their appearances as models for the noise term in the *signal + noise* models of deterministic signals perturbed by an additive noise and in this case the mean zero assumption is quite natural.

2.1.1 Comparison with Wiener's Theory

It is instructive to compare the elements of spectral theory introduced so far in the present approach to Wiener's spectral theory as presented in Section 1.3. It is the content of the classical ergodic theorem (which should be viewed as a mere generalization of the law of large numbers) that the limit (1.26) exists if f is a stationary process. Moreover, if the process f is ergodic (that is, if the shift-invariant functions of the process are almost surely constant), this limit takes place in the sense of almost sure convergence, it is deterministic, and it is equal to the autocovariance $C_f(\tau)$ defined by (2.4) and (2.5) when $y = x + \tau$. In other words, the spectral theory of stationary processes contains Wiener's theory when the process in question is ergodic. Moreover, the sample estimates computed from one realization of the series, for example, the sample mean $\overline{f}$ defined in (1.34) or the sample autocovariance function c_f defined in (1.33) and (1.34), or even the sample autocorrelation function ρ_f defined in (1.36), appear as estimates of their theoretical counterparts m, C_f, and ρ_f given earlier.

2.1.2 A First Example: White Noise

Let $\epsilon = \{\epsilon_0, \epsilon_1, \ldots, \epsilon_{N-1}\}$ be a (finite) sequence of uncorrelated mean zero random variables with common variance σ^2. In other words

$$\mathbb{E}\{\epsilon_j\} = 0 \qquad \text{and} \qquad \mathbb{E}\{\epsilon_j \epsilon_k\} = \begin{cases} \sigma^2 & \text{if } j = k, \\ 0 & \text{if } j \neq k. \end{cases} \tag{2.8}$$

Such a sequence is weakly stationary since its autocovariance function C_ϵ is given by

$$C_\epsilon(j) = \begin{cases} \sigma^2 & \text{if } j = 0, \\ 0 & \text{if } j \neq 0. \end{cases} \tag{2.9}$$

Consequently, the spectral function (recall that this is the sequence defined as the Fourier transform of the autocovariance function) is constant. More precisely

$$\nu_\epsilon(k) = \sum_{j=0}^{N-1} C_\epsilon(j) e^{-2i\pi kj/N} = \sigma^2, \qquad k = 0, 1, \cdots, N-1. \tag{2.10}$$

Recall formula (1.38) giving the definition of the finite discrete Fourier transform of a finite sequence. Note that the constant value of the white-noise sequence should be $10 \log_{10} \sigma^2$ if expressed in decibels and 0 if $\sigma = 1$. Let us also consider the case of discrete but infinite white-noise sequences. So let us assume now that ϵ is a doubly infinite sequence of mean zero uncorrelated random variables ϵ_k for $k = 0, \pm 1, \pm 2, \cdots$ having the same variance σ^2. Formulae (2.8), (2.9), and (2.10) still hold, except for the fact that the spectral function $\nu_\epsilon(\omega)$ is computed for all $\omega \in [0, 1)$, whereas it was previously computed only for the natural frequencies $\omega_k = k/N$. Indeed, recalling the definition (1.11) of the discrete Fourier transform of a doubly infinite sequence, the spectral function $\nu_\epsilon(\omega)$ is given by:

$$\nu_\epsilon(\omega) = \widehat{C}_\epsilon(\omega) = \sum_{j=-\infty}^{+\infty} e^{-2i\pi j\omega} C_\epsilon(j)$$

for all the values of the frequency variable $\omega \in [0, 1]$. With this definition it is possible to prove that, since $C_\epsilon(j)$ is zero unless $j = 0$, in which case it is equal to σ^2, the spectral measure ν_ϵ of the doubly infinite discrete white noise ϵ is absolutely continuous and that the spectral density $\nu_\epsilon(\omega)$ is equal to the constant σ^2.

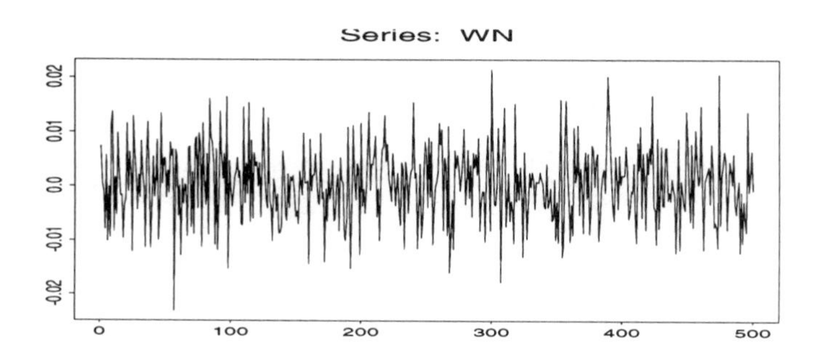

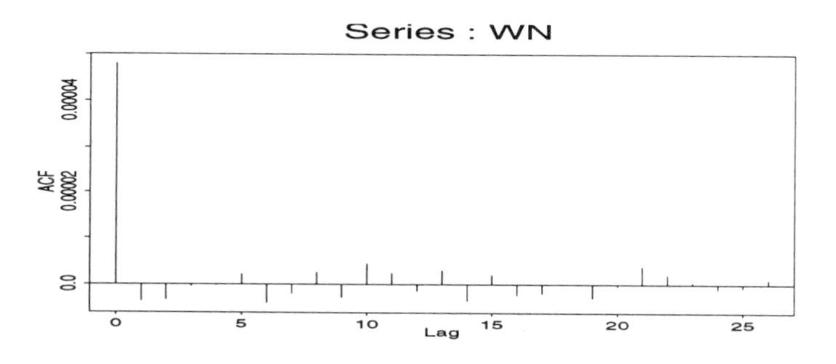

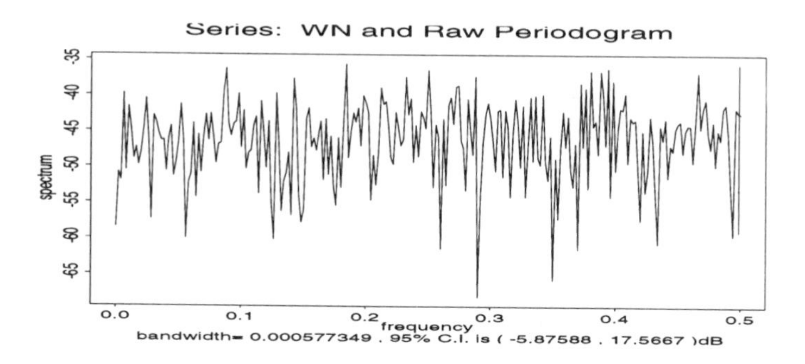

Figure 2.1. White noise: The top plot shows a typical sample from a Gaussian white noise series with length $N = 512$. The middle plot shows the values of the sample autocorrelation function for lags up to 26, and the bottom plot shows the periodogram of the series. The statistical properties of the periodogram are discussed later in the chapter.

After the cases of the finite and doubly infinite white noise sequences, the next step is to define and analyze the continuous time generalization of the white noise sequences defined earlier. It is natural to define such a white noise as a continuous family $\epsilon = \{\epsilon(x);\ x \in \mathbb{R}\}$ of mean zero uncorrelated random variables $\epsilon(x)$ of order 2 satisfying (2.8) and consequently (2.9). Unfortunately, such a mathematical object is very singular: it is not possible to define it in such a way that, when the random parameter is fixed, the sample path $x \hookrightarrow \epsilon(x)$ is a function. Indeed, the fact that the autocovariance function C_ϵ is a delta function at 0 (which was a bona fide function on the set of integers but which is not a function on the set $\mathbb{R}$ of real numbers) forces these sample paths to be Schwartz distributions. There is nevertheless a way to give a rigorous mathematical meaning to the notion of continuous-time white-noise process. It involves the notion of orthogonal L^2 measure. We shall use this concept later to define stochastic integrals and the general form of the spectral representation of stationary processes. But for the time being, we want to hang on to the notion of white noise given by a uncorrelated family of mean zero random variables with the same variance. We shall keep this notion at the intuitive level and, when we simulate such a noise, since we can only deal with a finite sample of such an object, we shall use the (rigorous) notion of white noise introduced for finite sequences!

2.1.3 A Second Example: Random Trigonometric Polynomials

Let $\omega_0, \omega_1, \cdots, \omega_m$ be distinct values in $[0, 1]$ and let $A_0, A_1, \cdots, A_m$ and $B_0, B_1, \cdots, B_m$ be uncorrelated mean zero random variables of order 2. We assume that for $j = 0, 1, \ldots, m$ the random variables A_j and B_j have the same variance. We denote by σ_j^2 this common variance and we set $\sigma^2 = \sigma_0^2 + \sigma_1^2 + \cdots + \sigma_m^2$. We now define the random process $f = \{f(x);\ x \in \mathbb{R}\}$ by

$$f(x) = \sum_{j=0}^{m} [A_j \cos \pi x \omega_j + B_j \sin \pi x \omega_j], \qquad x \in \mathbb{R}. \tag{2.11}$$

The mean is independent of the time variable x, since $\mathbb{E}\{f(x)\} = 0$, obviously. Moreover, using elementary trigonometry and $\mathbb{E}\{A_j\} = \mathbb{E}\{B_j\} = 0$ and $\mathbb{E}\{A_j^2\} = \mathbb{E}\{B_j^2\} = \sigma_j^2$ for $j = 0, 1, \cdots, m$, and the fact that $\mathbb{E}\{A_j B_k\} = 0$ for all values of j and k and $\mathbb{E}\{A_j A_k\} = \mathbb{E}\{B_j B_k\} = 0$ whenever $j \neq k$, we easily get

$$\mathbb{E}\{f(x+h)f(x)\} = \sum_{j=0}^{m} \sigma_j^2 \cos(\pi h \omega_j),$$

which shows that the process is wide-sense stationary. The process is in fact (strictly) stationary if the random variables A_j and B_j are jointly Gaussian. Notice also that

$$C_f(h) = \int \cos(\pi h\omega) d\nu(\omega) \tag{2.12}$$

if ν is the point measure putting mass σ_j^2 on ω_j for $j = 0, 1, \dots, m$. We derived directly in this simple case the existence of the spectral measure provided in general by Bochner's theorem.

We shall see in the next subsection that the representations (2.12) and (2.11) of the autocovariance function and of the process itself are not specific to the particular class of random trigonometric polynomials, but that they hold in the full generality of wide-sense stationary processes of second order. Formula (2.11) can be viewed as still another instance of the spectral representation of stationary stochastic processes and as explained earlier, this representation of the original signal as a random weighted sum of complex exponentials is what is missing in Wiener's approach to spectral theory.

2.2 Spectral Representations

After all this build-up, we finally come to the derivation of the spectral representation of a general stationary stochastic process. For pedagogical reasons we first consider the case of a finite (mean zero) random sequence $f = (f_0, f_1, \cdots, f_{N-1})$ of the second order (i.e., each of the random variables f_j has a moment of order 2) and we denote by $\hat{f}$ its (discrete) Fourier transform. The sequence $\hat{f}$ is a complex random sequence; its mean is constant and equal to 0, and it is also of order 2. Note that by inversion of the Fourier transform one can recover the original sequence f from its Fourier transform $\hat{f}$. Indeed,

$$f_j = \frac{1}{N} \sum_{k=0}^{N-1} \hat{f}_k e^{2\pi jk/N}, \qquad j = 0, 1, \cdots, N-1. \tag{2.13}$$

If we compute the autocovariance function of the sequence $\hat{f}$, we get

$$\mathbb{E}\{\hat{f}_h \overline{\hat{f}_k}\} = \sum_{j,j'=0}^{N-1} \mathbb{E}\{f_j \overline{f_{j'}}\} e^{-2i\pi(hj-kj')/N},$$

and if the random sequence f is wide-sense stationary, i.e., if its autocovariance $\mathbb{E}\{f_j \overline{f_{j'}}\}$ is a function $C_f(j'-j)$ of the difference $j'-j$, then we have

$$\mathbb{E}\{\hat{f}_h \overline{\hat{f}_k}\} \quad = \quad \sum_{j,j''=0}^{N-1} C_f(j'') e^{2i\pi kj''/N} e^{2i\pi(k-h)j/N},$$

which is equal to $N\nu_f(k)$ when $k = h$ (recall formula (1.43) in Chapter 1) and 0 otherwise if we use the notation ν_f for the (discrete) Fourier transform of the autocovariance function C_f of f, i.e., its spectral function according to our terminology. Consequently, the sequence $W = \{W_0, \dots, W_{N-1}\}$ defined by

$$W_k = \frac{1}{\sqrt{N}\sqrt{\nu_f(\omega_k)}} \hat{f}_k$$

is a finite (discrete) standard white noise in the sense given earlier. Here standard simply means that the common variance is equal to 1. Notice that f has the representation

$$f_j = \frac{1}{N} \sum_{k=0}^{N-1} e^{2i\pi kj/N} \sqrt{\mathcal{E}(k)} W_k \tag{2.14}$$

in terms of the white noise W provided we set $\mathcal{E}(k) = N\nu_f(k)$. The square root is well defined, since the Fourier transform $\nu_f(k)$ is nonnegative because the autocovariance function C_f is itself nonnegative definite. Formula (2.14) is not the first instance of the *spectral representation* formula of a stationary stochastic process, but it is our first general statement of it.

2.2.1 Application to Monte Carlo Simulations

Let us remark that the sequence f is (jointly) Gaussian if and only if the sequence W is (complex) Gaussian, and since the statistical distribution of a Gaussian process is entirely determined by its first two statistical moments, the spectral representation provides us with an answer to the following question:

Given a numerical sequence $C = \{C(j)\}_j$ having the properties of an autocovariance function, how can we simulate samples of a mean zero Gaussian process $f = \{f_j\}_j$ whose autocovariance function C_f is equal to the given sequence C?

The spectral representation formula provides a simple answer to this question. Indeed, given C one can compute its Fourier transform, say, ν. This function is nonnegative and we can compute its square root. Now, to create a realization of a sample $f = \{f_0, \cdots, f_{N-1}\}$ of length N with the prescribed distribution, it is enough to create a sample of size N of standard $N(0,1)$ random variables and use them to create a complex white noise $W = \{W_k\}$ of length $N/2$, but because we are looking for a real sequence f this is all we need to complete the simulation. Indeed, one multiplies this complex white noise W by the square root of $\mathcal{E} = N\nu$ and takes the inverse Fourier transform to obtain the desired sample from formula (2.14). Recall that computing the Fourier and inverse Fourier transforms is very fast if N is a power of 2 and one uses the fast Fourier transform algorithm.

Such a simulation procedure is of great practical importance because it generalizes without any difficulty to random fields (instead of random sequences) for which the practical applications are abundant. Indeed, the numerical simulation of Gaussian random fields with a prescribed autocovariance structure (or equivalently with a given spectrum) has been used successfully in domains as diverse as fluid mechanics and turbulence simulations, ground motion modeling for earthquake prediction, texture simulation for image analysis, physical oceanography, ...

2.2.2 The General Case

It is tempting to try to extend the spectral representation formula to doubly infinite stationary sequences. Indeed, using the duality given by the Fourier transform between the Fourier series and the functions on the interval $[0, 2\pi)$, we easily guess the generalization

$$f_j = \frac{1}{\sqrt{2\pi}} \int_0^{2\pi} e^{ij\xi} \sqrt{\mathcal{E}(\xi)} W(\xi) d\xi \tag{2.15}$$

of the spectral representation (2.14) derived for finite time series. But the main problem is to give a meaning to the random process $W = \{W(\xi);\ \xi \in [0, 2\pi)\}$. Indeed, one would like this process to be a white-noise process indexed by the continuous variable ξ, and as we already mentioned, this is a very singular object, and certainly not a function as we would like. The main difficulty in extending the spectral representation (2.14) so that it holds in greater generality (for all wide-sense stationary processes would be great) is to give a meaning to this integration with respect to the white noise. This is done by combining the terms $W(\xi)$ and $d\xi$ into a random measure $W(d\xi)$ whose definition makes sense. For the sake of completeness we present a derivation of this general representation formula directly in the case of the processes $f = \{f(x);\ x \in \mathbb{R}\}$ on the whole real line, the case of doubly infinite sequence being argued in the same way. The reader is warned that this subsection can be skipped in a first reading, especially if he is not interested in mathematical rigor and generality.

We consider a mean zero wide-sense stationary stochastic process of order 2, $f = \{f(x);\ x \in \mathbb{R}\}$, and we denote by C_f its autocovariance function, i.e., $C_f(x) = \mathbb{E}\{f(x)f(0)\}$, and by $\sigma^2 = C_f(0)$ the common variance. We assume that the autocovariance function goes to 0 as $x \to \infty$ fast enough. Let us assume, for example, that C_f is integrable. In this case the spectral measure $\nu_f(d\omega)$ is absolutely continuous in the sense that

$$\nu_f(d\omega) = \mathcal{E}_f(\omega)d\omega ,$$

and its density $\mathcal{E}_f(\omega)$ is the Fourier transform of the autocovariance function, namely

$$\mathcal{E}_f(\omega) = \int_{-\infty}^{+\infty} e^{-i\omega x} C_f(x)dx.$$

This (nonnegative) function is often called the *power spectrum*, or *spectral density*. The continuous time version of the spectral representation formula (2.14) takes the form

$$f(x) = \frac{1}{\sqrt{2\pi}} \int_{-\infty}^{+\infty} e^{i\omega x} \sqrt{\mathcal{E}(\omega)} W(d\omega) \tag{2.16}$$

for some *white noise random measure* $W(d\omega)$. A modicum of care is needed to give a precise meaning to this notion of white noise random measure and to define the integral. We try to limit the technicalities to a minimum but we nevertheless proceed to the discussion of some of the gory details of the stochastic integration theory needed for our purposes. A family $W = \{W(A);\ A \in \mathcal{B}_b(\mathbb{R})\}$ of mean zero random variables parameterized by the bounded Borel sets of the real line is called a white noise random measure if it satisfies:

$$\mathbb{E}\{W(A)\overline{W(A')}\} = |A \cap A'|, \tag{2.17}$$

where we used the notation $|A|$ for the Lebesgue measure of the Borel set A. If φ is a elementary step function of the form

$$\varphi = \sum_{j=1}^{k} \varphi_j \mathbf{1}_{A_j} \tag{2.18}$$

for some finite collection of (possibly complex) numbers $\varphi_1, \cdots, \varphi_k$ and disjoint bounded Borel sets $A_1, \cdots, A_k$, then it is natural to define its *stochastic integral* with respect to W as

$$I_W(\varphi) = \int \varphi(\omega) W(d\omega) = \sum_{j=1}^{k} \varphi_j W(A_j).$$

Here and throughout the book we use the notation $\mathbf{1}_A$ for the indicator function of the set A, i.e., the function which is equal to 1 on A and 0 outside. With this definition it is easy to check that

$$\mathbb{E}\{|I_W(\varphi)|^2\} = \int |\varphi(\omega)|^2 d\omega,$$

which shows that this integration can be used to define a norm-preserving unitary operator I_W from the space of square integrable functions on the real line into the space of square integrable random variables. This makes it possible to define $I_W(\varphi)$ for any square integrable function φ. One first approximates φ by a sequence of elementary step functions φ_n of the form (2.18), and then one defines $I_W(\varphi)$ as the limit in quadratic mean (i.e. in the L^2 space over the probability space on which the white noise measure is defined) of the random variables $I_W(\varphi_n)$. We shall still use the integral notation

$$I_W(\varphi) = \int \varphi(\omega) W(d\omega)$$

and think of W as of a random measure. Note that the random variables $W(A)$ and $W(A')$ are uncorrelated (and consequently orthogonal in the Hilbert space L^2 over the probability space) when the Borel sets A and A' are disjoint. For this reason white noise families such as W are often called *orthogonal L^2-measures*. We would like to think of (2.17) in the more suggestive form

$$\mathbb{E}\{W(d\omega)\overline{W(d\omega')}\} = \delta(\omega - \omega') d\omega d\omega',$$

which will be very useful for computational purposes.

We close this subsection with a brief discussion of the (even more) general case of a spectral measure $\nu_f(d\omega)$ which is not necessarily assumed to be absolutely continuous.

As we already noticed, this happens typically when the autocovariance function does not go to zero fast enough. This is indeed the case when the autocovariance function has an almost periodic component due to the presence of a component of the signal given by a random trigonometric polynomial. In this case the spectral measure ν_f has a nontrivial pure point component and the preceding discussion does not apply as such. Nevertheless, it is possible to reconcile the spectral decomposition formula for random trigonometric polynomials (2.11) and the spectral representation (2.16) for processes with a spectral density into a single spectral representation formula. Moreover, this formula also applies to the pathological cases of processes with a nontrivial singular continuous spectral measure $\nu_{f,sc}$. Such a spectral representation formula reads

$$f(x) = \frac{1}{\sqrt{2\pi}} \int e^{i\omega x} \widetilde{W}(d\omega), \tag{2.19}$$

where $\widetilde{W}$ is now a family of mean zero random variables $\widetilde{W}(A)$ of order 2 satisfying

$$\mathbb{E}\{\widetilde{W}(A)\overline{\widetilde{W}(A')}\} = \nu_f(A \cap A').$$

As before, $\widetilde{W}$ can be viewed as an orthogonal L^2 measure and the notion of stochastic integral introduced earlier can still be used to give a meaning to the representation (2.19). In the absolute continuous case, one merely has $\widetilde{W}(d\omega) = \sqrt{\mathcal{E}(\omega)}W(d\omega)$.

We do not give the details of a rigorous mathematical derivation of the spectral representation formula (2.16) and its variant (2.19). We rely on the intuition developed in the case of the finite sequences and we merely refer the interested to the standard textbooks on spectral theory of time series. See, for example, [58], [195], [235], [239], or [248] for complete proofs.

2.3 Nonparametric Spectral Estimation

For a mean zero wide-sense stationary process of order 2 the knowledge of the second-order statistics (i.e., the autocovariance function) is equivalent to the knowledge of the spectrum. Moreover, this knowledge becomes equivalent to the knowledge of the entire distribution of the process in the case of Gaussian processes. In any case, the estimation of the second-order statistics (or equivalently the spectrum) is an important problem and the spectral representation formulas appears to be a very useful tool in this respect.

A good part of the statistical literature on spectral estimation is devoted to parametric models. Auto regressive (AR) models, moving average (MA) models, auto regressive moving average (ARMA) models, auto regressive integrated moving average (ARIMA) models, $\cdots$ are among the most popular. These models are determined by a finite number of parameters, and spectral estimation reduces to the estimation of these parameters. The contention of the book is that a nonparametric approach is less restrictive. As a consequence we try to stay away from these models and we restrict ourselves to a model-independent approach to spectral estimation much in the spirit of the time-frequency analysis of signals. We nevertheless use a specific AR model as illustration of spectral smoothing at the end of the chapter.

2.3.1 The Periodogram Revisited

The periodogram was introduced in the context of finite (not necessarily random) time series $f = \{f_0, f_1, \cdots, f_{N-1}\}$. It was defined as

$$\hat{\nu}_f(\omega) = \frac{1}{N}\left|\sum_{j=0}^{N-1} f_j e^{-2i\pi j\omega}\right|^2$$

and shown to be equal to the Fourier transform of the sample autocovariance function appropriately defined as in (1.33). The purpose of this section is to study the statistical properties of these objects when the f_j's are sample observations of a stationary stochastic process and when the sample autocovariance function and the periodogram (i.e., the sample spectral function) are regarded as estimates. In particular, we assume that the numbers $f_0, f_1, \cdots, f_{N-1}$ are observations of mean zero random variables from a doubly infinite stationary process $f = \{f_j\}_{j=-\infty,\cdots,+\infty}$, and we denote by $C_f(j) = \mathbb{E}\{f_j f_0\}$ the autocovariance function of this time series. The sequence C_f is assumed to decay fast enough at ∞ for f to have a spectral density which we will denote ν_f. Typically, we shall assume that the sequence $\{C_f(j)\}_j$ is summable, in which case the spectral density ν_f is nothing but the Fourier transform of C_f.

Warning

> It is important to emphasize the meaning of the hat $\hat{}$ which appears on some of the notation. We follow the convention used in the statistical practice of using a hat $\hat{}$ to distinguish between the value of an estimator (obtained from sample observations) and the value of the corresponding object/parameter of the model.

Notice that this use of the hat $\hat{}$ can come in conflict with its use for the Fourier transform. This double use of the same notation is very unfortunate, but we hope that the context will help clarify its meaning. We tried very hard to resolve this dilemma and after numerous unsuccessful attempts we decided to follow the standard practice of signal/functional analysis of using a hat $\hat{}$ for the Fourier transform and of statistics of using hat $\hat{}$ for estimators. The case of the spectral function is a typical example. We introduced the notation $\nu_f(\omega)$ in Chapter 1; recall formulae (1.41) or (1.43), for example. We now use the same definition formulae, but instead we use the notation $\hat{\nu}_f(\omega)$ because the elements entering in the formulae are realizations/observations of random variables and the resulting quantity $\hat{\nu}_f(\omega)$ is the corresponding estimate of the theoretical value $\nu_f(\omega)$.

If one computes the expected value of the sample spectral density one gets (recall formula (1.33))

$$\begin{aligned}
\mathbb{E}\{\hat{\nu}_f(\omega)\} &= \sum_{j=-(N-1)}^{N-1} \left(1 - \frac{|j|}{N}\right) C_f(j)\cos(2\pi j\omega) \\
&= \int_0^1 F_N(\omega - r)\nu_f(r)dr, \qquad (2.20)
\end{aligned}$$

where F_N is the Fejér kernel

$$F_N(\omega) = \frac{1}{N}\left(\frac{\sin(N\pi\omega)}{\sin(\pi\omega)}\right)^2, \qquad \omega \in [0,1]. \qquad (2.21)$$

Formula (2.20) shows that for each fixed $\omega \in [0,1]$ the sample spectral density $\hat{\nu}_f(\omega)$ is a *biased* estimator of the spectral density $\nu_f(\omega)$. Indeed, its expected value $\mathbb{E}\{\hat{\nu}_f(\omega)\}$ is not equal to its true value $\nu_f(\omega)$. It is merely a weighted average of the values of the "true" spectral density in a neighborhood of the desired frequency. But because of the factor N^{-1} in the definition (2.21) of the Fejer kernel, the bias disappears as the sample size N grows to ∞ (at least when the spectral density is continuous). So this shortcoming should not be too serious for large samples. Unfortunately, one can also show that

$$\lim_{N\to\infty} \mathrm{Var}\{\hat{\nu}_f(\omega)\} = \begin{cases} \nu_f(\omega)^2, & \text{if} \quad \omega \neq 0.5 \,, \\ 2\nu_f(\omega)^2, & \text{if} \quad \omega = 0.5 \,. \end{cases} \qquad (2.22)$$

The fact that the variance of $\hat{\nu}_f(\omega)$ is proportional to $\nu_f(\omega)^2$ is another reason to plot the logarithm of the periodogram instead of the periodogram itself. Indeed the variance of the logarithm of $\hat{\nu}_f(\omega)$ converges toward a constant independent of the frequency ω. In any case, the result given by formula (2.22) is pretty bad because it says that, even though the bias disappears when the sample size increases, the variance of the sample spectral density does not go to zero and consequently, this estimate cannot converge to the true value of the spectral density. In other words, it is *not a consistent* estimator! Even worse, if one considers two different frequencies ω and ω', then

$$\lim_{N\to\infty} \mathrm{Cov}\{\hat{\nu}_f(\omega)\hat{\nu}_f(\omega')\} = 0,$$

which shows that the values of the sample spectral density at two different frequencies are asymptotically uncorrelated. In fact, if the original sequence is a Gaussian white noise, one can prove that the values of the sample spectral density at the natural frequencies (i.e., the values of the periodogram) are *independent* and distributed according to χ^2 distributions in this asymptotic regime of large sample size (i.e., when $N \to \infty$).

In particular, in the large sample limit the logarithm of the periodogram looks very much like a white noise itself. This fact is confirmed by the plot given as the bottom plot of Figure 2.1 of the periodogram of a Gaussian white noise time series.

2.3.2 Sampling and Aliasing

We now consider the effect of sampling on the spectrum of a stationary stochastic process. The present discussion parallels the discussion of the aliasing phenomenon of Chapter 1, and we shall use similar notation. We assume that $f^{(d)} = \{f_j\}_{j=0,\pm1,\cdots}$ is a (doubly infinite) mean zero stationary sequence of order 2 of real-valued random variables. We use the superscript $^{(d)}$ to emphasize the discretization produced by the sampling of the continuous time process. We shall consider the effect of working with a finite subset of the sequence in the later subsection devoted to data tapering. As before, we assume that these random variables are observations of a continuous time stationary (mean zero) process $f = \{f(x);\ x \in \mathbb{R}\}$ of order 2 and we assume that for some time origin x_0 and some sampling interval Δx we have

$$f_j = f(x_0 + j\Delta x), \qquad j = 0, 1, \cdots, N-1.$$

The autocovariance function $C^{(d)}$ of the sequence $f^{(d)}$ appears as a sampling of the autocovariance $C_f(x) = \mathbb{E}\{f(x)f(0)\}$ of the process f. Indeed,

$$C^{(d)}(j) = \mathbb{E}\{f_j f_0\} = \mathbb{E}\{f(x_0 + j\Delta x)f(x_0)\} = C_f(j\Delta x).$$

Consequently, the discussion which we had in the first chapter to compare the Fourier transform of a function and the discrete Fourier transform of a sequence of its samples applies to the present situation. In particular, if we denote by $\nu^{(d)}(\omega)$ the spectral function of the discrete sequence $f^{(d)}$, then the "folding of the frequency interval" takes the form

$$\nu^{(d)}(\omega) = \sum_{k=-\infty}^{+\infty} \nu_f(\omega + \frac{k}{\Delta x}), \qquad |\omega| \le \omega_{(N)} \equiv \frac{\pi}{\Delta x}. \tag{2.23}$$

In fact, as we now show, formula (2.23) can be recovered directly from the spectral representation formula. Indeed,

$$\begin{aligned} f_j &= f(x_0 + j\Delta x) \\ &= \frac{1}{\sqrt{2\pi}} \int e^{i\omega(x_0+j\Delta x)} \tilde{W}(d\omega) \\ &= \frac{1}{\sqrt{2\pi}} \sum_{k=-\infty}^{+\infty} \int_{(2k-1)\pi/\Delta x}^{(2k+1)\pi/\Delta x} e^{i\omega x_0} e^{i\omega j\Delta x} \tilde{W}(d\omega) \\ &= \frac{1}{\sqrt{2\pi}} \int_{-\pi/\Delta x}^{\pi/\Delta x} e^{i\omega j\Delta x} W^{(d)}(d\omega), \end{aligned} \tag{2.24}$$

provided we set

$$W^{(d)}(d\omega) = \sum_{k=-\infty}^{+\infty} e^{i(\omega+k\pi/\Delta x)x_0} W\left(d(\omega + \frac{k\pi}{\Delta x})\right).$$

Notice that, using the properties of the orthogonal measure $W(d\omega)$, one can easily check that $\mathbb{E}\{W^{(d)}(d\omega)\} = 0$ whenever $|\omega| \le \omega_{(N)}$. Moreover, if $-\omega_{(N)} \le \omega' < \omega \le \omega_{(N)}$, one can also show that

$$\begin{aligned} \mathbb{E}\{W^{(d)}(d\omega')W^{(d)}(d\omega)\} &= \sum_{k=-\infty}^{+\infty} \sum_{l=-\infty}^{+\infty} e^{i(\omega+k\pi/\Delta x)x_0} e^{-i(\omega+l\pi/\Delta x)x_0} \\ &\quad \times \mathbb{E}\{\tilde{W}(d(\omega' + l\pi/\Delta x))\overline{\tilde{W}(d(\omega + k\pi/\Delta x))}\} \\ &= 0, \end{aligned}$$

which shows that $W^{(d)}(d\omega)$ is an orthogonal L^2 white noise measure and consequently that (2.24) gives the spectral representation of the stationary sequence $f^{(d)}$. Its power spectrum is then given by

$$\begin{aligned} \nu^{(d)}(\omega)d\omega &= \mathbb{E}\{|W^{(d)}(d\omega)|^2\} \\ &= \sum_{k=-\infty}^{+\infty} \mathbb{E}\{|\tilde{W}\left(d(\omega + \frac{k\pi}{\Delta x})\right)|^2\} \\ &= \sum_{k=-\infty}^{+\infty} \nu_f(\omega + k\pi/\Delta x)d\omega, \qquad |\omega| \le \omega_{(N)}, \end{aligned}$$

which (modulo differentiations) is exactly formula (2.23).

The moral of this theoretical exercise is the following. If the spectral density $\nu_f(\omega)$ of the continuous time process $f = \{f(x);\, x \in \mathbb{R}\}$ vanishes beyond the Nyquist frequency (i.e., for $|\omega| > \omega_{(N)}$), then one can expect very good agreement between $\nu_f(\omega)$ and $\nu^{(d)}(\omega)$ and one can estimate the spectral density of the sampled series in lieu of the (theoretical) spectral density of the continuous time signal. On the other hand, if the spectral density $\nu_f(\omega)$ of the continuous signal is significantly different from 0 beyond the Nyquist frequency, then using an estimate of $\nu^{(d)}(\omega)$ to estimate $\nu_f(\omega)$ will give poor results. Given our discussion of spectrum folding and aliasing phenomenon given in Chapter 1, it is not surprising that this phenomenon is known under the name of *aliasing* in the statistical community. It occurs when there is a significant spectral energy leak between the spectral bands appearing in the computation of formula (2.24).

2.4 Spectral Estimation in Practice

Despite the alarming news of the previous subsections concerning the dangers of aliasing and the fact that the sample spectral density is a biased inconsistent estimate of the true spectral density, we shall see that one can nevertheless use this very sample spectral density (i.e., the periodogram) as a spectral estimator.

> Spectral analysis is a very touchy business: subtle choices are needed to get the right results!

It is clear from the examples (and the theoretical results presented earlier) that the sample spectral densities $\hat{\nu}_f(\omega)$ are too wiggly to be faithful estimates of the true spectral densities $\nu_f(\omega)$, which are usually smooth functions of ω. The naive remedy is to smooth the spectral estimate, replacing $\hat{\nu}_f(\omega)$ by an estimate of the form

$$\tilde{\nu}(\omega) = \int_0^1 K_N(\omega - r)\hat{\nu}_f(r)\, dr \ , \tag{2.25}$$

for some kernel function K_N called *spectral window* for the circumstance. We use a subscript N to emphasize the fact that the choice of the spectral window is made as a function of the length of the sample time series. Notice that this smoothing operation appears as a convolution in the Fourier domain, so it can be rewritten in the time domain as an operation of multiplication. Indeed, the smoothing of formula (2.25) is equivalent to replacing the sample autocovariance function $\hat{c}_f(j)$ by something of the form $k_N(j)\hat{c}_f(j)$ for a specific sequence $k_N = \{k_N(j)\}_{j=0,\pm1,\dots}$ (i.e., the sequence whose Fourier transform is the spectral window K_N).

The effect of the smoothing by the spectral window K_N is twofold:

- On one hand, the bias is increased because the values of the sample spectral density are smeared by neighboring values, but it seems reasonable to expect that, when the sample size N goes to infinity, this bias will still disappear if the *width* of the spectral window goes to zero with N.
- On the other hand, the averaging inherent to the smoothing by K_N is lowering the variance of the sample spectral density, and it is reasonable to expect that the variance could go to 0 when N goes to ∞ if the width of the spectral window does not go to zero too fast.

In fact, it is possible to strike a just balance and find asymptotic rates of decay for the spectral window such that, in the large sample limit, the periodogram and the sample spectral density become unbiased and consistent. We give more details later. There are two standard ways to lessen the bias of the periodogram: data tapering and prewhitening. We discuss them separately.

2.4.1 Data Tapering

For the purpose of the present discussion we assume that $f = \{f_j\}_{j=0,\pm1,\cdots}$ is a mean zero stationary sequence of order 2 real-valued random variables and the spectral estimation which we perform is from the observed finite sequence $f_0, f_1, \cdots, f_{N-1}$, which we rewrite as the infinite sequence

$$bf = \{b_j f_j\}_{j=0,\pm1,\cdots}, \tag{2.26}$$

where the doubly infinite sequence $b = \{b_j\}_{j=0,\pm1,\ldots}$ is defined by

$$b_j = \begin{cases} 1 & \text{if } j = 0, 1, \cdots, N-1, \\ 0 & \text{otherwise.} \end{cases} \tag{2.27}$$

The sequence b appearing in (2.26) is called a *data window* or *taper*. Figure 2.2 gives the plot of a *cosine taper* as defined in formula (2.28) with parameter $p = .2$. This parameter p should be a number between 0 and .5. It gives the proportion of points to be tapered at each end of the data stream. If M is an integer and if we set $p = M/N$, then such a cosine taper is defined by the formula

$$b_j = \begin{cases} (1 - \cos(\pi(j - .5)/M))/2 & \text{if } \ j = 1, \cdots, M, \\ 1 & \text{if } \ j = M+1, \cdots, N-M, \\ (1 - \cos(\pi(N - j + .5)/M))/2 & \text{if } \ j = N-M+1, \cdots, N. \end{cases} \tag{2.28}$$

The second plot of Figure 2.2 show the result of the tapering of the white noise series used earlier in this chapter. For the purpose of the present

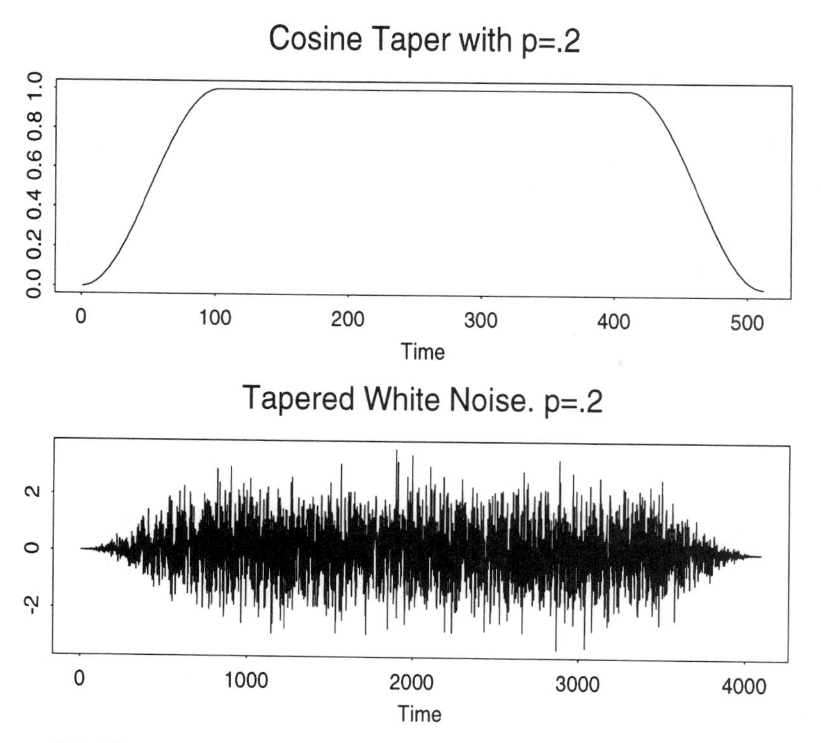

Figure 2.2. Top: cosine taper. Bottom: white noise time series with 20% of the points tapered at each end and tapered white noise.

discussion we assume momentarily that b is arbitrary and we denote by $\hat{b}$ its Fourier transform,

$$\hat{b}(\omega) = \sum_{j=-\infty}^{+\infty} b_j e^{-2i\pi j\omega}, \qquad \omega \in [0,1].$$

In the presence of such a tapering, the sample spectral density or periodogram becomes

$$\nu_f^b(\omega) = \frac{1}{N} \left| \sum_{j=0}^{N-1} b_j f_j e^{-2i\pi j\omega} \right|^2 ,$$

and the bias can be computed in exactly the same way provided one replaces the Fejer kernel by the Fourier transform $\hat{b}$. The idea behind tapering is to select the data window so that the kernel $\hat{b}(\omega)$ has much smaller side lobes that the Fejer kernel. This is usually achieved by making sure that the b_j go to 0 in a smoother fashion than the abrupt change in the rectangular data taper (corresponding to no taper at all).

2.4.2 Prewhitening

This method is based on an a priori knowledge of the form of the theoretical spectral density. This is particularly useful in parametric spectral estimation when a model (AR, ARMA, ARIMA, ...) is assumed. As explained earlier, except for the discussion of an AR example at the end of this chapter, we shall refrain from using parametric models.

For the sake of the present discussion we assume that the sequence $f = \{f_0, f_1, \cdots, f_{N-1}\}$ is a sample from a stationary time series whose spectrum is to be estimated. Let us choose positive integers K and L, let us set $M = N-(K+L)$, and let us define the new series $g = \{g_0, g_1, \cdots, g_{M-1}\}$ by

$$g_j = \sum_{k=-L}^{K} a_k f_{j+K-k}, j = 0, 1, \cdots, M-1 ,$$

where the coefficients a_k of the convolution filter are to be determined. Computing the sample spectral density of g, we get

$$\nu_g(\omega) = \left| \sum_{k=-L}^{K} a_k e^{-2i\pi k\omega} \right|^2 \nu_f(\omega). \tag{2.29}$$

So, even if ν_f has a wide dynamic range, it might be possible to choose the filter coefficients a_k so that the dynamic range of ν_g is much smaller.

Consequently, one can consider the following procedure to construct an estimator of ν_f with low bias:

• Use the filtered data to produce a spectral estimator for g in the form

$$\hat{\nu}_g(\omega) = \left| \sum_{j=0}^{M-1} b_j g_j e^{-2i\pi j\omega} \right|^2,$$

where $b = \{b_j\}_j$ is a data taper of our choice.

• If we succeeded in forcing a small dynamic range on g, we should have

$$\mathbb{E}\left\{ \frac{\hat{\nu}_g(\omega)}{\left|\sum_{k=-L}^{K} a_k e^{-2i\pi k\omega}\right|^2} \right\} \approx \frac{\nu_g(\omega)}{\left|\sum_{k=-L}^{K} a_k e^{-2i\pi k\omega}\right|^2} = \nu_f(\omega),$$

and we can use this formula to produce an estimate of ν_f.

In the ideal case the filter $a = \{a_k\}_k$ reduces the series f to a white noise, hence its name of a prewhitening filter. Prewhitening is not a panacea. It has the following obvious shortcomings:

• The reduction of the length of the original series can have a serious effect on other estimates.

• There is a "chicken-and-egg" problem, since constructing a prewhitening filter requires some a priori knowledge of the spectral function which we are supposed to estimate. This last difficulty is often resolved by using an estimate from an a priori model (AR models are very popular in this respect).

2.4.3 Block Averaging

A natural way to reduce the variance of the spectral estimates was proposed by Bartlett. He suggested to break the time series into non-overlapping blocks, to compute a periodogram based on the data in each block alone, and then to average the individual periodograms together to form an overall spectral estimate. Such an estimate is not exactly of the form of a lag window spectral estimate, but because of the variance reduction, block averaging is regarded as an alternative. Notice that this procedure requires rather long time series. Indeed, for the average to make sense, a significant number of blocks is needed and all these blocks must have the same spectral characteristics. *This cannot be the case if the stationarity assumption is not satisfied.*

This naive block averaging procedure can be improved in two ways. First, a taper can be used in each block to reduce the bias, and second, we can have the blocks overlap. Indeed, Welsh showed that such an overlap can produce estimates with a better variance. The spectral estimate based on the averaging of the periodograms in overlapping intervals is called the Welsh-Bartlett estimate. We shall see later in Chapter 6 similar estimators in the context of time-frequency analysis.

The examples discussed here use a variant of the original form of the smoothing by running averages. This variant is based on repeated passes on the raw periodogram. It is explained in detail in [52]. We use it because it is implemented in Splus, which we used to produce the examples. See the end of the chapter for the details of the S commands.

Figure 2.3 shows the original raw periodogram of the white noise series already used earlier in the chapter, together with two other spectral estimates obtained by two different smoothing attempts. See the last section for the values of the parameters used. The theoretical spectrum of a white noise is a constant function. Because of the logarithmic scale, this function should be equal to 0 if we use the standard unit of decibels to measure the spectral power. If we pay attention to the change in scale on the vertical axis, one notices that the smoothing improves the estimates dramatically. The latter goes from a white-noise-looking object to a smoother function. We can use the values of the means and the variances of the estimates plotted in this figure to assess the quality of the estimations. The means are −2.98451, −0.9564203, and −0.7168848, respectively. These values trying to approach 0 give an indication of the improvement in the bias. The corresponding values of the variances are 30.04696, 4.012902, and 1.712197, and their decreasing values illustrate the usefulness of the smoothing of the periodogram.

2.5 Examples and S-Commands

We give the details of the S-commands used to produce the results presented above and we present two extra analyses. All of the S-commands used in this chapter are from the standard S language and we shall also use one of the data set included in the Splus package for the purpose of illustration.

2.5.1 White Noise Spectrum

The white noise example was produced as follows.

Generation of a white noise series of length 512*:*

```
> WN <- rnorm(512,0,1)
> tsplot(WN)
```

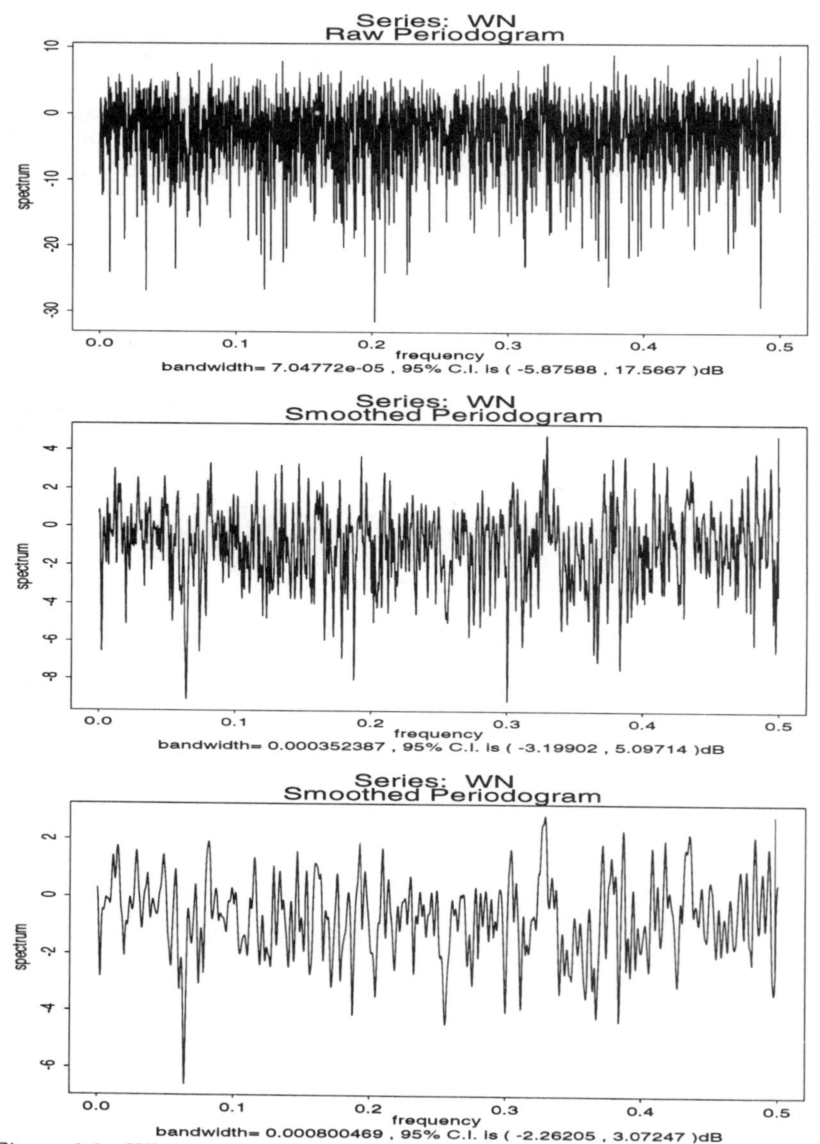

Figure 2.3. White noise: The top plot reproduces the raw periodogram of the white noise series used throughout this chapter. The other plots give smoothed versions of the periodogram. The second plot was obtained with two smoothing passes, whereas the third plot uses four passes.

```
> acf(WN)
> spec.pgram(WN,plot=T)
```

The function `acf` by default computes and plots the sample autocorrelation function of the series given as argument. On the other hand the function `spec.pgram` produces a plot only if the parameter `plot` is set to `T` (i.e., true).

Computation of smoothed spectra:

```
> PWN <- spec.pgram(WN,plot=T)
> P2WN <- spec.pgram(WN,spans=c(3,5),plot=T)
> P4WN <- spec.pgram(WN,spans=c(3,5,7,9),plot=T)
```

The plots produced by these commands are included in Figure 2.3. The following commands are used to compute the values of the means and the variances used in the text. The function `spec.pgram` returns a list and one of the elements of this list, namely `$spec`, contains the values of the spectral power in decibels.

Numerical values of the means and the variances:

```
> mean(PWN$spec)
-2.98451
> mean(P2WN$spec)
-0.9564203
> mean(P4WN$spec)
-0.7168848
> var(PWN$spec)
30.04696
> var(P2WN$spec)
4.012902
> var(P4WN$spec)
1.712197
```

2.5.2 Auto-regressive Models

An auto-regressive model of order p is a sequence $f = \{f_n\}$ of random variables satisfying

$$f_n = \phi_1 f_{n-1} + \cdots + \phi_p f_{n-p} + \epsilon_n \ , \ n \in \mathbb{Z}, \tag{2.30}$$

where the coefficients ϕ_j of the model are deterministic constants and where the sequence $\epsilon = \{\epsilon_n\}_n$ is a white noise with unknown variance σ^2. The recursion formula (which essentially say that f_n regresses itself linearly on its

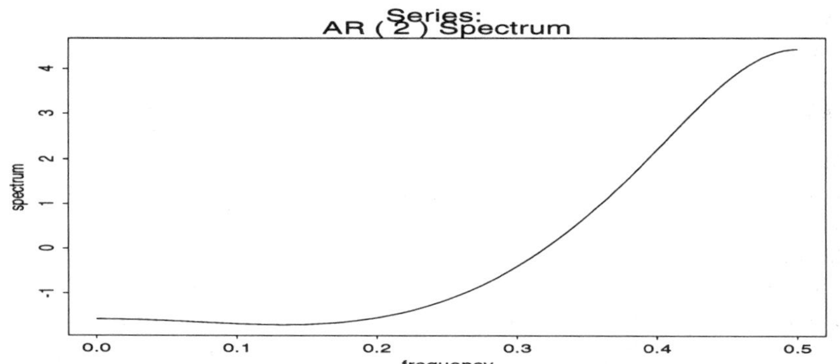

Figure 2.4. AR process: Spectral density for the AR(2) with coefficients −0.3 and 0.1.

past values) can be used to show that the autocorrelation function decays exponentially. Moreover, writing down the first values of the autocovariance function in terms of the parameters ϕ_j's and σ^2 of the model gives a system of equations (called *Yule-Walker equations*) which can be used to estimate the parameters of the model from the values of the sample autocovariance function. This is the standard way to fit an auto-regressive model to sample data. (See the following example.) From its definition (2.30) it is easy to show that the spectral density of an AR(p) process f is given by the formula

$$\nu_f(\omega) = \frac{\sigma^2}{2\pi} \frac{1}{|1 - \phi_1 e^{-2\pi i\omega} - \cdots - \phi_p e^{-2\pi i p\omega}|^2} . \tag{2.31}$$

We can use the following S-commands to compute and plot such a spectral density.

Auto-regressive spectrum:

```
> ARMODEL <- list(ar=c(-.3,.1),order=2,var.pred=1.0)
> MODELSP <- spec.ar(ARMODEL,plot=T)
```

Figure 2.4 gives the plot of the spectral density for the AR(2) model with coefficients −0.3 and 0.1 as computed from the formula (2.31). We now simulate a sample realization of length 1024 of such a AR model, and we estimate its spectrum by first fitting an AR model to the time series so obtained and by using the estimated parameters to derive the spectrum.

AR spectral estimation from a sample:

```
> AR <- arima.sim(model=list(order=c(2,0,0),ar=c(-0.3,0.1)),
n=1024)
```

```
> tsplot(AR)
> spec.pgram(AR,plot=T)
> ARR <- ar(AR)
> ARRSP <- spec.ar(ARR,plot=T)
```

Figure 2.5 shows the original signal on the top, the raw periodogram of the sample AR time series in the middle, and the spectral estimate obtained by formula (2.31) from the estimated values of the coefficients at the bottom. The first obvious remark is that the raw periodogram does much worse at estimating the true spectrum (recall that the latter is shown in Figure 2.4.) So it is worth using a parametric model when we have one. Indeed, even though a model of order $p = 7$ (instead of $p = 2$) was fitted to the sample series, the resulting spectrum is nevertheless much closer to the true one.

2.5.3 Monthly CO_2 Concentrations

We conclude this chapter with an example from the Splus package. The monthly concentrations of CO_2 at Mauna Loa, Hawaii from January 1958 to December 1975 are contained in a data set co2 included with the Splus package. Figure 2.6 gives a plot of these values. Figure 2.7 gives the raw periodogram and two other spectral estimates obtained by smoothing.

Plot of the CO_2 concentrations:

```
> tsplot(co2)
```

This plot reveals a linear upward trend and a cyclic behavior whose period seems to be of 12 months. None of these remarks should come as a surprise. The spectral analysis is performed via the S-commands:

Spectral analysis of the monthly CO_2 concentrations:

```
> SPCO <- spec.pgram(co2,plot=T)
> SP2CO <- spec.pgram(co2,spans=c(3,5),plot=T)
> SP4CO <- spec.pgram(co2,spans=c(3,5,7,9),plot=T)
```

The plots are reproduced in Figure 2.7. One sees clearly the peak corresponding to the period of 1 year and to its harmonics. By default, the function `spec.pgram` removes the linear trend before computing the spectral estimates, but it does not remove the seasonal effect. As we can see from Figure 2.7, it builds a peak at the frequency of the effect and secondary peaks at the harmonics. This can be a shortcoming if these large peaks mask smaller spectral features of interest. In such cases it might be desirable to remove the seasonal component prior to the spectral analysis especially if such a component can be identified without ambiguity. Splus provides a function to do just that. It is called `stl`, and we illustrate its use

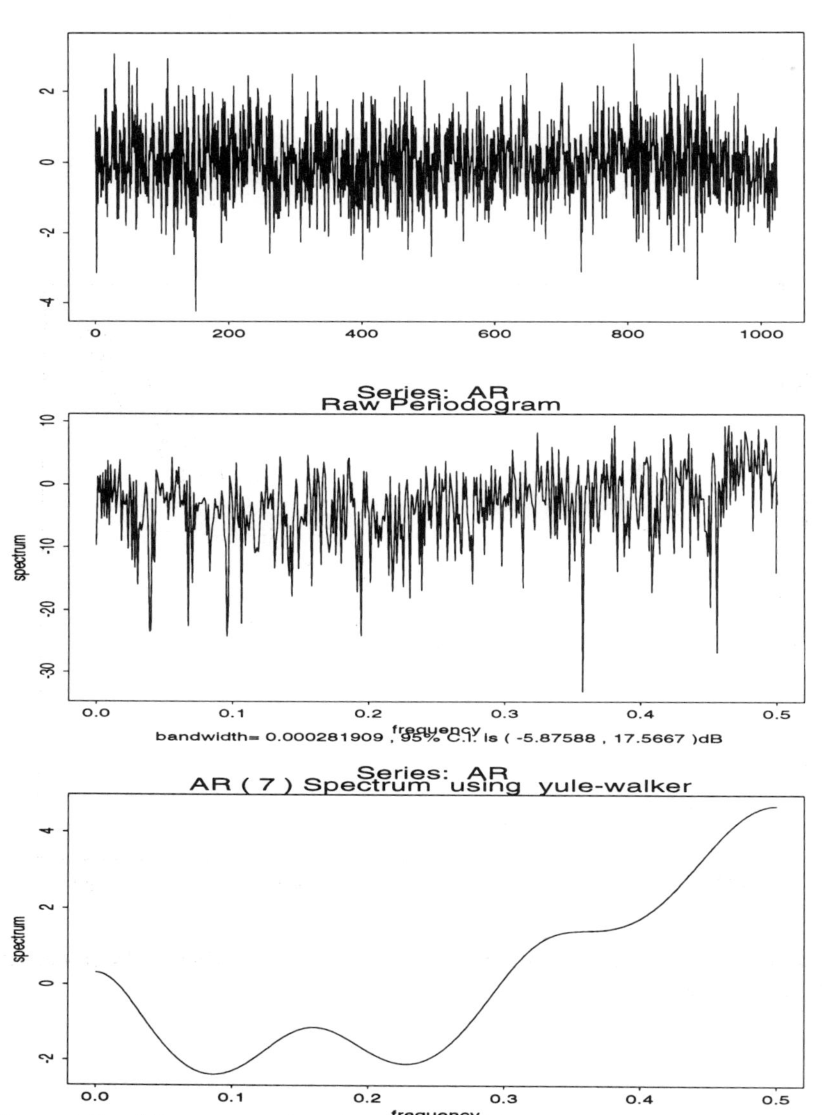

Figure 2.5. AR process: Sample of an AR(2) with coefficients -0.3 and 0.1 (Top). Its raw periodogram is given in the plot in the middle while the spectral estimate obtained by fitting first an auto-regressive model to the data and using the spectral form of auto-regressive spectra is given in the bottom plot.

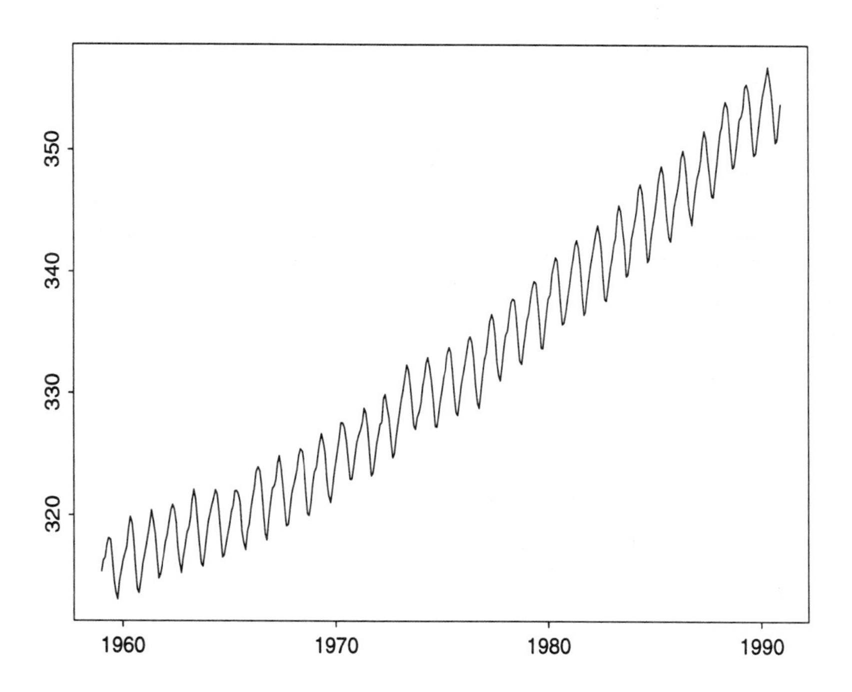

Figure 2.6. Monthly concentrations of CO_2 at Mauna Loa, Hawaii, from January 1958 to December 1975.

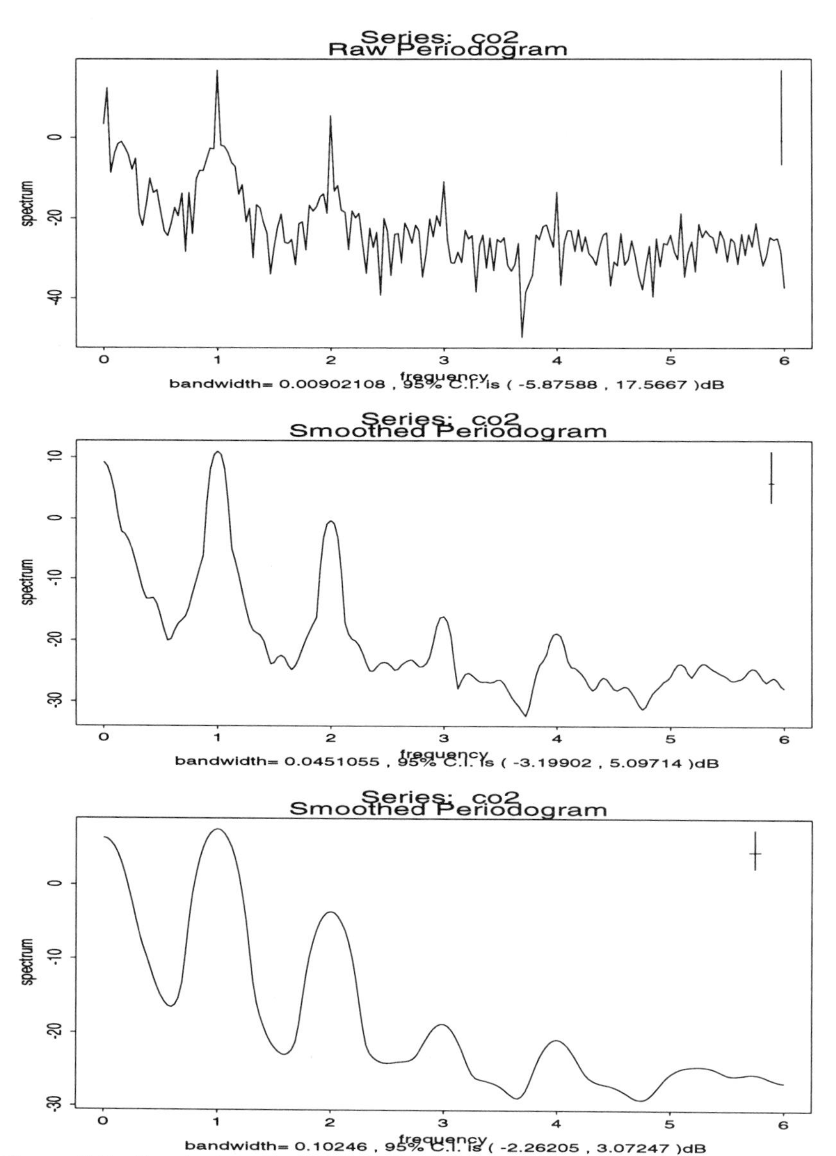

Figure 2.7. Spectral analysis of the monthly concentrations of CO_2. The raw periodogram (top) and the smoothed periodograms (middle and bottom) contain the peaks indicative of the yearly period and of its harmonics.

in the case of the CO_2 monthly concentrations. The following S-commands produce (among other things) the plots contained in Figure 2.8.

Removing a seasonal component:

```
> co2.stl <- stl(co2,"periodic")
> tsplot(co2.stl$sea)
> tsplot(co2,co2.co2.stl$rem)
> spec.pgram(co2.stl$sea,plot=T)
> spec.pgram(co2.stl$rem,plot=T)
> spec.pgram(co2.stl$sea,spans=c(3,5,7),plot=T)
> spec.pgram(co2,stl$rem,spans=c(3,5,7),plot=T)
```

2.6 Notes and Complements

The spectral theory of stationary stochastic processes is a classical chapter of *the book of statistics*. We refer to reader to the classical texts of D. Brillinger [58], Bloomfield [52], and Koopmans [195] for the general theory and to the texts of Newton [235] and Percival and Walden [239] for the same theory with a strong emphasis on the applications.

As mentioned in the text, the examples in the last section were borrowed from the Splus manual.

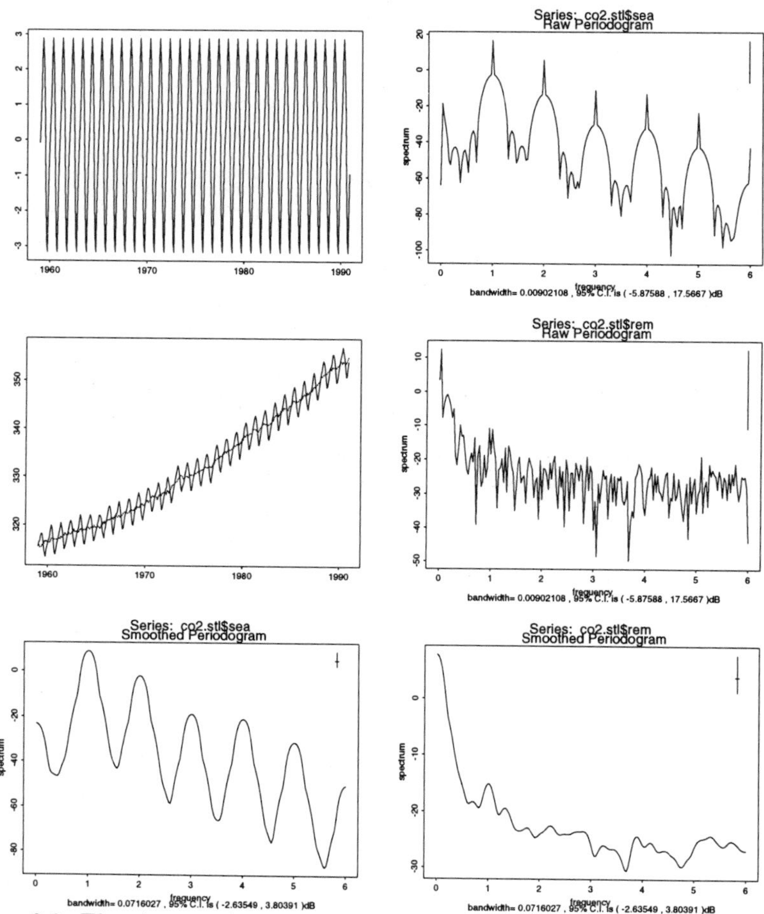

Figure 2.8. The plots on the top row give the seasonal component as identified by the function **slt** of Splus and its raw periodogram. The middle row contains the plots of the remainder (i.e., what is left once the seasonal component has been removed from the original data) and its raw periodogram. Finally, the bottom row contains two smoothed versions of the raw periodogram given above.

Part II

Gabor and Wavelet Transforms

Chapter 3

The Continuous Gabor Transform

We begin our presentation of the time-frequency signal transformations with the case of the continuous Gabor transform (CGT for short). The continuous Gabor transform is nothing but a simple localization of the Fourier transform via the introduction of a sliding window function. The existence of this window makes this transform into a function of two parameters: a time parameter giving the location of the center of the window and a frequency parameter for the computation of the Fourier transform of the windowed signal. Let us stress that although Gabor's original representation was discrete, we use the terminology "continuous Gabor transform" to conform with the widespread practice. The Gabor transform is based upon time and frequency translations, as opposed to time translations and scalings as in the case of the wavelet transform which we consider in the next chapter.

3.1 Definitions and First Properties

We start with the basic definitions and properties of the continuous Gabor transform. It is a particular case of the general scheme described in the previous chapter.

3.1.1 Basic Definitions

Given a function $g \in L^2(\mathbb{R})$, we construct the corresponding family $\{g_{(b,\omega)};\ b \in \mathbb{R}, \omega \in \mathbb{R}\}$ of Gabor functions obtained by shifting and modulating copies of g:

$$g_{(b,\omega)}(x) = e^{i\omega(x-b)} g(x-b) \ . \tag{3.1}$$

The function g is used to concentrate the analysis near specific points in the time domain. It will be called the window function of the analysis. It plays the role of the taper in standard spectral theory (see Chapter 2). Once the window g is fixed, the associated continuous Gabor transform is defined as follows.

Definition 3.1 *Let $g \in L^2(\mathbb{R})$ be a window. The continuous Gabor transform of a finite-energy signal $f \in L^2(\mathbb{R})$ is defined by the integral transform*

$$G_f(b,\omega) = \langle f, g_{(b,\omega)} \rangle = \int f(x)\, \overline{g(x-b)e^{-i\omega(x-b)}}\, dx\ . \tag{3.2}$$

Contrary to the theory of the Fourier transform of square integrable functions, this defining integral does not need to be interpreted in an L^2 sense. Indeed, since $f(\,\cdot\,)$ and $\overline{g(\,\cdot\,-b)}$ are both square integrable, their product is integrable and the integral makes perfectly good sense for all the values of b and of ω. This localized Fourier transform has received many names in the signal analysis literature. It is often referred to as the time-dependent Fourier transform, or the windowed Fourier transform, or even the short-term Fourier transform. For the sake of definiteness we shall use the terminology Gabor transform. The set of coefficients $G_f(b,\omega)$ is sometimes called the Gabor representation of f. The Gabor representation is complete. Indeed, the continuous Gabor transform is invertible on its range:

Theorem 3.1 *Let $g \in L^2(\mathbb{R})$ be a nontrivial window (i.e., $||g|| \neq 0$.) Then every $f \in L^2(\mathbb{R})$ admits the decomposition*

$$f(x) = \frac{1}{2\pi||g||^2} \int_{-\infty}^{\infty} \int_{-\infty}^{\infty} G_f(b,\omega) g_{(b,\omega)}(x)\, db\, d\omega\ , \tag{3.3}$$

where equality holds in the weak $L^2(\mathbb{R})$ sense.

A simple polarization argument shows that the representation formula (3.3) is in fact equivalent to a norm-preserving property of the CGT. The latter is known as the "Gabor-Plancherel formula." It reads

$$\int_{\mathbb{R}^2} |G_f(b,\omega)|^2 db\, d\omega = 2\pi ||g||^2\ ||f||^2 \tag{3.4}$$

and results from a simple application of Fubini's lemma. More precisely, it follows from Fubini's lemma if one assumes that $g \in L^1(\mathbb{R}) \cap L^2(\mathbb{R})$. The general case is obtained through classical limit arguments. The reader is referred to [161] for a complete proof.

In other words, the mapping

$$L^2(\mathbb{R}) \ni f \hookrightarrow \frac{1}{||g||\sqrt{2\pi}} G_f \in L^2(\mathbb{R}^2)$$

is norm-preserving. This formula gives a precise form to the notion of conservation of energy between the time domain and the time-frequency domain. As such, it is the basis for the interpretation of $|G_f(b,\omega)|^2$ as a time-frequency energy density (up to the multiplicative factor $||g||\sqrt{2\pi}$.) In the notations of Chapter 1 we have $\lambda = (b,\omega)$, $\Lambda = \mathbb{R} \times \mathbb{R}$, and $\mathcal{H} = L^2(\mathbb{R} \times \mathbb{R}, d\omega db)$ for the target space.

It is interesting to note the symmetry of the CGT. Indeed, by Plancherel's formula, we may also write

$$G_f(b,\omega) = \frac{1}{2\pi}\langle \hat{f}, \widehat{g_{(b,\omega)}} \rangle = \frac{1}{2\pi}\int \hat{f}(\xi)\, \overline{\hat{g}(\xi-\omega)e^{i\xi b}}\, d\xi \ , \tag{3.5}$$

so that $G_f(b,\omega)$ may also be thought of as a Gabor transform of the Fourier transform of f (up to a phase factor).

3.1.2 Redundancy and Its Consequences

The Gabor transform maps functions of one variable into functions of two variables. The CGT of a signal $f \in L^2(\mathbb{R})$ contains highly redundant information. This fact is reflected in the existence of a (smooth) *reproducing kernel*:

Proposition 3.1 *Given a window function $g \in L^2(\mathbb{R})$, the image of $L^2(\mathbb{R})$ by the corresponding continuous Gabor transform is the reproducing kernel Hilbert space:*

$$\mathcal{H}_g = \left\{F \in L^2(\mathbb{R}^2), P_g F = F\right\} \ . \tag{3.6}$$

where P_g is the integral operator with kernel $\mathcal{K}_g$ (the reproducing kernel) is defined by

$$\mathcal{K}_g(b,\omega;b',\omega') = \frac{1}{2\pi||g||^2}\langle g_{(b',\omega')}, g_{(b,\omega)}\rangle \ . \tag{3.7}$$

The operator P_g is nothing but the orthogonal projection onto $\mathcal{H}_g$, i.e., the self-adjoint operator satisfying

$$P_g^2 = P_g^* = P_g \ .$$

We use the notation P^2 for the composition $P \circ P$ and the exponent * for the (complex) adjoint of an operator or a matrix. The proposition is proved by considering the inner products of the left- and right-hand sides of equation (3.3) with Gabor functions. This leads to the formula

$$G_f(b,\omega) = \int \mathcal{K}_g(b,\omega;b',\omega')\, G_f(b',\omega')db'd\omega',$$

from which all the claims of the proposition follow easily.

Remark 3.1 It is interesting to notice that the action of the orthogonal projection P_g is simply an inverse CGT (3.3) followed by a CGT (3.2).

Remark 3.2 The redundancy of the CGT has the following interesting consequence. *It is possible to use in the inverse CGT a window which is different from the one used in the direct transform.* The formulas are modified in the following way. Let $h \in L^2(\mathbb{R})$ be the reconstruction window. Then equation (3.3) has to be replaced by

$$f(x) = \frac{1}{2\pi\langle h,g\rangle}\int_{-\infty}^{\infty}\int_{-\infty}^{\infty} G_f(b,\omega)h_{(b,\omega)}(x)dbd\omega \ . \qquad (3.8)$$

An immediate consequence is the existence of associated reproducing kernels. Indeed, let $P_{g,h}$ be the integral operator defined by the kernel

$$\mathcal{K}_{g,h}(b,\omega;b,\omega') = \frac{1}{2\pi\langle h,g\rangle}\langle h_{(b',\omega')}, g_{(b,\omega)}\rangle \ .$$

One easily shows that $P_{g,h}$ is another (non-orthogonal) projection onto $\mathcal{H}_g$. Again, the action of $P_{g,h}$ may be seen as an inverse CGT followed by a CGT, but now with different window functions. The new kernels correspond to *oblique* (i.e., non-orthogonal) projections.

3.1.3 Invariance Properties

The Gabor transform of a shifted copy of a signal f is the shifted copy of the Gabor transform of f. In the same way, the Gabor transform of a modulated copy of a signal f is the corresponding frequency-shifted copy of the Gabor transform of f.

Translations and modulations become (two-dimensional) transformations in the time-frequency plane and the proposition below gives the expected invariance properties of the Gabor transform. They have important consequences for some applications such as detection problems, since matched filtering type techniques may be extended to more general situations. Let us quote, for instance, the case of narrow-band radar analysis of Doppler effect, where the goal is to measure a relative speed through the frequency shift of a reflected signal. This type of analysis is conveniently carried out with Gabor transform.

More generally, if one puts together time shifts and modulations (i.e., frequency shifts), the result is slightly more complicated

Proposition 3.2 *Let $f \in L^2(\mathbb{R})$, and set*

$$\tilde{f}(x) = e^{i\omega_0(x-x_0)} f(x - x_0) \ .$$

Then

$$G_{\tilde{f}}(b, \omega) = e^{i\omega_0(b-x_0)} G_f(b - x_0, \omega - \omega_0) \ . \tag{3.9}$$

Although the proof of the proposition is immediate, we reproduce it in order to emphasize the origin of the phase factor. It is convenient to introduce the translation and modulation operators. For $b \in \mathbb{R}$ and $\omega \in \mathbb{R}$, we set

$$T_b f(x) = f(x - b) \, , \quad E_\omega f(x) = e^{i\omega x} f(x) \ . \tag{3.10}$$

Then we have $\tilde{f}(x) = T_{x_0} E_{\omega_0} f(x)$, $g_{(b,\omega)}(x) = T_b E_\omega g(x)$, and

$$G_{\tilde{f}}(b, \omega) = \langle f, E_{-\omega_0} T_{-x_0} T_b E_\omega g \rangle = \langle f, E_{-\omega_0} T_{b-x_0} E_\omega g \rangle \ .$$

Now, the proposition follows from the relation

$$E_\omega T_b = e^{i\omega b} T_b E_\omega \ . \tag{3.11}$$

This formula is slightly different from the previous invariance formulas because of the presence of the phase factor. Thanks to equation (3.11), we see that this phase factor reflects the non-commutativity of the translation and modulation operators, in other words, the fact that these two families of operators actually generate a three-parameter group, the so-called *Weyl-Heisenberg group*. The Gabor functions are generated from the action of the Weyl-Heisenberg group on the window g. Such aspects have been analyzed in particular in [153, 154, 124, 125]. We shall not elaborate on these points, but it is worth mentioning that geometrical and group theoretical approaches constitute an important part of time-frequency representations. We give a slightly more precise description of these points in the following remark.

Remark 3.3 *The Weyl-Heisenberg group:* G_{WH} is the group generated by elements of the form $(b, \omega, \varphi) \in \mathbb{R}^2 \times [0, 2\pi]$ with group multiplication

$$(b, \omega, \varphi) \cdot (b', \omega', \varphi') = (b + b', \omega + \omega', \varphi + \varphi' + \omega b' \mod 2\pi) \ .$$

It may, for example, be realized as the group of matrices

$$(b, \omega, \varphi) = \begin{pmatrix} 1 & \omega & \varphi \\ 0 & 1 & b \\ 0 & 0 & 1 \end{pmatrix} \ .$$

The measure $d\mu(b,\omega,\varphi) = db\, d\omega\, d\varphi$ is invariant in the sense that

$$\mu((b',\omega',\varphi')\cdot A) = \mu(A)$$

for all $(b',\omega',\varphi') \in G_{WH}$ and all bounded Borel subsets A of G_{WH}. In addition, G_{WH} acts naturally on $L^2(G_{WH}, d\mu)$, as follows. For each $(b',\omega',\varphi') \in G_{WH}$, one defines the unitary operator $\lambda_{(b',\omega',\varphi')}$ which acts on any $F \in L^2(G_{WH}, d\mu)$ according to

$$[\lambda_{(b',\omega',\varphi')}F](b,\omega,\varphi) = F((b',\omega',\varphi')^{-1}\cdot(b,\omega,\varphi)), \qquad (b',\omega',\varphi') \in G_{WH}\ .$$

We can also use the action of G_{WH} on $L^2(\mathbb{R})$ defined in a similar way. For each $(b',\omega',\varphi') \in G_{WH}$ is associated the unitary operator $\pi_{(b',\omega',\varphi')}$ acting on $L^2(\mathbb{R})$ according to

$$[\pi_{(b',\omega',\varphi')}f](x) = e^{i[\varphi'+\omega'(x-b')]}f(x-b')$$

if $f \in L^2(\mathbb{R})$. Both λ and π are *representations* of G_{WH}, in the sense that they preserve the group multiplication. For example, we have

$$\pi_{((b,\omega,\varphi)\cdot(b',\omega',\varphi'))} = \pi_{(b,\omega,\varphi)}\pi_{(b',\omega',\varphi')}$$

for all $(b,\omega,\varphi), (b',\omega',\varphi') \in G_{WH}$. λ is called the (left) *regular* representation of G_{WH}. Once these notations have been introduced, the invariance expressed in Proposition 3.2 may be rephrased as follows. Let us use the notation G for the Gabor transform viewed as a mapping from $L^2(\mathbb{R})$ into $L^2(\mathbb{R}^2)$. Then

$$\lambda \circ G = G \circ \pi\ .$$

This expresses the fact that $L^2(\mathbb{R})$ is isomorphic to a subspace of $L^2(\mathbb{R}^2)$, and that the actions of the group on these two subspaces are isomorphic. The operator $G : L^2(\mathbb{R}) \to L^2(\mathbb{R}^2)$ is called an *intertwining operator* between π and λ.

Nevertheless, since the departure from perfect invariance is mild, the modulus of the Gabor transform (sometimes referred to as the *spectrogram*) is still invariant.

3.2 Commonly Used Windows

Before discussing examples of Gabor transforms, it is useful to discuss the choice of the analyzing window g. As we have seen, any square-integrable window would do the job, but some choices are more convenient than others. The main property for being a "good window" for CGT is to be "well

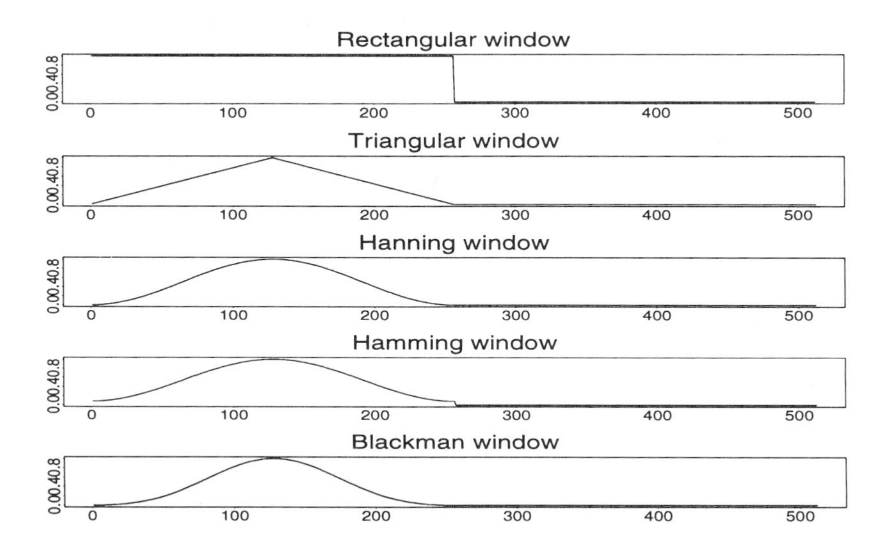

Figure 3.1. Graphs of commonly used window functions g. Notice that the label of the horizontal axis should be divided by 512 to correspond to the defining formulas given in the text.

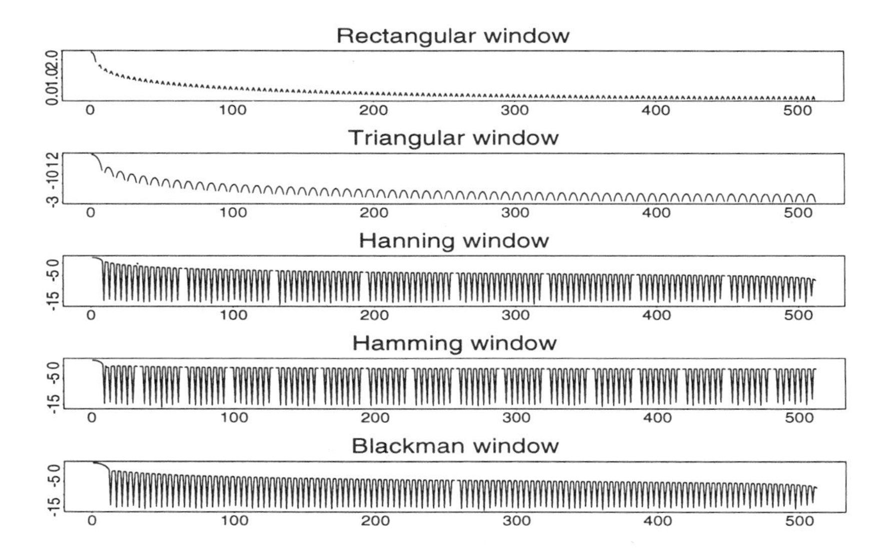

Figure 3.2. Plots of the logarithm $\omega \hookrightarrow \log_{10} |\hat{g}(\omega)|$ of the modulus of the commonly used window functions g given in Figure 3.1.

localized" in both the time and frequency domains. For this, there are standard choices, inherited from the spectral estimation literature. Some of the most commonly used windows are shown in Figure 3.1 and the logarithms of their Fourier transforms are displayed in Figure 3.2. The main reason for the use of these window functions is that they have simple functional forms and their Fourier transforms are essentially concentrated around the origin $\omega = 0$ as can be seen from Figure 3.2. The mathematical formulas defining these window functions are as follows. Notice that we often use scaled and shifted versions of those windows given by the following formulas. For example, we have chosen to write the formulas so that all the windows are supported in the interval $[0, 1]$, even though one will most of the time require that the window be "centered" around 0.

- *Rectangular*

$$g(x) = \begin{cases} 1, & 0 \le x \le 1 , \\ 0, & \text{otherwise} . \end{cases} \tag{3.12}$$

- *Bartlett (triangular)*

$$g(x) = \begin{cases} 2x, & 0 \le x \le 1/2 , \\ 2(1-x), & 1/2 < x \le 1 , \\ 0, & \text{otherwise} . \end{cases} \tag{3.13}$$

- *Hanning*

$$g(x) = \begin{cases} 1(1-\cos(2\pi x))/2, & 0 \le x \le 1 , \\ 0, & \text{otherwise} . \end{cases} \tag{3.14}$$

- *Hamming*

$$g(x) = \begin{cases} .54 - .46\cos(2\pi x), & 0 \le x \le 1 , \\ 0, & \text{otherwise} . \end{cases} \tag{3.15}$$

- *Blackman*

$$g(x) = \begin{cases} .42 - .5\cos(2\pi x) + .08\cos(4\pi x), & 0 \le x \le 1 , \\ 0, & \text{otherwise} . \end{cases} \tag{3.16}$$

Which window to choose depends on several objective and subjective reasons. For example, the Hamming's window seems to be the favorite one in speech analysis. But the most commonly used window function is the Gaussian window:

$$g(x) = e^{-x^2/2\sigma^2} . \tag{3.17}$$

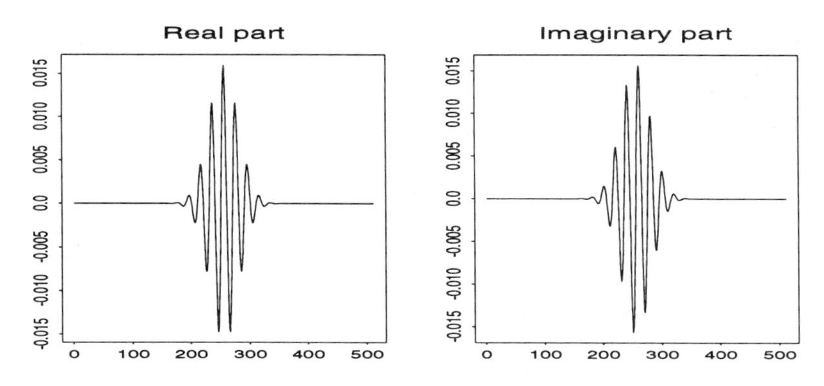

Figure 3.3. Real and imaginary parts of the Gabor function generated in Example 3.1.

Indeed, the associated Gabor functions minimize the time-frequency spread, i.e., they reach the bound given by Heisenberg's inequality. All the Gabor transform computations reported in this monograph use the Gaussian window. Examples of Gabor functions with Gaussian window generated in Example 3.1 are displayed in Figure 3.3.

3.3 Examples

For the sake of illustration we consider first simple academic models in order to illustrate the behavior of the CGT. We look at examples that may be completely understood with simple numerical calculations. We shall see later how to perform Gabor analysis of real life signals and especially speech signals, and to use such time-frequency representations to extract salient features with a clear physical interpretation.

3.3.1 Academic Signals

Sine Waves

It is easy to check that the continuous Gabor transform of a sine wave $f(x) = \sin(\lambda x)$ is given by

$$G_f(b,\omega) = \frac{1}{2i}\left[e^{i\lambda b}\overline{\hat{g}(\lambda-\omega)} - e^{-i\lambda b}\overline{\hat{g}(\lambda+\omega)}\right] .$$

Thus, if we assume that the window function g has been chosen well localized near the origin in the frequency domain, the CGT should be localized

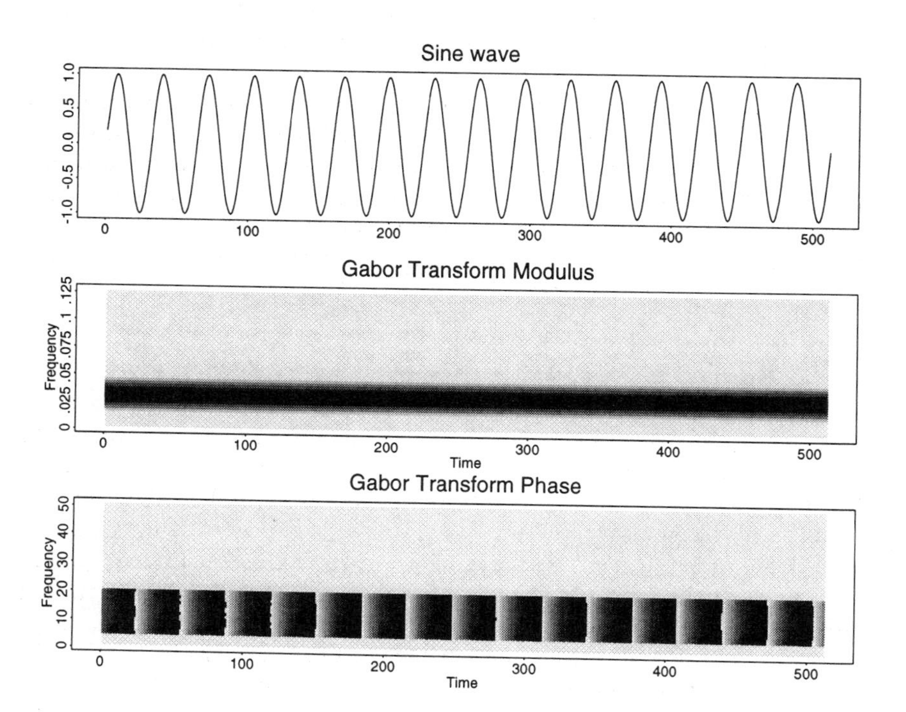

Figure 3.4. Plots of the modulus and the phase of the CGT of a sine wave of frequency 1/32 Hz, generated in Example 3.3.

near the lines $\omega = \pm\lambda$, as can be seen in Figure 3.4 (where only the positive values of the frequency variable ω are displayed). There, the sampling frequency is assumed to be 1 Hz, i.e., one sample per second, and the frequency of the sine wave was set to 1/32 Hz. The phase of the Gabor transform has also an interesting behavior. In the zone of interest of the time-frequency plane, i.e., for values of ω close to the frequency of the sine wave (outside this zone, the phase is meaningless because the modulus is too small and it has been discarded), the phase of the CGT behaves like that of the signal (same frequency and no phase shift). As we shall see in Chapter 7, such an information may be turned into algorithms for determining the frequency of a signal.

It is important to understand that in such a case, the localization is not perfect as it would be in the case of the Wigner-Ville transform of such a signal: using a window has somewhat lowered the resolution of the time-frequency representation.[1] On the other hand, the window avoids producing a "ghost term" near the zero frequency that would have been produced by the Wigner-Ville transform, for instance, because of interferences between positive and negative frequencies.[2]

Chirps

Our first example only intended to show the behavior of the CGT in a simple situation. Indeed, this signal is stationary and Fourier analysis would have been more appropriate. We now consider our first example of a model of a genuinely nonstationary signal. Since there is no universally accepted definition of a chirp, we shall not attempt to give a precise definition. For us a chirp will essentially be a wave with a time-dependent frequency, and more precisely, by chirp we shall mean a sine of a polynomial function of time. Figure 3.5 gives the plots of the modulus and the phase of the CGT of the specific example of the function $f(x) = \sin(2\pi[x + (x - 256)^2/8000])$. Chirp models are a generalization of the sine wave models. The representation of a periodic function as a superposition of elementary waves with a specific frequencies (see also the spectral representation formula for the stationary random signals given in Chapter 2) indicate that the classical tools of standard Fourier analysis can be very appropriate. On the other hand, if instead of working with superpositions of elementary waves with a specific frequency we work with a single wave, but the frequency is allowed to vary with time, the example of the chirp which we just considered shows that the tools of time-frequency analysis (as provided by the continuous

[1] We shall see when describing the "squeezing" or "reassignment" algorithms how the frequency resolution may be restored in some cases.

[2] In order to overcome this problem, the Wigner-Ville distribution is generally computed on the analytic signal of the original signal.

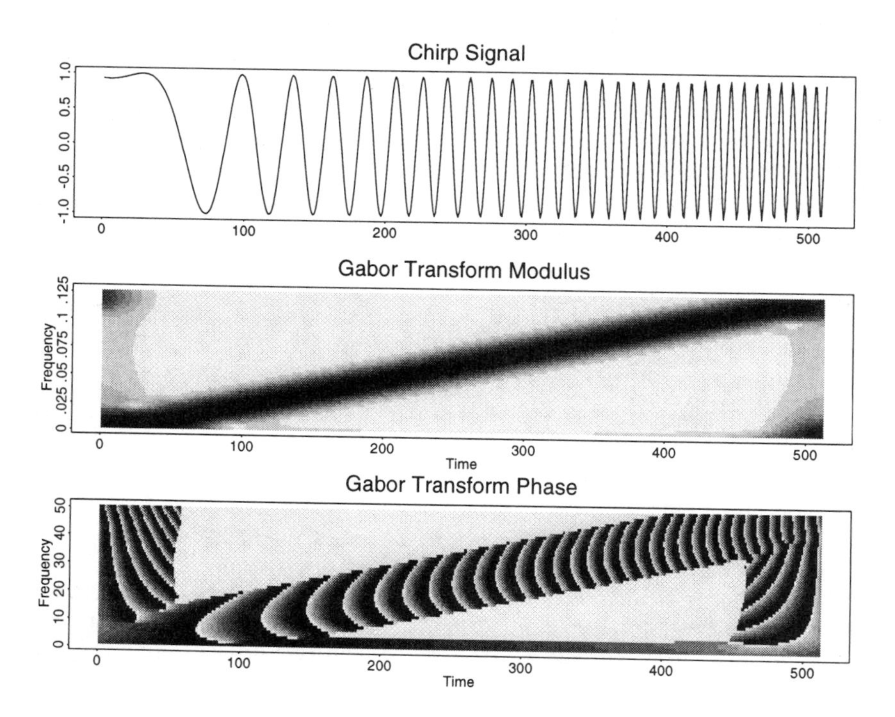

Figure 3.5. Plots of the modulus and the phase of the CGT of a chirp signal, as generated in Example 3.4.

Gabor transform, for example) are more appropriate.

Figure 3.5 gives an example of "linear chirp" (i.e., a chirp whose frequency depends linearly on time). See Example 3.4 for the Splus commands used to generate the figure. The sampling frequency was set to 1, and the chirp's frequency runs from 0 Hz to 0.125 Hz, i.e., the whole frequency range of the figure. As we just explained, we see that the modulus of the CGT is concentrated near the straight line representing the frequency, as in equation (3.21). As in the previous case, the phase information is also interesting. Looking at the behavior of the phase near the ridge, it can be seen that on the ridge, the phase of the CGT behaves exactly in the same way as that of the original signal. Observe also the curvature of the lines of constant phase. These observations can be turned into algorithms for determining the local frequencies of specific classes of signals.

Finally, note the boundary effects at the upper left and lower right parts of the figure. They are due to the fact that the CGT has been computed on a periodized version of the signal.

Frequency-Modulated Signals

We now consider signals which are more complex than the elementary chirps and whose analysis will is more difficult with the tools of classical Fourier analysis. As a typical example we mention the following simple model:

$$f(x) = \sum_{\ell=1}^{L} A_\ell(x) \cos \phi_\ell(x) \ . \tag{3.18}$$

Here, $A_\ell(x) \geq 0$ and $\phi_\ell(x)$ are smooth functions. Despite its simple form, this model is of crucial importance for the analysis of the so-called *frequency-modulated signals*. Successive integration by parts and a series of technical estimates from the asymptotic theory of oscillating integrals can be used to show that, at least for sufficiently large values of ω, the Gabor transform of a signal of the form given in (3.18) is given by

$$\begin{aligned} G_f(b,\omega) &= \frac{1}{2}\sum_{\ell=1}^{L} A_\ell(b) \left(e^{i\phi_\ell(b)}\overline{\hat{g}(\phi'_\ell(b)-\omega)} + e^{-i\phi_\ell(b)}\overline{\hat{g}(-\phi'_\ell(b)-\omega)}\right) \\ &\quad + r(b,\omega), \end{aligned} \tag{3.19}$$

where the remainder $r(b,\omega)$, whose size depends upon the derivatives and second derivatives of the amplitude and phase functions, can be shown to be negligible compared to the sum as long as the amplitudes and frequencies are slowly varying. If we assume that the phase functions $\phi_\ell(x)$ have been chosen in such a way that their derivatives are positive, and if we restrict ourselves to positive values of ω, the continuous Gabor transform of (3.18) is essentially equal to the sum of components of the form

$$A_\ell(b)e^{i\phi_\ell(b)}\overline{\hat{g}(\phi'_\ell(b)-\omega)}\ . \tag{3.20}$$

Assuming as usual that the window function is well localized near the origin of the frequency domain (which is the case for the Gaussian window which we use), each such component localizes in the neighborhood of a curve,

$$\omega = \varphi_\ell(b) = \phi'_\ell(b), \tag{3.21}$$

in the time-frequency (b,ω)-plane. Such curves are named *ridges*. The characterization of frequency-modulated signals may be achieved by detecting the ridges. We shall describe these points in great detail in a later chapter of this text.

We do not give examples of the Gabor transform of specific academic examples of frequency-modulated signals with more than one component, for, looking at Figure 3.5, one imagines easily the result in the case $L > 1$. Furthermore, we shall see that the Gabor transforms of (real-life) speech signals have exactly the same features as the transforms of frequency-modulated signals!

The Wigner-Ville representation of signals of the form (3.18) would be more difficult to interpret because of the presence of the "ghost" cross terms mentioned in the previous chapter. However, if only one single component is present (i.e., if $L = 1$) and if its frequency is a linear function of time, it may be shown that the Wigner-Ville representation is optimally localized on the instantaneous frequency line (i.e., the ridge.) As we already mentioned, the optimality of the Wigner-Ville representation for this class of linear chirps was used as propaganda for this type of bilinear representation.

Remark 3.4 *Importance of the window size:* Up to now, we have not discussed the problem of the specific choice of the window. The functional form of the window has its importance, but this point should be considered as minor. The most important point to be stressed is the importance of the bandwidth of the window (which is fixed once for all, since Gabor functions have constant bandwidth), as is well known to speech specialists for example. The choice of the bandwidth is not innocuous. It actually depends on the processing one has in mind. Clearly, if the window's bandwidth exceeds the "distance" between two harmonic components in a signal, the CGT cannot discriminate between them and interferences between neighboring components create undesirable artifacts. We shall see several examples illustrating this remark, in particular in Figure 3.11 and Example 3.5, later.

Transients

Throughout the book, we shall discuss several examples of acoustic transients. Since real-life data is usually of a sensitive nature and often classified, we shall rely on artificialy academic examples presented to us as realistic recreations of the *real thing*. Let us stress one more time that once a window has been chosen, all the Gabor functions have the same time resolution. It is essentially given by the size of the window, as may be seen in Figure 3.6. The first plot of this figure gives the example of transient signal which we consider in this section. It is a *train* of 10 repetitions of an elementary peak. It will be referred to as $A0$. See the discussion in Example 3.2 at the end of the chapter for details. The signal is assumed to be sampled at unit frequency for simplicity, and the Gabor transform is computed for frequencies ranging from 0 Hz to the Nyquist limit, i.e., 0.5 Hz. The CGT is computed with Gaussian windows of different sizes. In the first case (second plot from the bottom) the window is small (i.e., broad-band) enough so that the transients are "well separated" by the windows and can be seen individually. For a larger (i.e., narrow-band) window (second plot from the top of the figure), the size of the time frequency signature of each transient is wider. Each individual window sees several transients at the same time and as a consequence, the peaks are smeared and the transform cannot "separate" them. The last plot gives the phase of the CGT.

We shall come back to this point in the discussion of a speech example, and in the next chapter devoted to continuous wavelet analysis (see, however, [137, 203] for Gabor-transform-based algorithms for transient detection.)

3.3.2 Discussion of Some "Real-Life" Examples

An Example from Earthquake Engineering

We now present one of the data sets which we will revisit over and over in order to illustrate the use of the various tools presented in the book.

Seismic damage assessment is an important problem in earthquake engineering. Properties of buildings and structures are monitored before, during, and after quakes based on the evolution over time of critical indices which are used in post-earthquake investigations to decide whether a structure that has experienced a strong motion event can still be considered as safe or if some service is required.

We consider records from model tests conducted at the University of Urbana - Champaign by Sozen *et al.* [160, 2, 70]. The tests modeled a scale version of a 10-story 3-bay reinforced concrete structure under sim-

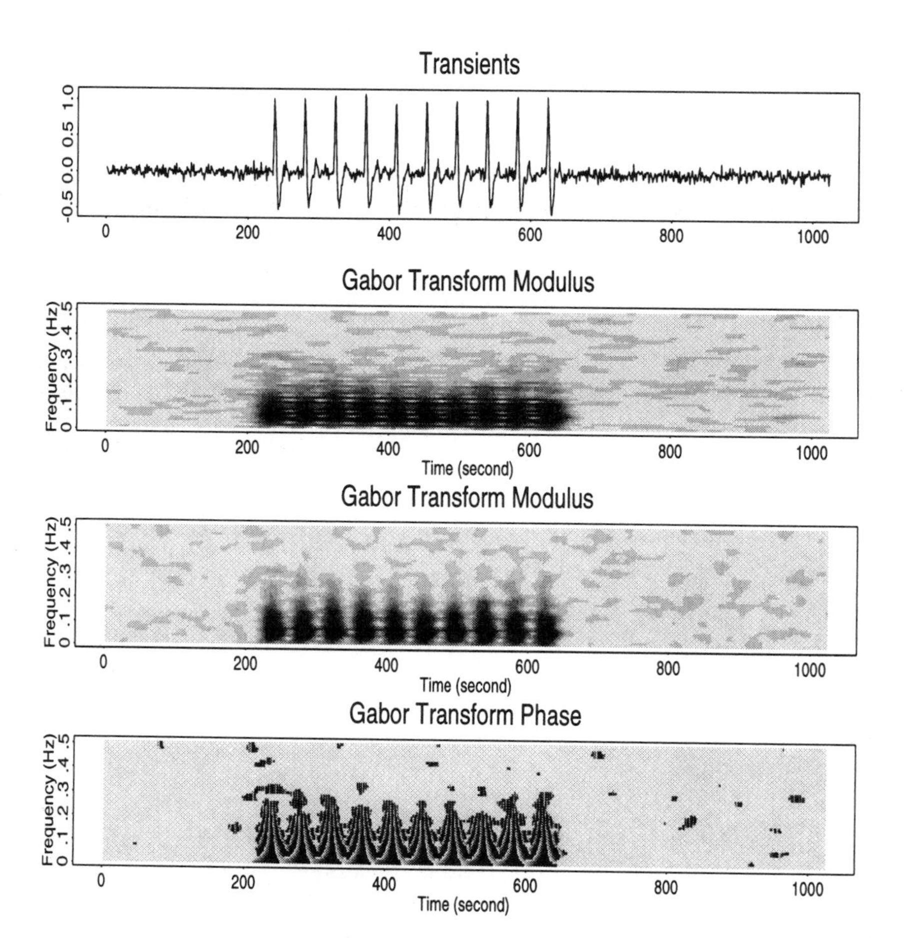

Figure 3.6. CGT of a transient signal: original signal, modulus for two different window sizes, and phase.

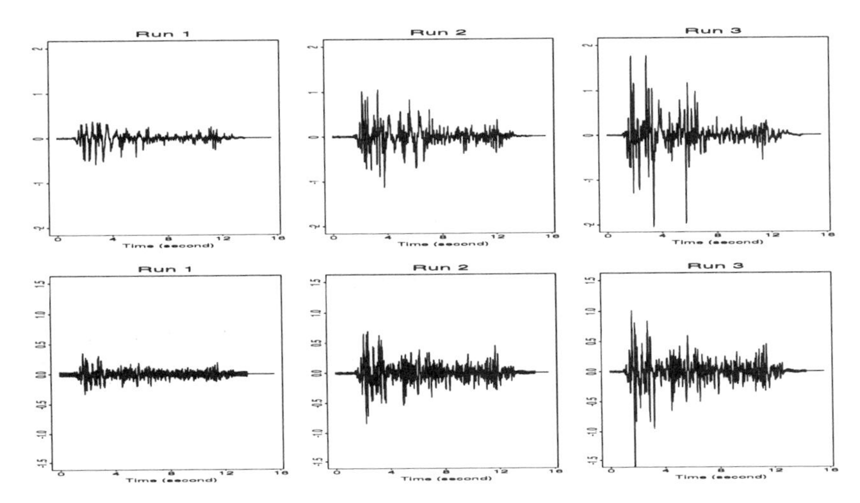

Figure 3.7. The time series on the top shows the recorded accelerations (in gs) on the 10th floor of the structure plotted against time (in s). The measurements of the three runs have been concatenated and are shown back to back. The time series at the bottom of the figure gives the accelerations at the base of the structure.

ulated seismic excitations. The base excitations were patterned after the 1940 El Centro earthquake NS acceleration time series and were scaled to increasingly higher PGA. The test frame specimen was excited by three consecutive horizontal acceleration processes at the ground surface with increasing intensity and no repair or strengthening between runs. The acceleration records from the base (labeled "B") and the 10th story (labeled "10") of the test structure are shown in Figure 3.7.

These data have been analyzed by Cakmak *et al.* in [107] and [230]. Using mathematical models and nonparametric spectral estimation techniques, they showed a significant increase throughout the 15-s spanned by the three runs of the fundamental period of the structure. In [230] the authors assume that the structure response can be modeled by a locally linear system and they use local spectral measures to estimate local (in time) transfer functions and various peak picking techniques to identify what they call a maximum softening damage index. They argue that this index is a reliable measure of damage, and they show that it increases over the time span of the three runs.

More precisely, considering the basement signal as the input and the 10th-floor one as the output of a linear system, the transfer function of the system is estimated as the ratio of the Fourier transforms of the output and input. Such a ratio exhibits a peak for a low frequency, say, ν_0. This

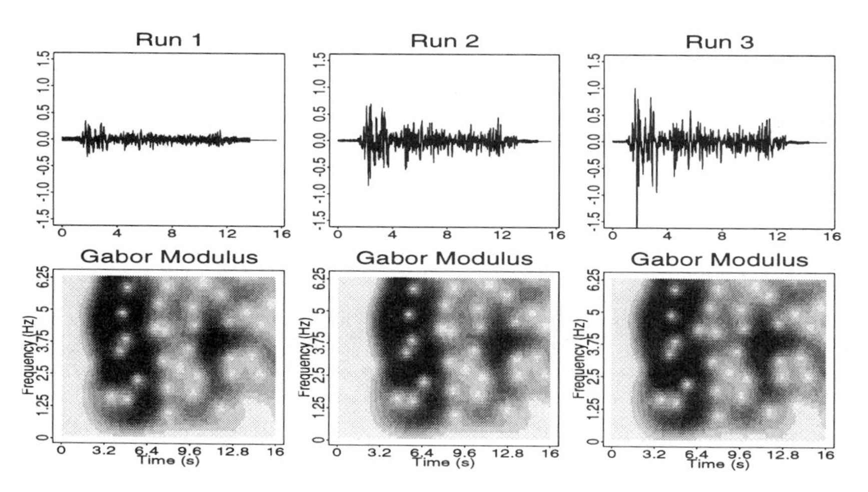

Figure 3.8. CGT of the three runs of the quake data. The modulus of the CGT is imaged for the accelerations measured in the basement (bottom) and the signals are reproduced on the top.

low frequency is seen to decrease when consecutive runs are done. The authors of [107, 230] use this fact to quantify the degree of softening of the structure.

These conclusions are confirmed by the time-frequency analysis of the data which we now perform. For each of the three runs we compute the continuous Gabor transform of the acceleration data for the basement and the 10th-story data. The results for the basement are shown in Figure 3.8 where the modulus of the transform is imaged over the time/frequency plane. We use a Gaussian window with sufficient width in order for the frequency resolution to pick up the features shown. See Example 3.7 later for details on the choices of the parameters. Notice that, even though the signals of the three runs seem significantly different in the time domain, the time/frequency signatures given by the Gabor transform are essentially identical (up to a normalization factor transparent to the observer.) For this reason, we shall try to visualize the damage to the structure via the changes in the time/frequency characteristics of the 10th-floor acceleration only. This is in contrast with the analysis done in [230] where a local transfer function is computed for every run by dividing the spectrogram of the upper level accelerations by the corresponding values for the basement.

Figure 3.9 shows the results of the similar computations for the accelerations on the upper level. We refrained from concatenating the data of the three runs and computing one single transform because of the differ-

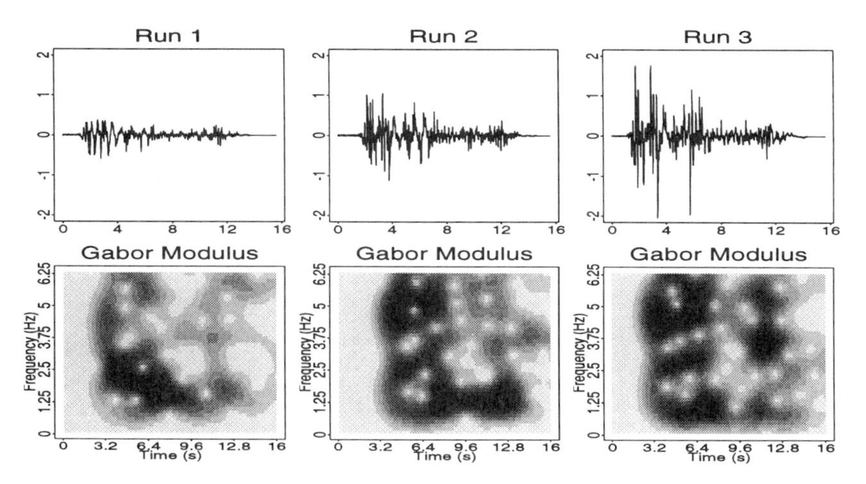

Figure 3.9. CGT of the three runs of the accelerations measured on the upper level of the frame.

ent ranges of the signals in the three runs. It is clearly seen from Figure 3.9 that the energy concentration changes from run to run. In particular, there is a significant contribution at low frequencies, which is gradually shifted from approximately 2 Hz (run 1) to 1 Hz (run 3.) This may be seen more quantitatively in Figure 3.10, which represents the "Gabor transfer function"

$$m(\omega) = \frac{V_{out}(\omega)}{V_{in}(\omega)} , \tag{3.22}$$

where $V(\omega)$ is the *Gabor spectrum*, to be studied later in Chapter 6,

$$V(\omega) = \int |G_f(b,\omega)|^2 db ,$$

and the label *out* (resp. *in*) refers to 10th-story (resp. floor) signal. There, the shift of the fundamental frequency appears clearly. Note that in addition, the peak gets gradually smaller.

However, there is more information to be obtained from Figures 3.8 and 3.9. Looking at the first run (before which the building was not damaged), we see that the distribution of energy in the time-frequency domain at the basement and at the 10th floor differ drastically. This means that the linear system modeling the building has filtered out most of the frequencies of the input signal (in particular the high frequencies) and concentrated essentially all the energy into a single "chirplike" component. This effect is less obvious in the case of run 2 and disappears completely for run 3. In the

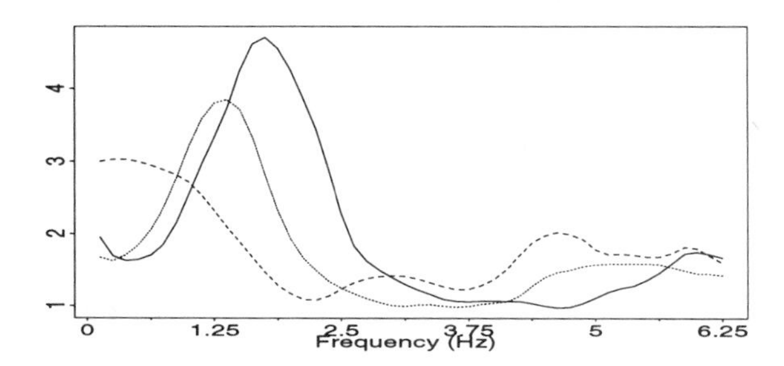

Figure 3.10. Gabor transfer function as defined in (3.22) for the three quake signals; solid line: run 1; dotted line: run 2; dashed line: run 3.

latter case, the time-frequency energy distribution for the basement and the 10th-floor signals are very similar. This means that the above-mentioned filtering done by the building was not done anymore (particularly at high frequencies). It is very tempting to interpret this fact as an indication of the damage.

We shall revisit these data in Section 4.5.3 for their wavelet analysis.

An Example from Speech Analysis

Let us consider the speech signal of Figure 3.11. It was provided to us by Stephane Maes. This example is one of those we shall come back to many times in this book. A possible model to describe such a signal (at least its central part) is given by equation (3.18) with frequencies approximately of the form $\phi'_\ell(x) \approx \ell\phi'_0(x) \approx \ell\omega_0$, where ω_0 is the *pitch* frequency. These models have been successfully applied to (voiced) speech in [222, 223, 252], for instance, in the context of speech compression. The rationale behind such an approach is the following. Whenever a model like (3.18) is used, the signal is described by a certain number of slowly varying functions (essentially the local amplitudes and frequencies), which are thus easily compressed (by plain subsampling in [222, 223], but any other compression scheme would do as well).

It is clear that if the bandwidth of the window used for the CGT computation exceeds ω_0, every Gabor function with frequency in the frequency band of the signal will "see" simultaneously several harmonic components of the signal, thus yielding interferences and corresponding beatings. This is clearly visible in the bottom image of Figure 3.11, where the Gabor transform modulus oscillates as a function of time with frequency equal to the pitch of the signal. Notice the regularity of the pattern that may be observed in that image.

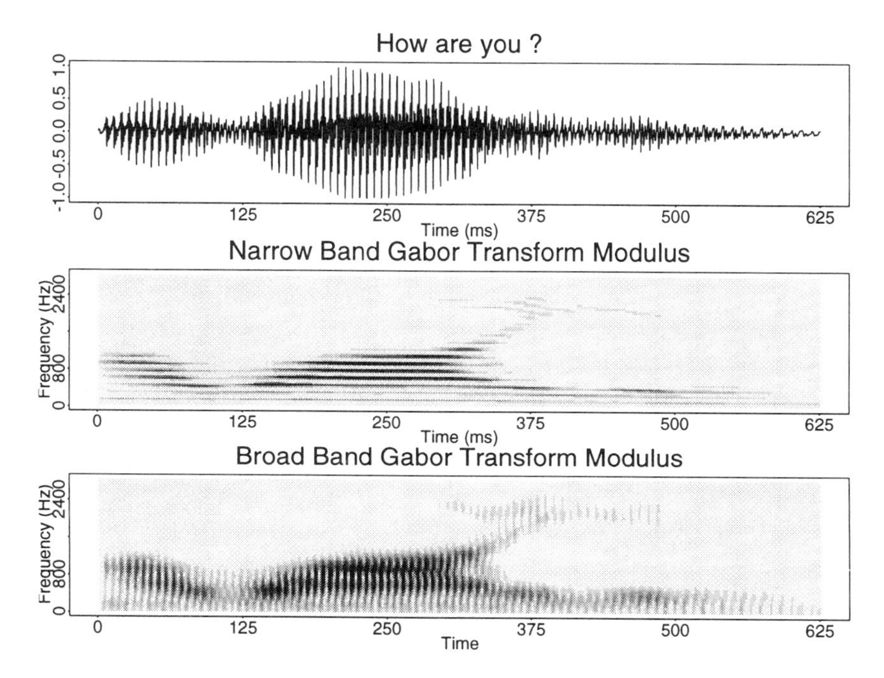

Figure 3.11. CGT of the speech signal /How are you/ with two different window sizes.

In contrast, when the bandwidth is smaller than the pitch frequency, there is no such interference. One says that the frequencies are *well separated* and the Gabor transform reveals the harmonic structure of the signal. This can easily be seen in the middle image of Figure 3.11 (see Example 3.5 for the S-commands used to generate the figure). We shall come back to this example later on. We shall show in Chapter 7 how the CGT may be systematically estimated, and in Chapter 8 how it can be used to reconstruct the harmonic components independently from each other, and then resynthesize the signal.

An Example from Underwater Acoustics

We now describe another class of signals which we are discussing in some detail throughout this book. This kind of signal may be conveniently described by means of time-frequency methods, since it exhibits spectral characteristics which vary as function of time.

We consider two time series from the data base prepared by Gerry

Dobeck *et al.* (see, for example, [273, 274]) at the Naval Surface Warfare Center in Panama City. Their goal was to test the performance of pattern recognition classifiers of acoustic backscatters. If dolphins can recognize underwater objects with their active sonar, humans should be able to detect metallic objects and discriminate them from natural clutter. One of the goals of [273, 274] was to propose a method of generation of synthetic clutter with a time-frequency signature similar to the one of natural clutter. The idea was to prepare surrogate data at a low cost. Even though we shall discuss the time-frequency characteristics of natural clutter for the sake of comparison with the characteristics of the objects which have to be identified, we shall not pursue any further the avenue of the generation of synthetic clutter. For the sake of illustration we consider only a time series giving the acoustic returns from a metallic object and a time series of the returns from natural clutter. Both returns were generated from frequency-modulated transmit signals.

We first consider the CGT of these signals. We show in Figures 3.12 and 3.13 the outputs of the CGT in the two cases (see Example 3.6 for the S commands). The first Gabor transform shows two distinct structures. They could come from the reflections from two different objects or from the main reflection and a secondary one for the same cylinder. Notice also that the appearance of the top plot in Figure 3.13 is different from what can be seen in Figure 3.12, as if there was a quantization anomaly. We believe that this is indeed the case, for a large scaling factor was applied to the values plotted in Figure 3.13 (notice that the scales on the vertical axes are different). The Gabor transform is computed for 50 different frequencies, between 0 and $\nu_s/4$. In this particular example the sampling frequency ν_s is $\nu_s = 2$kHz. The two time-frequency representations are clearly different from each other. In particular, the first one (corresponding to the presence of a metallic object) exhibits two chirps. We shall see more on this signal in Chapter 7 when we discuss the Crazy Climbers ridge detection method.

3.4 Examples and S-Commands

This section is devoted to the description of the S-commands used to produce the results presented earlier. Some of these commands are from the standard S language, but most are specific to the time-frequency Swave package presented in Part II of this monograph.

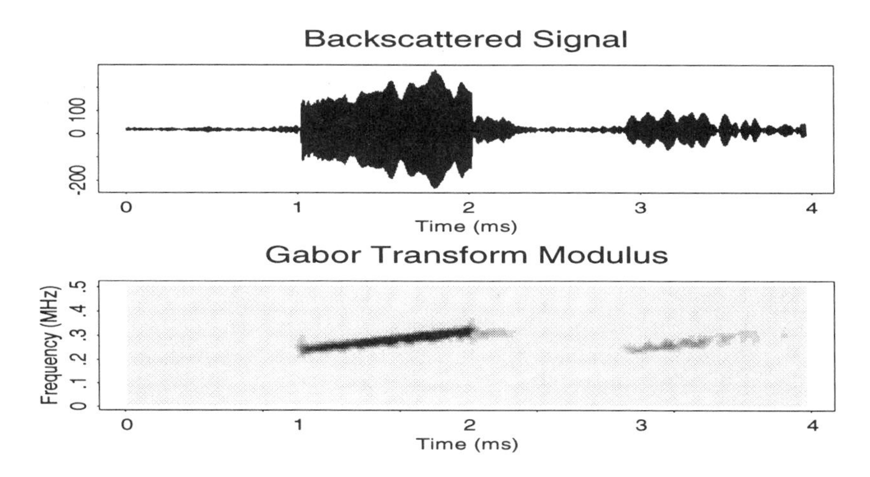

Figure 3.12. Continuous Gabor transform of the acoustic returns from an underwater metallic object.

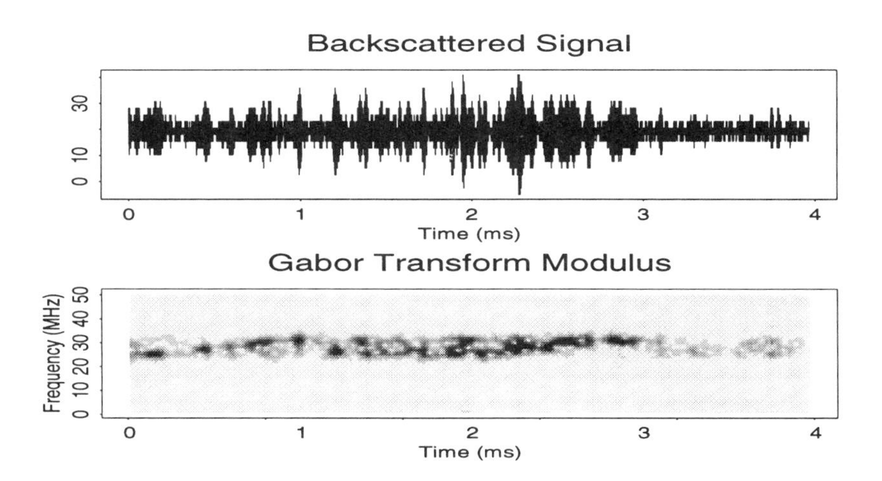

Figure 3.13. Continuous Gabor transform of the acoustic returns from natural underwater clutter.

3.4.1 Gabor Functions

The Gabor functions used throughout this text are based on the Gaussian window

$$g(x) = e^{-x^2/2\sigma} \ , \tag{3.23}$$

where σ is a fixed `scale` parameter. The default value $\sigma = 1$ is often too small, i.e., yields Gabor functions with poor frequency resolution. It has to be readjusted frequently. Examples of Gabor functions may be generated by using the Splus command `gabor`, as follows:

Example 3.1 Gabor Functions
Generate a Gabor function at given frequency (here 0.1 *Hz), for a sampling frequency (here* 1 *Hz), location (here* 256*) and scale (here* 25*):*

```
> gab <- gabor(512,256,.1,25)
> par(mfrow=c(1,2))
> tsplot(Re(gab))
> title(''Real part'')
> tsplot(Im(gab))
> title(''Imaginary part'')
```

3.4.2 CGT of Simple (Deterministic) Signals

The computation of the (continuous) Gabor transform relies on the use of the Swave function `cgt`. It implements the continuous Gabor transform, with the Gaussian window (3.23). The function `cgt` takes three arguments: the input signal, the number of frequencies n_f, and the sampling rate for the frequency axis δ_f. By convention, all the functions of the package assume a sampling frequency equal to 1. In addition, the code has been written in such a way that the Nyquist limit corresponds to $n_f\delta_f = 1$. Then setting δ_f to a value smaller than $1/n_f$ would produce aliasing. We actually suggest doing this once on purpose in order to understand the phenomenon and to recognize it when it is produced by accident. The fourth argument of the function sets the size of the window. When this last parameter is set to 1, the size of the window is approximately 10 samples.

In addition, the command `cleanph` allows the representation of the phase of the CGT (between $-\pi$ and π) and forces it to be equal to $-\pi$ when the modulus is less than a certain threshold (chosen by the user) and considered as insignificant.

Let us start with the simple example of transients:

Example 3.2 Transients. Suppose we are given a transient signal, here named $nA0$. We compute here its CGT with Gaussian windows with two different sizes.

Compute Gabor transform of transient signal, here with $n_f = 50$, $\delta_f = 0.02$ *and two different values for the scale parameter, namely* `scale` = *10,18:*

```
> par(mfrow=c(1,4))
> tsplot(nA0)
> title(''Transients'')
> cgtA0 <- cgt(nA0,50,.02,10)
> cgtA0 <- cgt(nA0,50,.02,18)
```

The values 50 for the parameter n_f and .02 for the parameter δ_f of the `cgt` S-function of the Swave package mean that the whole frequency domain will be represented (the sampling frequency corresponding to 1 Hz), and that 50 different values of the frequency will be displayed. scale=25 corresponds to a window of approximately 250 points.

To display the phase of the Gabor transform:

```
> tmp<- cleanph(cgtA0,.1)
> title(''Phase of Gabor transform'')
```

Clearly, the phase is irrelevant when the modulus is smaller than a certain limit. We fix here 10% as a threshold on the modulus. The result is displayed in Figure 3.6.

Notice that the lines of constant phase tend to "converge" toward the singularities. However, the Gabor functions being of constant size, the time-resolution cannot be better than the window's size.

Example 3.3 Sine wave

Generate the sequence, here a sine wave of frequency 1/32 *Hz sampled with unit sampling frequency:*

```
> x <- 1:512
> sinwave <- sin(2*pi*x/32)
> par(mfrow=c(1,3))
> tsplot(sinwave)
> title(''Sine wave'')
```

Compute the Gabor transform with $n_f = 50$, $\delta_f = .005$ *and scale* $\sigma = 25$. *This corresponds to frequencies ranging from* 0 *to* 0.125 *Hz. Display the phase:*

```
> cgtsinwave <- cgt(sinwave,50,.005,25)
> tmp <- cleanph(cgtsinwave,.01)
> title(''Gabor Transform Phase'')
```

Let us now describe the example of a chirp signal, here a linear chirp.

Example 3.4 Chirp.
Generate the chirp and compute the Gabor transform between frequencies 0 *and* 0.125 *Hz:*

```
> x<- 1:512
> chirp <- sin(2*pi*(x + 0.002*(x-256)*(x-256))/16)
> tsplot(chirp)
> title(''Chirp signal'')
> cgtchirp <- cgt(chirp,50,.005,25)
> tmp <- cleanph(cgtchirp,.01)
```

The result is displayed in Figure 3.5.

3.4.3 "Real-Life" Examples

We now consider the example of speech signal discussed in the text. The signal (provided with the Swave package) is named HOWAREYOU. The Gabor transform is computed for 70 different values of the frequency variable, between 0 and 2800 Hz (35% of the sampling frequency), with two different bandwidths for the window. The results are displayed in Figure 3.11.

Example 3.5 How are you?
Read signal from disk, display it, and compute its Gabor transform with two different sets of parameters:

```
> HOWAREYOU <- scan(''signals/HOWAREYOU'')
> sinwave <- sin(2*pi*x/32)
> par(mfrow=c(1,3))
> tsplot(HOWAREYOU)
> cgtHOWAREYOU<- cgt(HOWAREYOU,70,0.01,60)
> cgtHOWAREYOU<- cgt(HOWAREYOU,70,0.01,10)
```

We now give the S commands used to analyze the backscattered acoustic signal described in Subsection 3.3.2. In the present example, the signal is first read from the disk (in a subdirectory *signals* provided with the Swave package), then analyzed. Since the signal has a nonzero DC component, it is convenient first to subtract the mean before computing the Gabor transform. The CGT is computed for 50 different values of the frequency, ranging from 0 to 250 kHz. This is the frequency range where most of the signal's energy is present.

We also give the commands needed for the example in Figure 3.12, corresponding to the file `backscatterer.1.220`. The commands for the other signal (file `backscatterer.1.000`) are the same.

Example 3.6 Backscattered Acoustic Signal.
Get the signal from disk, display it and compute its CGT:

```
> back1 <- scan("signals/backscatter.1.220")
> tsplot(back1)
> gb1 <- cgt(back1 - mean(back1), 50, 0.005, 10)
> gb <- cgt(back - mean(back), 50, 0.005, 50)
```

Finally, we give the commands used to analyze the quake data described in Subsection 3.3.2. In this example, there are six signals, respectively `H1.B`, `H2.B`, `H3.B`, which represent the acceleration signal measured at the basement for the three runs, and three other signals, `H1.10`, `H2.10`, and `H3.10`, giving the acceleration signal measured at the 10th floor. All these have the form of two columns arrays, with time and acceleration. The Gabor transform was computed with narrow-band Gabor functions (`scale` =100) for frequencies ranging from 0 to 6.25 Hz (with a sampling frequency $\nu_s = 250Hz$). From the Gabor transforms of the six signals, corresponding "transfer functions" may be estimated. This is done using the utility `tfmean`, which returns the average of a time-frequency representation as a function of the frequency.

Example 3.7 Quake data.
Compute Gabor transform:

```
> cgtH1.B <- cgt(H1.B[2,], 50, 0.001, 100)
> cgtH1.10 <- cgt(H1.10[2,], 50, 0.001, 100)
> cgtH2.B <- cgt(H2.B[2,], 50, 0.001, 100)
> cgtH2.10 <- cgt(H2.10[2,], 50, 0.001, 100)
> cgtH3.B <- cgt(H3.B[2,], 50, 0.001, 100)
> cgtH3.10 <- cgt(H3.10[2,], 50, 0.001, 100)
```

Estimations of the transfer functions and plots:

```
> tsplot(tfmean(cgtH1.10)/tfmean(cgtH1.B),
> tfmean(cgtH2.10)/tfmean(cgtH2.B),
> tfmean(cgtH3.10)/tfmean(cgtH3.B))
```

3.5 Notes and Complements

Continuous and discrete Gabor transforms have been used for quite a long time in the engineering literature. The first systematic works in the signal analysis and information theory context go back to Gabor himself [139] in the mid-1940s. Actually, it is worth emphasizing that Gabor analysis appeared in a different form in early works of Schrödinger on the harmonic oscillator [262]. Indeed, the usual Gabor functions are nothing but the so-called *canonical coherent states*, introduced in quantum mechanics for quantizing the classical harmonic oscillator. Quite a long time after that, canonical coherent states were rediscovered by several authors (among them Glauber [144], Klauder [193], and Sudarshan [271]) in the context of the quantum description of coherent optics.

About approximately the same time, similar tools were introduced in signal processing, and in particular in speech processing with the *sonagraph* developed in the 1940s by the Bell Labs for time-frequency acoustic signal analysis. See, for example, [245, 246]. Since then, techniques based on various forms of Gabor representations have become a standard tool in the speech processing community (with applications in speech analysis, but also in speech compression and speaker identification), and more generally in signal processing.

We owe to A. Grossmann the reinterpretation of the Gabor transform in a group-theoretical framework. Together with Morlet and Paul [153, 154], they realized that the theory of square-integrable group representations can provide a unifying framework for many time-frequency representations (see also [124, 125]). A detailed treatment of the group-theoretical approach to the Gabor transform may be found in [261].

There was a renewal of interest in Gabor-type representations due to the discovery of wavelets (see next chapter), and also to the emergence of the theory of frames, mainly developed by Daubechies [96, 93] (to be described in Chapter 5), which helped solve the problem of the discretization of the continuously-defined Gabor transforms. These issues will be addressed in detail in the subsequent chapters.

Chapter 4

The Continuous Wavelet Transform

We now present the wavelet transform as an alternative to the Gabor transform for time-frequency analysis. We revisit some of the examples presented in the previous chapter. However, time-frequency is not the only aspect of wavelet analysis, and we emphasize an important feature of wavelets, namely their ability to perform very local *time-scale analysis*.

4.1 Definitions and Basic Properties

We begin with some of the basics of the continuous wavelet transform.

4.1.1 Basic Definitions

Let $\psi \in L^1(\mathbb{R}) \cap L^2(\mathbb{R})$ be a fixed function.[1] From now on it will be called the *analyzing wavelet*. It is also sometimes called the *mother wavelet* of the analysis. The corresponding family of wavelets is the family $\{\psi_{(b,a)};\ b \in \mathbb{R}, a \in \mathbb{R}_+^*\}$ of shifted and scaled copies of ψ defined as follows. If $b \in \mathbb{R}$ and $a \in \mathbb{R}_+^*$, we set

$$\psi_{(b,a)}(x) = \frac{1}{a}\,\psi\left(\frac{x-b}{a}\right), \quad x \in \mathbb{R}\ . \tag{4.1}$$

The wavelet $\psi_{(b,a)}$ can be viewed as a copy of the original wavelet ψ rescaled by a and centered around the "time" b. Given an analyzing wavelet ψ, the associated continuous wavelet transform is defined as follows.

[1] In fact, it is sufficient to assume $\psi \in L^1(\mathbb{R})$, but for convenience also assume that $\psi \in L^2(\mathbb{R})$. This extra assumption ensures the boundedness of the wavelet transform.

Definition 4.1 *Let $\psi \in L^1(\mathbb{R}) \cap L^2(\mathbb{R})$ be an analyzing wavelet. The continuous wavelet transform (CWT for short) of a finite-energy signal f is defined by the integral*

$$T_f(b,a) = \langle f, \psi_{(b,a)} \rangle = \frac{1}{a} \int f(x) \overline{\psi\left(\frac{x-b}{a}\right)} dx \,. \tag{4.2}$$

The value $T_f(b,a)$ contains information concerning the signal f at the scale a around the point b. In the notation of Chapter 1 we have $\lambda = (b,a)$ and $\Lambda = \mathbb{R} \times \mathbb{R}^*_+$ and $\mathcal{H} = L^2(\mathbb{R} \times \mathbb{R}^*_+, a^{-1}dbda)$. $\lambda = (b,a)$ is often called the time-scale variable of the analysis.

Remark 4.1 Let us stress our choice for normalization of wavelets: the wavelets $\psi_{(b,a)}$ have been normalized in such a way that $||\psi_{(b,a)}||_1 = ||\psi||_1$ where we use the standard notation $|| \cdot ||_1$ for the norm of the Banach space $L^1(\mathbb{R}, dx)$ of absolutely integrable functions on the real line:

$$||f||_1 = \int |f(x)| \, dx.$$

Different normalizations may be found in the literature, in particular the L^2 normalization used by Mallat and his collaborators, who use the definition

$$\psi_{b,a}(x) = \frac{1}{\sqrt{a}} \psi\left(\frac{x-b}{a}\right). \tag{4.3}$$

We shall also find other normalizations convenient when studying fractional Brownian motions in Chapter 6. As a rule we allow ourselves to change the normalization when needed, but we make sure each time that the reader is warned of the change.

Remark 4.2 A simple application of Parseval's relation gives the wavelet coefficients in terms of the Fourier transforms of the signal and the analyzing wavelet. Indeed, since $\widehat{\psi_{(b,a)}}(\xi) = e^{ib\xi} a^{-1} \hat{\psi}(\xi)$, we have that

$$T_f(b,a) = fracla \int \hat{f}(\xi) e^{-ib\xi} \overline{\hat{\psi}(\xi)} \, d\xi \,. \tag{4.4}$$

The following definition will play an important role in the sequel.

Definition 4.2 *We say that the wavelet ψ is* progressive, *when $\psi \in H^2(\mathbb{R})$.*

Using the notation Θ introduced earlier for the Heaviside function, this definition can be restated by saying that

$$\hat{\psi}(\xi) = \Theta(\xi) \hat{\psi}(\xi)$$

whenever ψ is progressive. From (4.4) one sees immediately that the wavelet transform of a real-valued signal is the same as the wavelet transform of the associated analytic signal when the wavelet is progressive, i.e.,

$$T_f(b,a) = \langle f, \psi_{(b,a)} \rangle = \frac{1}{2} \langle Z_f, \psi_{(b,a)} \rangle$$

(recall formula (1.21) linking the Fourier transform of a real-valued signal to the Fourier transform of its associated analytic signal).

A crucial property of the continuous wavelet transform is that, under a mild condition on the analyzing wavelet (see equation (4.5)), the transform is invertible on its range. But different situations need to be considered.

Wavelet Analysis of the Hilbert Space $L^2(\mathbb{R})$

We have the following.

Theorem 4.1 *Let $\psi \in L^1(\mathbb{R}) \cap L^2(\mathbb{R})$, be such that the number c_ψ defined by*

$$c_\psi = \int_0^\infty |\hat{\psi}(a\xi)|^2 \frac{da}{a} \tag{4.5}$$

is finite, nonzero, and independent of $\xi \in \mathbb{R}$. Then every $f \in L^2(\mathbb{R})$ admits the decomposition

$$f(x) = \frac{1}{c_\psi} \int_{-\infty}^{\infty} \int_0^\infty T_f(b,a) \psi_{(b,a)}(x) \frac{da}{a} db \,, \tag{4.6}$$

where the convergence holds in the strong $L^2(\mathbb{R}, dx)$ sense.

This result was first proven in this form in [151], but it goes back to earlier works of Calderón. The proof (which can now be found in all textbooks discussing wavelet analysis, for example [Chui92a, Daubechies92a]), goes as follows. First consider $d_a(x) = \int T_f(b,a)\psi_{(b,a)}(x)db$. Clearly, $d_a \in L^2(\mathbb{R}, dx)$, $\hat{d}_a(\xi) = \hat{f}(\xi)|\hat{\psi}(a\xi)|^2$, and the result follows from the scaling invariance of the measure da/a and standard limiting arguments.

The assumption appearing in the statement of the preceding theorem is called the condition of admissibility , and an analyzing wavelet ψ satisfying such a condition is called an admissible wavelet.

Remark 4.3 Assuming the independence of c_ψ as defined in equation (4.5) with respect to ξ actually amounts to assuming the independence with respect to $\mathrm{sgn}(\xi)$. In particular, such a property is automatically fulfilled if the wavelet ψ is real. An immediate consequence of the assumption of finiteness of c_ψ is the fact that the integral of the wavelet ψ has to vanish, i.e.,

$$\int \psi(x)dx = 0.$$

In signal processing parlance, an admissible wavelet is a band-pass filter.

Therefore, we have the following partial isometry between $L^2(\mathbb{R}, dx)$ and the target space of the transform, namely $\mathcal{H} = L^2(\mathbb{R} \times \mathbb{R}_+^*, a^{-1}dadb)$:

$$||f||^2 = \frac{1}{c_\psi} \int_{-\infty}^{\infty} \int_0^{\infty} |T_f(b,a)|^2 \frac{da}{a} db \tag{4.7}$$

for all $f \in L^2(\mathbb{R})$. This allows for the interpretation of the square of the modulus of the wavelet transform (suitably normalized) as a time-frequency or more precisely a time-scale energy density.

Progressive Wavelets for $H^2(\mathbb{R})$ and $L^2_{\mathbb{R}}(\mathbb{R})$

Here and in the following we use the notation $L^2_H(\mathbb{R})$ for the space of H-valued square integrable functions on the real line. We also use the notation $H^2(\mathbb{R})$ for the Hardy space of square integrable functions $f \in L^2(\mathbb{R})$ whose Fourier transform $\hat{f}(\xi)$ vanishes for $\xi \leq 0$ (see Chapter 1 for a definition). Then one can use Remark 4.2 to ignore the negative frequencies. This shows that the above symmetry assumption on ψ is not needed when the wavelet is progressive. Then it is sufficient to assume that the constant c_ψ is finite and nonzero. Its behavior for negative ξ does not play any role.

Let us now consider the case of real signals. We saw in Remark 4.2 that the wavelet transform of such a signal was depending only on the restriction of the Fourier transform to the positive frequencies. Moreover, since the Fourier transform $\hat{f}$ of $f \in L^2_{\mathbb{R}}(\mathbb{R})$ satisfies the Hermitian symmetry $\hat{f}(-\xi) = \overline{\hat{f}(\xi)}$, in order to reconstruct (synthesize) the signal f it is enough to reconstruct the part which corresponds to the positive frequencies, and this is why it may be convenient to analyze real-valued signals f using progressive wavelets, i.e., functions $\psi \in H^2(\mathbb{R})$. Of course, the $H^2(\mathbb{R})$ inversion formula for the wavelet transform yields a complex-valued reconstructed signal, from which the original one is easily recovered by taking the real part: let $\psi \in L^1(\mathbb{R}) \cap H^2(\mathbb{R})$, and define the wavelet transform of $f \in L^2_{\mathbb{R}}(\mathbb{R})$ as in (4.2). Then we have

$$f(x) = 2\Re \frac{1}{c_\psi} \int_{-\infty}^{\infty} \int_0^{\infty} T_f(b,a)\psi_{(b,a)}(x) \frac{da}{a} db\,. \tag{4.8}$$

Choosing progressive wavelets instead of real ones may be justified as follows. The wavelet transform of real signals with real wavelets obviously yields real wavelet coefficients. And even after squaring them, there is no natural way of making the connection with some "local spectrum" one would like to associate with a given signal (see, however, ([234], where this problem is attacked with smoothing techniques). In this respect, complex progressive wavelets are somewhat more natural. We shall see more on this aspect in the chapter devoted to the analysis of frequency-modulated signals.

4.1.2 Redundancy

For a given admissible wavelet ψ the image of $L^2(\mathbb{R})$ by the wavelet transform is a closed subspace $\mathcal{H}_\psi$ of $L^2(\mathbb{R} \times \mathbb{R}_+^*, a^{-1}dadb)$. This space is called the *reproducing kernel Hilbert space* of the analyzing wavelet ψ. It is the space of solutions $F(b,a)$ of the integral equation

$$F(b',a') = P_\psi F(b',a') = \int_{-\infty}^{\infty} \int_0^{\infty} \mathcal{K}_\psi(b',a';b,a) F(b,a) \frac{da}{a} db \ , \tag{4.9}$$

where the reproducing kernel $\mathcal{K}_\psi$ is given by

$$\mathcal{K}_\psi(b',a';b,a) = \frac{1}{c_\psi} \langle \psi_{(b,a)}, \psi_{(b',a')} \rangle \ . \tag{4.10}$$

This fact is readily proved by taking the inner product of both sides of equation (4.6) with the wavelet $\psi_{(b',a')}$. The corresponding integral operator P_ψ defined in equation (4.9) is easily shown to be an orthogonal projection on the space $\mathcal{H}_\psi$ (i.e., $P_\psi^* = P_\psi^2 = P_\psi$).

Remark 4.4 Equation (4.9) expresses the redundancy of the CWT. As before, a consequence of this redundancy is the existence of many different inversion formulas for the CWT or, otherwise stated, the possibility of using in the inversion formula (4.6) a reconstruction wavelet different from the analyzing wavelet ψ: if the function $\chi \in L^1(\mathbb{R}) \cap L^2(\mathbb{R})$ is such that $\hat{\psi}(\xi)\hat{\chi}(\xi)/\xi \in L^1(\mathbb{R}_+^*, d\xi)$ and that in addition the number

$$c_{\psi\chi} = \int_0^\infty \overline{\hat{\psi}(a\xi)} \hat{\chi}(a\xi) \frac{da}{a} \tag{4.11}$$

is finite, nonzero, and independent of ξ, then (4.6) may be replaced by

$$f(x) = \frac{1}{c_{\psi\chi}} \int_{-\infty}^{\infty} \int_0^{\infty} T_f(b,a) \chi_{(b,a)}(x) \frac{da}{a} db \ , \tag{4.12}$$

where the wavelet coefficients $T_f(b,a)$ are still defined by (4.2). Such a remark turns out to be of practical interest, since some analyzing wavelets may not be admissible in the sense of the preceding definition, but nevertheless be such that there exists a reconstruction wavelet χ satisfying the given condition (see [171] for an example where such a situation is encountered).

Remark 4.5 As we can already see from the preceding discussion, the continuous wavelet and Gabor representations may be considered as similar tools, in the sense that they may both be seen as time-frequency representations and as such they have very similar properties. However, there is an important difference which can be seen easily from the *filtering point of view.* Both representations may be understood as the outputs of a filter bank. More on that in Section 5.6. The difference lies in the bandwidth of the filters. The filters associated with the Gabor transform are constant bandwidth filters. At the opposite, the wavelets may be seen as constant *relative* bandwidth filters. This essentially means that the bandwidth depends linearly on the frequency.

4.1.3 Invariance

Like the Gabor transform, the wavelet transform possesses built-in invariance properties. For example, the CWT of a shifted copy of the signal f equals the corresponding time-shifted copy of the CWT of f. A similar property holds with dilation. More generally, we have the following.

Proposition 4.1 *Let $f \in L^2(\mathbb{R})$ and let us set*

$$\tilde{f}(x) = f\left(\frac{x - x_0}{\alpha}\right)$$

for some $\alpha \in \mathbb{R}_+^*$. Then

$$T_{\tilde{f}}(b, a) = \alpha T_f\left(\frac{b - x_0}{\alpha}, \frac{a}{\alpha}\right) . \tag{4.13}$$

This proposition may be given a proof following the lines of that of Proposition 3.2. Let us consider the translations T_b (see the previous chapter) and the dilation D_a:

$$D_a f(x) = \frac{1}{a} f\left(\frac{x}{a}\right) .$$

Then, we have for any $x_0 \in \mathbb{R}, \alpha \in \mathbb{R}_+^*$,

$$T_{\tilde{f}}(b, a) = \langle T_{x_0} D_\alpha f, T_b D_a \psi \rangle = \alpha \langle f, D_{1/\alpha} T_{b - x_0} D_a \psi \rangle = \alpha \langle f, T_{\frac{b - x_0}{\alpha}} D_{\frac{a}{\alpha}} \psi \rangle ,$$

which yields the result.

Notice that unlike the CGT (which was covariant up to a phase factor), the CWT is fully covariant.

Remark 4.6 Group theoretical aspects. The continuous wavelet transform may be given the same geometrical interpretation as the Gabor transform, in terms of the action of the affine group G_{aff} on $L^2(\mathbb{R})$. The affine group is generated by elements of the form $(b,a) \in \mathbb{R} \times \mathbb{R}_+^*$, with group law $(b,a) \cdot (b',a') = (b + ab', aa')$. G_{aff} has a left-invariant measure $d\mu_L(b,a) = dadb/a^2$ and a right invariant measure $d\mu_R(b,a) = dadb/a$. Again, the wavelet transform (with a different normalization) appears as intertwining operator between the left regular representation λ of G_{aff} on $L^2(G_{aff}, d\mu_L)$ defined by

$$[\lambda_{(b',a')}F](b,a) = F((b',a')^{-1} \cdot (b,a))$$

and the representation π on $H^2(\mathbb{R})$ defined by

$$\pi_{(b,a)}f(x) = \frac{1}{\sqrt{a}} f\left(\frac{x-b}{a}\right) .$$

Denoting by T the mapping $T : L^2(\mathbb{R}) \ni f \hookrightarrow T_f \in \mathcal{H}_\psi \subset L^2(\mathbb{R} \times \mathbb{R}_+^*)$, we have as in the Gabor case

$$T \circ \pi = \lambda \circ T .$$

The invariance of the transform is simply a rephrasing of this fact. We will not go into details on this group theoretical aspect of the theory. Instead we refer the interested reader to [11, 43, 153] for a more complete analysis.

4.1.4 A Simple Reconstruction Formula

To end this section on generalities, we give another (simpler) form of continuous wavelet decompositions, the so-called continuous *Littlewood-Paley decompositions.* Given a wavelet $\psi \in L^1(\mathbb{R}) \cap L^2(\mathbb{R})$ such that $\hat{\psi} \in L^1(\mathbb{R}_+^*, d\xi/|\xi|)$ and such that the quantity

$$k_\psi = \int_0^\infty \overline{\hat{\psi}(a\xi)} \frac{da}{a} \tag{4.14}$$

makes sense and is finite, nonzero, and independent[2] of ξ, we have the following simple inversion formula for the corresponding continuous wavelet transform:

$$f(x) = \frac{1}{k_\psi} \int_0^\infty T_f(x,a) \frac{da}{a} . \tag{4.15}$$

[2]Again, the fact that k_ψ is independent of ξ reduces to assuming that k_ψ is independent of the sign of ξ, because of the invariance properties of the measure in (4.15).

Indeed, $2\pi T_f(x,a) = \int \hat{f}(\xi)\overline{\hat{\psi}(a\xi)}e^{i\xi x}d\xi$, so that by Fubini's lemma,

$$\int T_f(x,a)\frac{da}{a} = \frac{1}{2\pi}\int \hat{f}(\xi)e^{i\xi x}d\xi \int \overline{\hat{\psi}(a\xi)}\frac{da}{a} ,$$

which yields (4.15).

The reconstruction formula (4.15) can be viewed formally as a particular case of (4.12) in which a Dirac mass would play the role of reconstruction wavelet. Such a formula is sometimes very convenient for numerical purpose, since it only involves a one-dimensional integral (instead of a two-dimensional one.) Equation (4.15) will sometimes be quoted as the *linear analysis-reconstruction scheme*, in contrast to equation (4.6), called the *bilinear scheme*. We consider later discrete analogues of such Littlewood-Paley decomposition that were used in approximation theory to characterize functional spaces.

4.2 Continuous Multiresolutions

An important feature of wavelet analysis is the existence of the so-called *multiresolution analysis*, responsible for fast algorithms and many other interesting properties and generalizations. Although the concept of multiresolution analysis is usually described in a purely discrete context, it is also interesting to describe it in a continuous situation. An immediate corollary of the decomposition formulas (4.6), (4.12), and (4.15) is the existence of associated approximate identities which take the forms of *scaling functions*. If we consider for example the bilinear analysis-reconstruction scheme described by equation (4.6), we have the following:

Remark 4.7 Let $\psi \in L^1(\mathbb{R}) \cap L^2(\mathbb{R})$ be such that $0 < c_\psi < \infty$, and let us assume that ϕ is such that

$$\left|\hat{\phi}(a\xi)\right|^2 = \int_a^\infty \left|\hat{\psi}(u\xi)\right|^2 \frac{du}{u} . \tag{4.16}$$

Assuming an obvious definition for $\phi_{(b,a)}(x)$, we set

$$S_f(b,a) = \langle f, \phi_{(b,a)}\rangle \tag{4.17}$$

for all $f \in L^2(\mathbb{R})$. Then equation (4.6) may be replaced by

$$\begin{aligned} f(x) &= \frac{1}{c_\psi}\lim_{a\to 0}\int S_f(b,a)\phi_{(b,a)}(x)db \\ &= \frac{1}{c_\psi}\left(\int S_f(b,a)\phi_{(b,a)}(x)db + \int_{-\infty}^{\infty}\int_0^a T_f(b,u)\psi_{(b,a)}(x)\frac{du}{u}db\right) \end{aligned} \tag{4.18}$$

(with the obvious generalization for the case of equation (4.12)). The function ϕ is called a (bilinear) *scaling function* associated with the wavelet ψ. We refer to [118] for more details.

In the case of the linear decomposition-reconstruction scheme of equation (4.15), we have, similarly:

Remark 4.8 Let $\psi \in L^1(\mathbb{R}) \cap L^2(\mathbb{R})$ be such that $0 < k_\psi < \infty$, and let us assume that φ is such that

$$\hat{\varphi}(\xi) = \int_a^\infty \hat{\psi}(u\xi)\frac{du}{u} . \tag{4.19}$$

Then if we set

$$S_f(b,a) = \langle f, \varphi_{(b,a)} \rangle \tag{4.20}$$

for all $f \in L^2(\mathbb{R})$ (with the obvious definition for $\varphi_{(b,a)}(x)$), equation (4.15) may be replaced by

$$\begin{aligned} f(x) &= \frac{1}{k_\psi} \lim_{a\to 0} S_f(x,a) \\ &= \frac{1}{k_\psi}\left(S_f(x,a) + \int_0^a T_f(x,u)\frac{du}{u}\right) . \end{aligned} \tag{4.21}$$

The function φ is called a *scaling function* of the analysis.

It is important to emphasize the difference between wavelet coefficients T_f and scaling function coefficients S_f. While the coefficients $S_f(b,a)$ provide an approximation of the function f at scale a, $T_f(b,a)$ give details at scale a, i.e., essentially differences between two consecutive approximations. This will appear clearly when we discuss pyramidal algorithms in Section 5.6.

4.3 Commonly Used Analyzing Wavelets

We shall discuss in this chapter a number of simple examples, illustrating the main features of continuous wavelet analysis. But before turning to the examples, let us just give some examples of wavelets ψ suitable for use in CWT. As we already said, we shall often concentrate on "progressive" wavelets in practice.

4.3.1 Complex-Valued Progressive Wavelets

Among the simplest examples are the so-called Cauchy wavelets, which appear naturally in the context of the theory of analytic functions. Indeed,

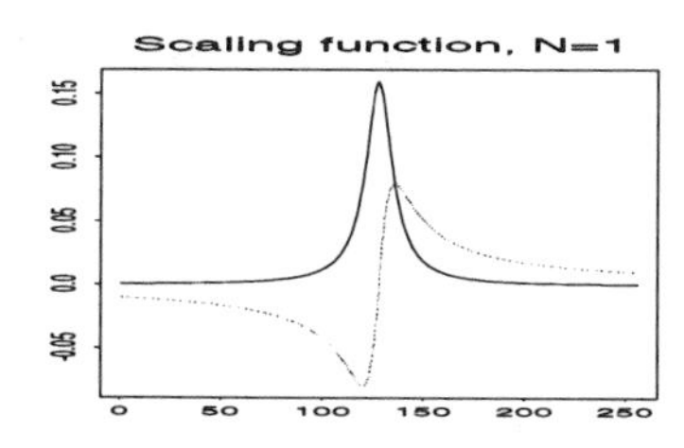

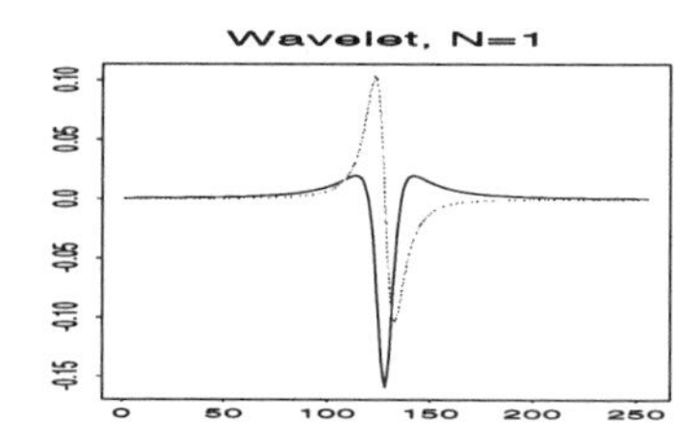

Figure 4.1. Cauchy wavelet and the corresponding scaling function (solid line: real part; dashed line: imaginary part).

given a function $f \in H^2(\mathbb{R})$, its analytic continuation to the upper half-plane $H_+ = \{z = b + ia;\ b \in \mathbb{R}, a > 0\}$ can be recovered from its boundary values on the real axis by the formula

$$f(z) = \frac{1}{2\pi i}\int_{-\infty}^{\infty}\frac{f(x)}{x-z}dx = \frac{1}{2\pi i}\frac{1}{a}\int_{-\infty}^{\infty}\frac{f(x)}{a^{-1}(x-b)-i}dx \ ,$$

i.e., as a scaling function transform, with a scaling function given by

$$\varphi_1^C(x) = \frac{1}{2\pi}\frac{1}{1-ix} \ . \tag{4.22}$$

According to equation (4.19), the corresponding wavelet in the continuous Littlewood-Paley approach reads

$$\psi_1^C(x) = \frac{1}{2\pi}\frac{1}{(1-ix)^2} \ . \tag{4.23}$$

These wavelet and scaling functions are displayed in Figure 4.1 (with a normalization chosen so that $\phi_1^C(0) = 1$).

They were intensively used in quantum mechanics by T. Paul (see [238]). The expression of $\phi_1^C(x)$ and $\psi_1^C(x)$ in the Fourier domain are

$$\widehat{\varphi_1^C}(\xi) = e^{-\xi}\Theta(\xi) \ , \tag{4.24}$$

where $\Theta(\xi)$ denotes the Heaviside step function and

$$\widehat{\psi_1^C}(\xi) = \xi e^{-\xi}\Theta(\xi) \ . \tag{4.25}$$

This last formula suggests the introduction of higher-order Cauchy wavelets ψ_N^C defined from their Fourier transforms via a straightforward generalization of formula (4.25):

$$\widehat{\psi_N^C}(\xi) = \frac{\xi^N}{(N-1)!}e^{-\xi}\Theta(\xi) \ , \tag{4.26}$$

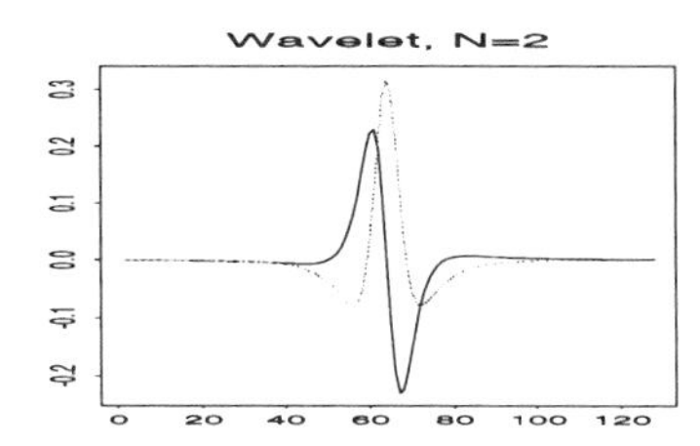

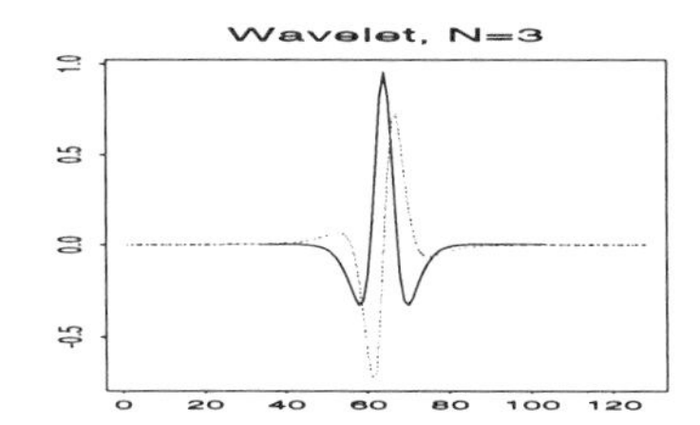

Figure 4.2. Second- and third-order Cauchy wavelets (solid line: real part; dashed line: imaginary part).

in which case

$$\widehat{\varphi_N^C}(\xi) = \left(1 + \xi + \frac{\xi^2}{2} + \cdots + \frac{\xi^{N-1}}{(N-1)!}\right) e^{-\xi}\Theta(\xi) \ . \tag{4.27}$$

The main difference between those higher-order Cauchy wavelets is that they have different numbers of vanishing moments. See Definition 4.3 for a rigorous definition. We shall see later, when studying singularity analysis with wavelets, what such a property may be very useful. Examples with $N = 2$ and $N = 3$ are displayed in Figure 4.2. Notice that the wavelets become more oscillatory as the number of vanishing moments increases.

Another very classical example is that of the so-called *Morlet wavelet*, which is simply a modulated Gaussian function

$$\psi(x) = e^{-\frac{x^2}{2\sigma^2}} e^{i\omega_0 x} \ , \tag{4.28}$$

where ω_0 is a frequency parameter controlling the number of oscillations of the wavelet within its Gaussian envelope, and σ controls the size of the envelope. The Morlet wavelet is not, strictly speaking, an admissible wavelet, since it is not of integral zero. However, for ω_0 large enough (larger than 5 in practice), the integral of ψ is small enough to ensure that for all practical purposes, it may be used numerically as if it were a wavelet. We use $\sigma = 1$ throughout the book.

4.3.2 Real-Valued Wavelets

Thus far we have given examples of complex-valued wavelets, more precisely analytic or progressive wavelets, i.e., wavelets belonging to $H^2(\mathbb{R})$. We shall see many examples of real-valued wavelets when discussing the discrete wavelets and the associated fast algorithms. But let us mention a few of them here.

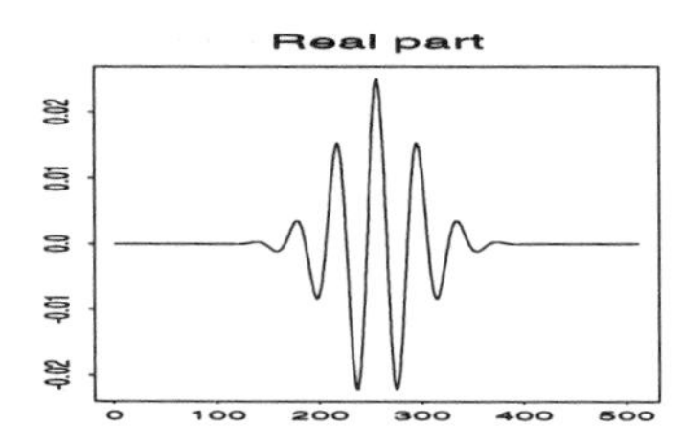

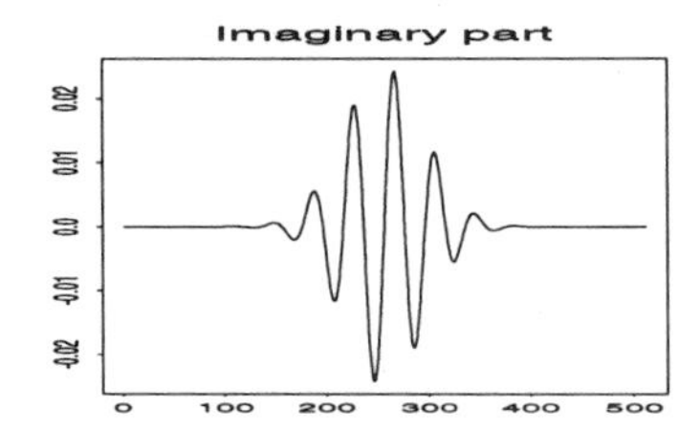

Figure 4.3. The Morlet wavelet (real and imaginary parts,) generated in Example 4.1.

The derivatives of Gaussians, and more precisely the Laplacians of Gaussian (the so-called LOG wavelet), have been very popular in the computer vision community. The LOG wavelet is defined as

$$\psi(x) = -\Delta e^{-x^2/2} = (1 - x^2)e^{-x^2/2} , \qquad (4.29)$$

where Δ is the usual Laplacian operator. An obvious generalization involves derivatives of any orders of Gaussian functions. Obviously, the wavelet

$$\psi_n(x) = -\frac{d^n}{dx^n} e^{-x^2/2}$$

has n vanishing moments. These wavelets can easily be turned into progressive wavelets by canceling their negative frequencies, i.e., by considering

$$\hat{\psi}(\xi) = K\xi^n e^{-\xi^2/2}\Theta(\xi) \qquad (4.30)$$

for some constant K.

The DOG (i.e., difference of Gaussians) was proposed as an alternative to the LOG wavelet in the computer vision literature. The DOG wavelet (displayed with the LOG wavelet in Figure 4.4) is simply obtained by considering the difference of two Gaussian functions at different scales, for example,

$$\psi(x) = 2e^{-2x^2} - e^{-x^2/2}. \qquad (4.31)$$

More generally, derivatives of smoothing functions were used thoroughly in the image processing literature. Let us consider, for example, the B-spline ϕ of degree 3 (or cubic spline,) obtained by convolving the characteristic function of the interval $[0, 1]$ with itself four times. Its Fourier transform is given by

$$\hat{\phi}(\xi) = \left(\frac{1 - e^{-i\xi}}{i\xi}\right)^4 . \qquad (4.32)$$

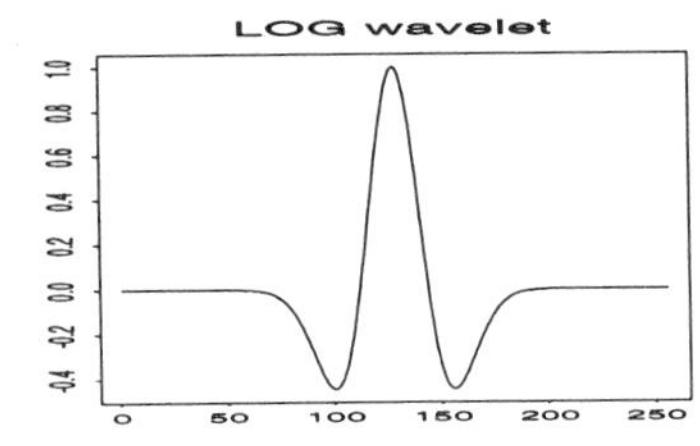

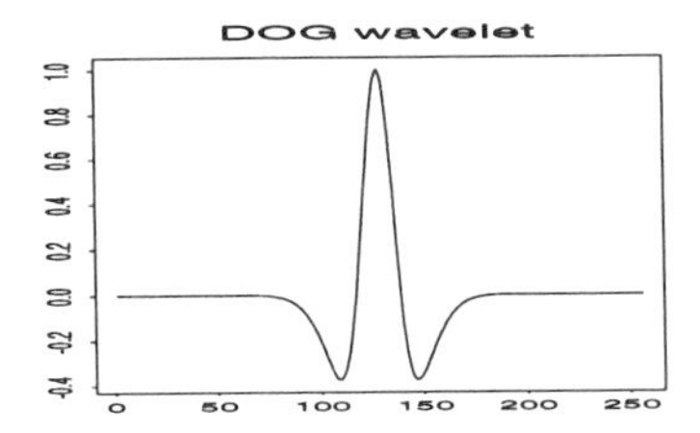

Figure 4.4. The LOG and DOG wavelets.

The derivative of such a cubic spline is given in the Fourier domain by

$$\hat{\psi}(\xi) = -i\xi \left(\frac{1 - e^{-i\xi}}{i\xi} \right)^4 . \tag{4.33}$$

We will see some applications of such a wavelet when studying local maxima of wavelet transforms.

4.4 Wavelet Singularity Analysis

One of the main features of wavelet analysis is its capacity of doing a very precise analysis of the regularity properties of functions. This is made possible by the scale variable. This fact actually goes back to the prehistory of wavelets (see, for example, [136]) and was intensively used by several authors in different contexts (see, e.g., [181, 182, 183, 171] and [Arneodo95]). In this section we give a brief account of some characteristic results of this theory and we discuss some of the illustrations of Section 4.5. Another goal of the present discussion is to show that for certain (singular) functions, essential information is carried by a limited number of wavelet coefficients. We shall see later in this volume how such an information may be taken advantage of in the design of numerical algorithms.

4.4.1 Hölder Regularity

We shall focus on the simplest case, namely the case of Hölder regularity. For this we first need to recall some definitions. For $0 < \alpha < 1$, let C^α denote the Banach space of functions f satisfying

$$|f(x) - f(y)| \leq C|x - y|^\alpha, \qquad x, y \in \mathbb{R}, \tag{4.34}$$

for some constant $0 < C < \infty$. Such functions are said to be Hölder continuous of order α. Then it is easily shown (see, e.g., [136, 171] for a proof) that

Theorem 4.2 *Let $\psi \in L^1(\mathbb{R})$ be such that*

$$c_\alpha = \int |x|^\alpha \, |\psi(x)| \, dx < \infty \ , \tag{4.35}$$

and

$$\int \sup_{|\delta|<1} |\psi'(u+\delta)| \, du < \infty \ . \tag{4.36}$$

Then $f \in C^\alpha$ if and only if

$$|T_f(b,a)| \leq C' a^\alpha, \qquad a \in \mathbb{R}, \tag{4.37}$$

for some finite positive constant C' independent of b.

To illustrate the properties of wavelets which are responsible for this characterization, we reproduce the proof of the first half of the theorem. Let ψ be a wavelet satisfying the above properties, and let $f \in C^\alpha$. Then since $\int \psi(x)dx = 0$, we have

$$\begin{aligned} |T_f(b,a)| &= \frac{1}{a}\left| \int [f(x) - f(b)] \, \overline{\psi\left(\frac{x-b}{a}\right)} dx \right| \\ &\leq \frac{C}{a} \int |x-b|^\alpha \left| \psi\left(\frac{x-b}{a}\right) \right| dx \\ &\leq C c_\alpha \, a^\alpha \ , \end{aligned}$$

which proves the first part of the theorem. The converse is proved in a similar way and makes use of the second assumption on the wavelet.

Remark 4.9 If one associates to any $f \in C^\alpha$ the infimum $\|f\|_\alpha$ of the constants C such that (4.34) holds, it may be proved that $\|\cdot\|_\alpha$ provides the quotient of C^α by the equivalence relation given by equality modulo constant functions, with a norm, which makes it a Banach space (notice that functions in this Banach space, which we will still denote by C^α, are only defined modulo additive constants). Then Theorem 4.2 expresses the fact that for a suitably chosen wavelet, the number $\inf a^{-\alpha}|T_f(b,a)|$ defines an equivalent norm on C^α. This is one of the simplest examples of the characterization of function spaces by mean of the wavelet transform. Elaborating on such arguments leads to the characterization of wider classes of functional spaces (see, e.g., [Meyer89a]or [136] for a review).

The previous result is global in the sense that it requires uniform regularity throughout the real line. Another classical example involves local regularity. For a given point x_0 we say that the function f is locally Hölder continuous of order α at x_0 if it satisfies

$$|f(x_0 + h) - f(x_0)| \le C|h|^\alpha \tag{4.38}$$

for some constant C and for $|h|$ small enough. Then we have the "local" counterpart of the previous result:

Theorem 4.3 *Let $\psi \in L^1(\mathbb{R})$ be such that condition* (4.35) *holds. If the function f is locally Hölder continuous of order α at x_0, then we have*

$$|T_f(b,a)| \le C'(a^\alpha + |b - x_0|^\alpha) \tag{4.39}$$

for some constant C'.

The proof of the theorem is also elementary and follows the lines of the previous argument. It is worth mentioning that there exists a converse to that result. It is more involved and it requires more sophisticated arguments. These aspects have been carefully analyzed by several authors [181, 171]; see in particular [183] and [212] for reviews especially tailored to mathematicians and signal analysts, respectively.

To this point, we have only studied the case of Hölder spaces of exponent $0 < \alpha < 1$. Let us now assume that $n < \alpha < n + 1$. Then the function f belongs to the Hölder space C^α if for any x there exists a polynomial P_n of order n and a constant C such that

$$|f(x) - P_n(h)| \le C|h|^\alpha \tag{4.40}$$

for $|h|$ small enough. Equivalently, f is C^α if its derivative of order n is $C^{\alpha-n}$. To study such functions, it is necessary to restrict further the class of wavelets to be used. To see this, let us start with a simple argument, and assume that the wavelet is the nth derivative of a function χ and let f be $n$times differentiable. Then repeated integrations by parts give

$$\langle f, \psi_{(b,a)}\rangle = (-a)^n \langle \frac{d^n f}{dx^n}, \chi_{(b,a)}\rangle \ .$$

Consequently, analyzing f with the wavelet ψ is equivalent to analyzing the nth derivative of f with the wavelet χ. To proceed further, it is convenient to introduce formally the notion of vanishing moments.

Definition 4.3 *Let $\psi \in L^1(\mathbb{R})$ be an analyzing wavelet. ψ is said to have M vanishing moments if*

$$\int_{-\infty}^{\infty} x^m \psi(x) dx = 0 \ , \qquad m = 0, \cdots M - 1 \ . \tag{4.41}$$

Saying that a wavelet ψ has M vanishing moments essentially amounts to saying that the Fourier transform $\hat{\psi}(\xi)$ of the wavelet behaves as $|\xi|^M$ for small frequencies.

Let us consider the following toy example. Assume that f is $M-1$ times continuously differentiable in a neighborhood U of a point $x = x_0$. Then for every $x \in U$ we may write

$$f(x) = f(x_0)+(x-x_0)f'(x_0)+\cdots+\frac{1}{(M-1)!}(x-x_0)^{M-1}f^{(M-1)}(x_0)+r(x) .$$

If the wavelet ψ (assumed to have compact support for the sake of the present argument) has M vanishing moments, then for sufficiently small values of the scale parameter a (i.e., such that $\text{supp}(\psi_{(b,a)}) \subset U$ for some b), we have

$$T_f(b,a) = \langle f, \psi_{(b,a)} \rangle = \langle r, \psi_{(b,a)} \rangle .$$

Then we can say that, at least in some sense, the wavelet does not "see" the regular behavior of $f(x)$ near $x = x_0$ and "focuses" on the potentially singular part $r(x)$. If in addition we assume that $f(x)$ is C^M near $x = x_0$, then we have $r(x) = (x-x_0)^M \rho(x)$, with $|\rho(x)| \le \sup|f^{(M)}|/M!$, and we immediately conclude that

$$T_f(b,a) \sim O(a^M) \quad \text{as} \quad a \to 0 .$$

More generally, Theorems 4.2 and 4.3 are still valid if one assumes in addition that the wavelet ψ has n vanishing moments (and a corresponding generalization of condition (4.36)). The proofs go along the same lines, except that instead of using the zero integral property of ψ to subtract $f(b)$ from $f(x)$, one uses all the vanishing moment assumption to subtract $P_n(x-b)$ from $f(x)$ inside the integral.

Elaborating on such arguments leads to the characterization of Hölder spaces of functions via wavelet transform (see, e.g., [Meyer89a, 136]) and the whole body of work on the characterization of singularities of a given signal [212]. We shall see a few examples of wavelet analysis of singularities in the next section.

Remark 4.10 It is instructive to examine the meaning of conditions of the type (4.39). Let us consider the cone in the (b,a) half-plane defined by the condition $|b-x_0| < a$. Within this cone we have $|T_f(b,a)| = O(a^\alpha)$ as $a \to 0$. Outside the cone, the behavior is governed by the distance of b to the point x_0. These two behaviors are generally different, and have to be studied independently. However, it is shown in [212] that non-oscillating singularities may be characterized by the behavior of their wavelet transform within the cone. Examples of non-oscillating singularities with different Hölder exponents are displayed later, in Figure 4.8.

Remark 4.11 It is also shown in [212] that rapidly oscillating singularities cannot be characterized by the behavior of their wavelet transform in the cone. A typical example is given by the function $f(x) = \sin 1/x$ whose instantaneous frequency tends to infinity as $x \to 0$. The wavelet transform modulus is maximum on a curve of equation $b = Ka^2$ for some constant K depending only on the wavelet, and this curve is not in the cone. In such a case, the oscillations have to be analyzed carefully (see, e.g., [183].) We shall come back to such examples in Section 4.4.2.

Remark 4.12 Let us end this discussion with a remark of a more "numerical" nature. In practice, all function values are only available for discrete values of the variable (with a fixed resolution given by the sampling frequency), so that the notions of singularities and Hölder exponents are meaningless (as is the limit $a \to 0$.) Nevertheless, one can say that the behavior of the wavelet coefficients across scales provides a good way of describing the regularity of functions whose samples coincide with the observations at a given resolution.

4.4.2 Oscillating Singularities and Trigonometric Chirps

Let us now slightly change our point of view and think again of functions in terms of amplitude and phase. The analysis outlined in Section 4.4.1 was essentially tailored to explore the singularities of the amplitude of functions, without taking oscillations into account. However, it is possible to proceed further and analyze singularities with oscillations. The study of such objects has led S. Jaffard and Y. Meyer [183], who introduced the notion of *trigonometric chirps*. By the definition given in [227] (see also [183] for a modification of this definition), a trigonometric chirp of type (α, β) at the point $x = x_0$ is a function of the form

$$f_{\alpha,\beta}(x) = |x - x_0|^\alpha g((x - x_0)^{-\beta}), \tag{4.42}$$

where $\alpha \in \mathbb{R}$, $\beta > 0$, and $g(x^{-\beta})$ means $g_+(x^{-\beta})$ if $x > 0$ and $g_-(|x|^{-\beta})$ if $x < 0$, and where g_+ and g_- belong to the Hölder class C^r, are 2π periodic, and satisfy

$$\int_0^{2\pi} g_\pm(x)dx = 0 \ .$$

The remarkable property of such trigonometric chirps is their behavior with respect to integration. Indeed, it follows from simple arguments that the antiderivatives

$$f^{(n)}_{\alpha,\beta}(x) = \int_{x_0}^x f_{\alpha,\beta}(y)dy$$

are trigonometric chirps of type $(\alpha + n(\beta+1), \beta)$: the joint action of integration and fast oscillations improves the regularity. Then it may be proved that the wavelet transform $T_f(b,a)$ of such a trigonometric chirp is "concentrated" around the ridge $a = a_r(b) = \beta |b|^{(\beta+1)}$. See [227] for details. In connection with Remark 4.10 and Remark 4.11, it is important to point out that such a curve lies (for scales small enough) outside the influence cone of the singularity. This explains in particular why in such cases it is not sufficient to study the behavior of the wavelet transform inside the cone to characterize the singularity.

4.5 First Examples of Wavelet Analyses

As before, we first consider a family of examples of deterministic signals, concentrating on the class of interest to us. Then we come back to "real signals," and in particular to the quake signal we already discussed in the previous chapter.

4.5.1 Academic Examples

Frequency Modulated Signals

In this subsection we consider only analytic (or progressive) wavelets, i.e., wavelets $\psi \in H^2(\mathbb{R})$. In the simplest case of a cosine wave, say, $f(x) = \cos\,\omega x$, it immediately follows that the wavelet transform is given by

$$T_f(b,a) = \frac{1}{2} e^{i\omega b} \overline{\hat{\psi}(a\omega)} \ .$$

If we assume that the Fourier transform $\hat{\psi}$ of the wavelet possesses good localization properties, and in particular that it is sharply peaked near a value $\xi = \omega_0$ of the frequency, the wavelet transform $T_f(b,a)$ is localized in the neighborhood of the horizontal line $a = \omega_0/\omega$. An example is given in Figure 4.5. Another important feature is the behavior of the phase of the CWT. In the region of interest, the CWT has the same phase behavior as the signal.

In the more general situation of time-varying amplitudes and frequencies, the preceding discussion can be generalized as follows:

Lemma 4.1 *Let*

$$f(x) = \sum_{\ell=1}^{L} A_\ell(x) \cos \phi_\ell(x) \ , \tag{4.43}$$

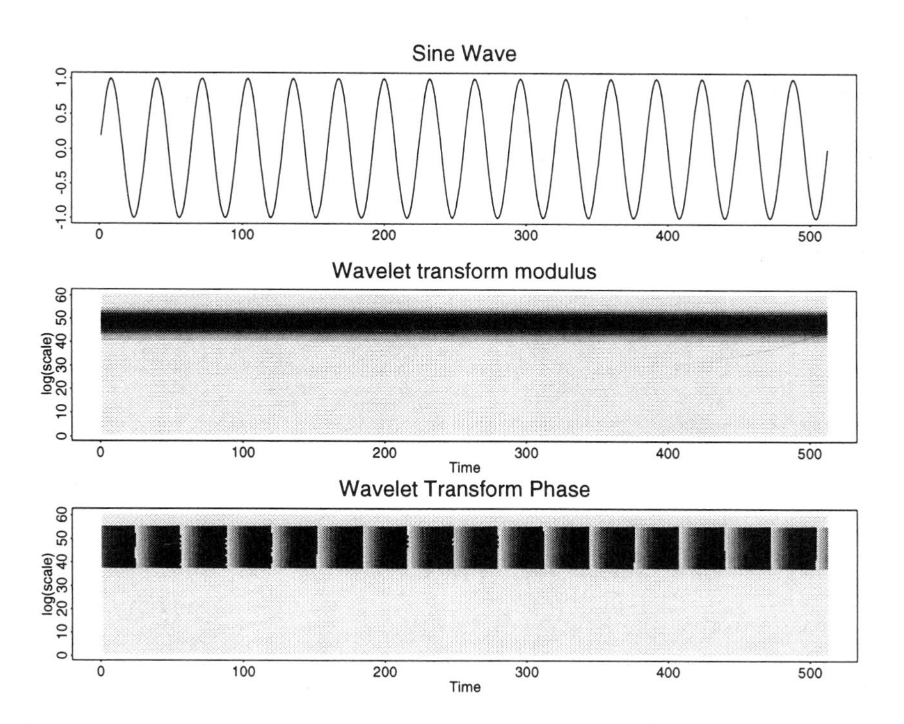

Figure 4.5. Plots of the modulus and the phase of the CWT of a sine wave, generated in Example 4.3.

where the amplitude and phase functions A_ℓ and ϕ_ℓ are twice continuously differentiable, slowly varying functions on $\mathbb{R}$. Then the corresponding wavelet transform may be written in the form

$$T_f(b,a) = \sum_{\ell=1}^{L} A_\ell(b)\cos\phi_\ell(b)\overline{\hat{\psi}(a\phi'_\ell(b))} + r(b,a) \ , \tag{4.44}$$

where the size of the remainder $r(b,a)$ depends on the first and second derivatives of the amplitude and phase functions in such a way that

$$r(b,a) \sim O\left(\frac{|A'_\ell|}{|A_\ell|}, \frac{|\phi''_\ell||\phi_\ell|}{|\phi'_\ell|^2}\right) \ . \tag{4.45}$$

The proof is based on a "term by term" control of the terms appearing in the Taylor's expansion of T_f (see [102] for details). This result shows that the continuous wavelet transform of functions of the form given in (3.18) will tend to "concentrate" in the neighborhood of curves of the form

$$a_\ell(b) = \frac{Cst}{\phi'_\ell(b)}$$

(the so-called *ridges* of the transform). This point will be discussed extensively in the part of Chapter 7 devoted to the *ridge-skeleton algorithm*, where more precise approximations will be discussed.

An example of linear chirp (generated in Example 4.5) is displayed in Figure 4.6. Observe that the modulus of the wavelet transform localizes in the neighborhood of a curve corresponding to the instantaneous frequency. It is not a straight line as in the Gabor case, since we are in the time-scale plane (recall that in some loose sense, the scale variable can be thought of as an inverse frequency) and the vertical axis represents the logarithm of the scale. As in the Gabor case, the phase behavior is also of interest, since on the ridge, the phase of the CWT coincides with the signal's phase.

It is worth emphasizing on this simple example the differences of behavior between the continuous wavelet transform and the continuous Gabor transform. To this end, let us consider a toy example consisting of a superposition of harmonic components:

$$f(x) = A_1\cos(\lambda x) + A_2\cos(2\lambda x) + A_3\cos(3\lambda x) + \dots .$$

If we consider the Gabor transform of such a signal and restrict our attention to positive values of ω for simplicity, the situation is quite simple if we assume that λ is large enough and g broad-band so that the terms involving negative frequencies may be neglected. We get

$$G_f(b,\omega) \approx A_1 e^{i\lambda b}\overline{\hat{g}(\omega-\lambda)} + A_2 e^{2i\lambda b}\overline{\hat{g}(\omega-2\lambda)} + A_3 e^{3i\lambda b}\overline{\hat{g}(\omega-3\lambda)} + \dots .$$

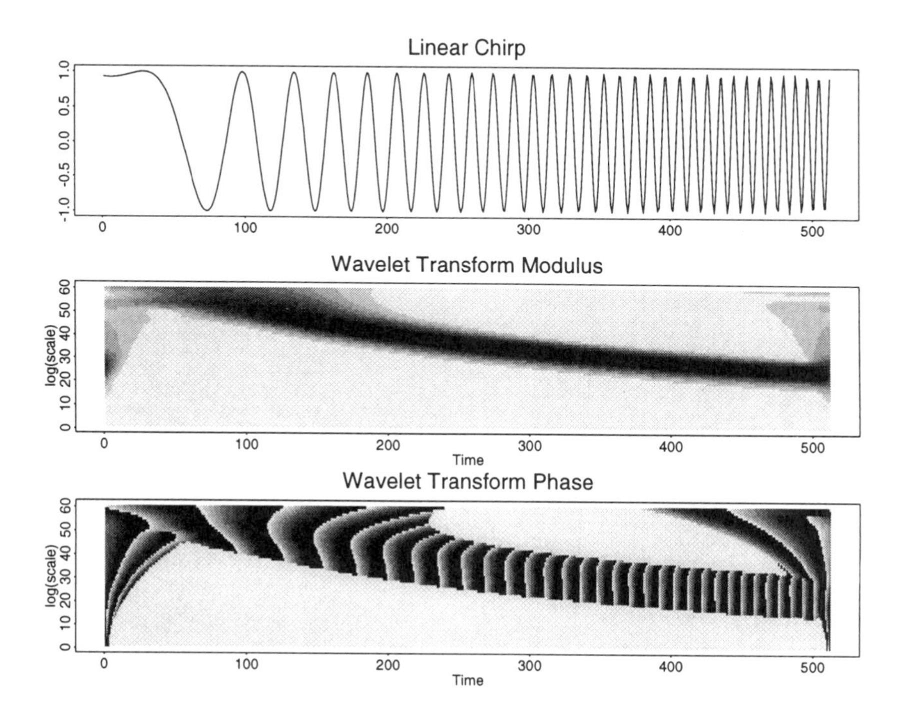

Figure 4.6. Plots of the modulus and the phase of the CWT of a chirp generated as in Example 4.5.

This formula gives contributions localized at frequencies $\omega = \lambda$, $\omega = 2\lambda$ and $\omega = 3\lambda$. If the bandwidth of the window g is small enough, then all the contributions will be neatly separated by the transform, i.e., the terms $\hat{g}(\omega-\lambda)$, $\hat{g}(\omega-2\lambda)$, $\hat{g}(\omega-3\lambda)$ will not overlap. More precisely, assuming for example that $\hat{g}(\xi)$ is localized in an interval $[-\gamma/2, \gamma/2]$ around the origin, then if $\gamma \leq \omega$ we easily see that whenever $\omega \approx k\lambda$,

$$G_f(b,\omega) \approx A_k e^{ik\lambda b} \overline{\hat{g}(\omega - k\lambda)} \ .$$

If we consider now the wavelet transform of the same example, with an analytic wavelet $\psi \in H^2(\mathbb{R})$,

$$T_f(b,a) \approx A_1 e^{i\lambda b} \overline{\hat{\psi}(a\lambda)} + A_2 e^{2i\lambda b} \overline{\hat{\psi}(2a\lambda)} + A_3 e^{3i\lambda b} \overline{\hat{\psi}(3a\lambda)} + \dots .$$

As before, we end up with a localization near the values $a = \omega_0/\lambda$, $a = \omega_0/2\lambda$, $a = \omega_0/3\lambda \cdots$. Assuming for simplicity that $\hat{\psi}(\xi)$ may be considered small outside an interval $[\gamma, 2\gamma]$, it is then easy to see that there exists a k such that for $a \approx \omega_0/(k\lambda)$, the term $\hat{\psi}(a(k+1)\lambda)$ may not be neglected any more compared with $\hat{\psi}(ak\lambda)$, so that we have for such values of the scale parameter a:

$$T_f(b,a) \approx A_k e^{ik\lambda b} \overline{\hat{\psi}(ka\lambda)} + A_{k+1} e^{i(k+1)\lambda b} \overline{\hat{\psi}((k+1)a\lambda)} + \dots \ .$$

Because of the presence of at least two oscillatory components, the wavelet transform will not be as simple as the Gabor transform. This is due to the interference phenomenon between the components.

As an illustration, we give in Figure 4.7 the case of a signal with three harmonic components.

4.5.2 Examples of Time-Scale Analysis

An important class of examples involves the use of the dilation variable a as a scale variable instead of a frequency or inverse frequency. This is the case when one is more interested in the local behavior of the function than in local periodicities.

In order to illustrate the behavior of the wavelet transform of singular functions, we display in Figure 4.8 the continuous wavelet transform of a function containing various kinds of singularities. As can be seen in Figure 4.8, and especially on the plot of the modulus of the wavelet transform, different types of singularities have different types of contributions to the wavelet transform modulus. The leftmost singularity in the signal is the weakest one, and as such, it has the smallest contribution at small scales. Then come the stronger singularities, and the behavior described in Theorem 4.3 may be seen in the perspective view. Essentially, the more singular

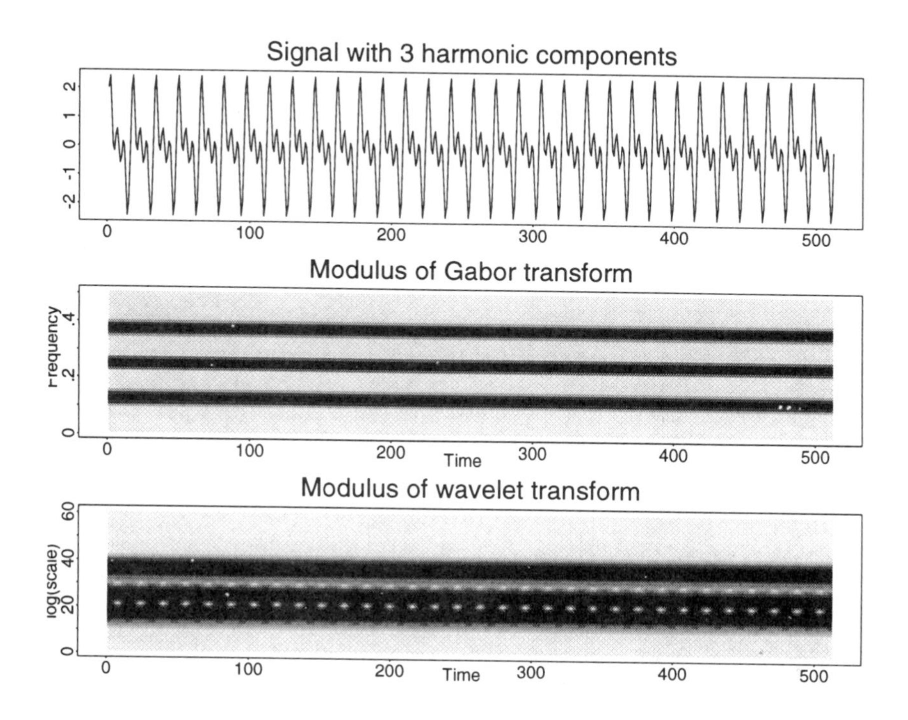

Figure 4.7. CGT and CWT of a signal with three harmonic components. The details are given in Example 4.4.

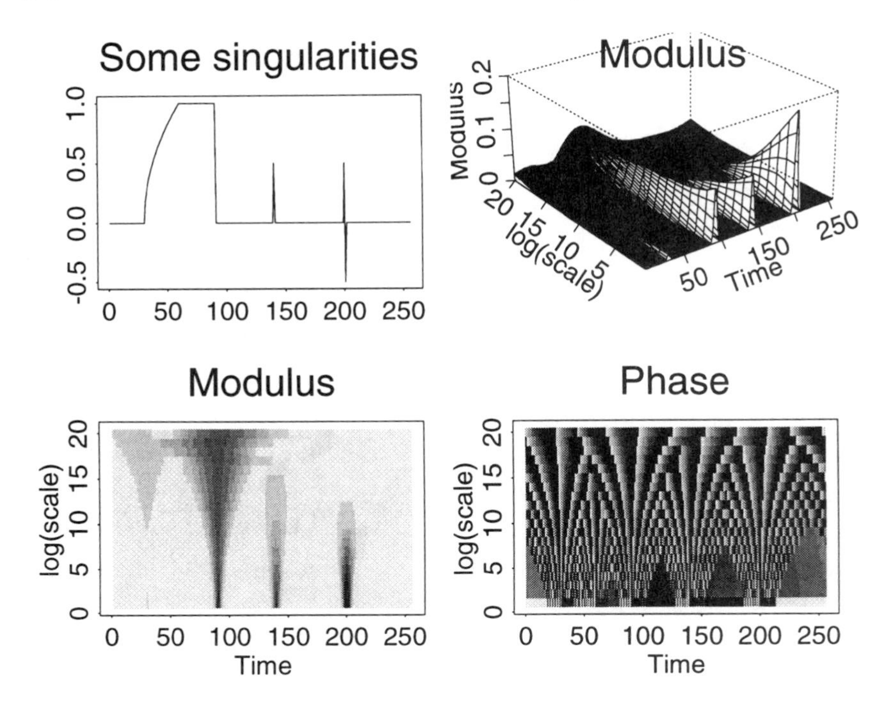

Figure 4.8. Plots of the CWT of a signal with singularities of various natures and strengths. Details on the signal can be found in the discussion of Example 4.2.

the function, the larger the contribution of the small scales to the modulus of the transform.

The behavior of the phase of the wavelet transform is also interesting to decipher: notice that the lines of constant phase "converge" toward the singularities as the scale decreases.

4.5.3 Non-academic Signals

We now return to the discussion of the trigonometric chirps introduced earlier. One may wonder if such signals do exist in nature. We shall see that they are pervasive in acoustics: bats and dolphins have known for quite a long time that these signals have optimality properties, and they took full advantage of these properties long before the mathematicians discovered time/frequency analysis. But we shall also consider another example with a potentially very exciting future.

Gravitational Waves Generated by Coalescing Binary Stars

There has been a recent renewal of interest and intense activity in the arena of gravitational wave detection [1, 56], and trigonometric chirps are bound to play a crucial role in this research. Gravitational waves are waves of fluctuation of the curvature of space-time. Their existence is predicted by general relativity, but so far they have not been observed, presumably because of their small amplitudes. According to the theory, when an object's curvature has fast variations, waves of curvature deformation propagating at the speed of light are emitted. Most gravitational wave detectors are based on the analysis of the relative motion of test masses. A gravitational wave that is adequately polarized pushes the test masses back and forth and modifies the distance between them, and the relative fluctuation of the distance $\Delta L/L$ is measured. To give an idea of the precision required for this kind of measurement, the order of magnitude of the expected relative distance fluctuation is of the order of 10^{-21} !!!

Among the candidates capable of generating gravitational waves, the collapse of a binary star is very popular because the expected signal $h(x) = (\Delta L/L)(x)$ may be computed explicitly (with acceptable precision) from various post-Newtonian models. In the simplest Newtonian approximation,[3] it has a simple functional form:

$$h(x) = A\Theta(x_0 - x)(x_0 - x)^{\lambda} \cos\left(\phi - \frac{2\pi F}{\mu + 1}(x_0 - x)^{\mu+1}\right) . \tag{4.46}$$

That is, it takes the form of a trigonometric chirp (recall that we use the notation Θ for the Heaviside step function). Here, $\lambda = -1/4$, $\mu = -3/8$, and A and F are amplitude and frequency parameters which depend on the source (distance, masses, ...). Strictly speaking, $h(x)$ is not a chirp in the sense of Section 4.4.2. Indeed, the value of μ is not in the range given there. Nevertheless, the corresponding ridge still lies within the cone alluded to in Remarks 4.10 and 4.11. Moreover, the signal is in the same "family," in the sense that it has a definite Hölder exponent λ, and it has oscillations that "accumulate" at the point x_0 (the corresponding local frequency diverges at $x = x_0$).

The main problem in detecting such gravitational waves is that they are supposed to be embedded into a strong noise, with input signal-to-noise ratio (SNR) smaller than -20 dB depending on the distance to the source. In addition, very few events are expected. The first problem to be solved is a detection problem: detect the presence of an event (if any ...), and only then can the problem of the estimation of the corresponding parameters x_0, A, F, ϕ be addressed.

[3] Higher order post-Newtonian approximations are also considered. They involve polynomial corrections of different orders to the amplitude and the phase of the signal.

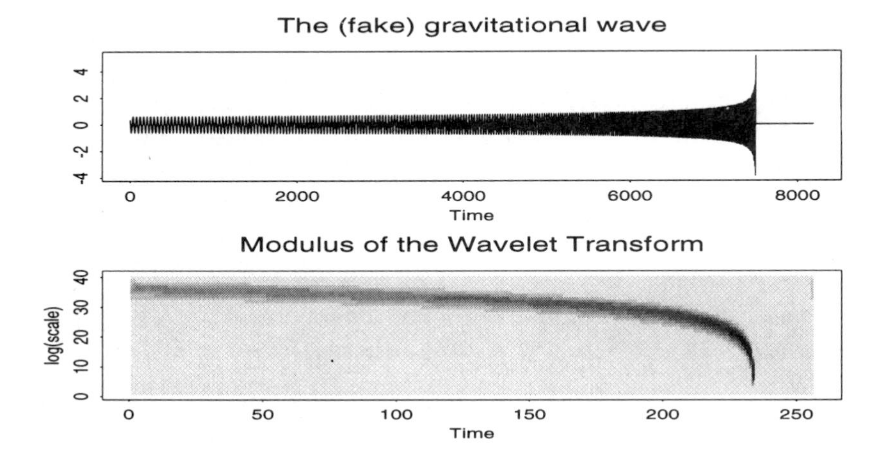

Figure 4.9. Pure gravitational wave and the modulus of its wavelet transform.

Wavelet analysis has been proposed [177, 178, 180] as a good candidate for on-line processing of the experimental signal. Indeed, if we forget for a while about the phase ϕ, and this can be done in a first analysis if a progressive wavelet is used, $h(x)$ appears as a shifted and dilated copy (up to a multiplicative factor) of a reference signal of the form $\Theta(-x)\,|x|^{\lambda}\cos|x|^{\mu+1}$. In view of Proposition 4.1, this is a typical problem to which wavelets are well adapted. In addition, such signals are naturally associated with ridges of the form

$$a_r(b) = \frac{1}{F}(x_0 - b)^{-\mu} \ ,$$

i.e., a two-dimensional family of ridges.

We show in Figure 4.9 a simulation of a gravitational wave together with its wavelet transform (five octaves and eight intermediate scales per octave). The wave was sampled at 1 kHz, and 1 s before collapse the frequency F was set to 50 Hz. As expected, we observe a significant concentration of energy near the ridge. Figure 4.10 represents the same analysis done on the signal buried in simulated experimental noise. For simplicity, we used Gaussian statistics for the noise. We have only taken into account the thermal noise (with spectral density $\mathcal{E}(k) \sim k^{-4}$) at low frequencies, mirror thermal noise (approximately white) at frequencies near 100 Hz, and shot noise at high frequencies, and we have neglected resonances of the experimental setup.

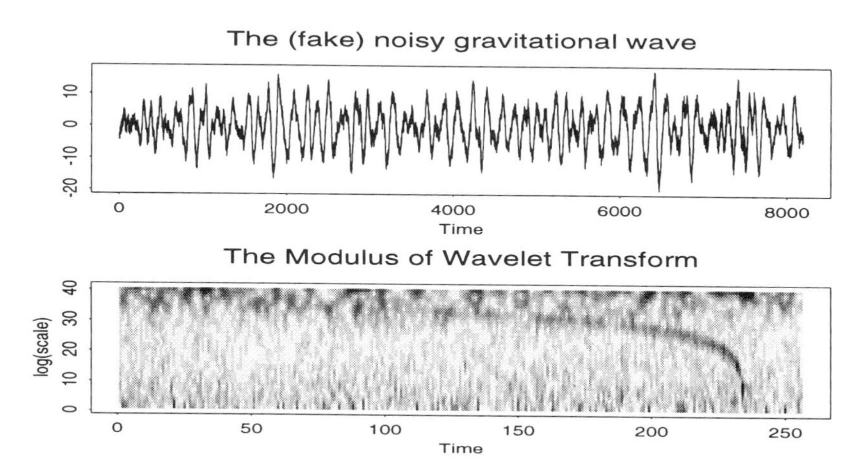

Figure 4.10. Noisy gravitational wave (in Gaussian noise) with the modulus of its wavelet transform.

The Quake Signals Revisited

The results of the analysis of the quake data (described in the previous chapter in Subsection 3.3.2) by the continuous wavelet transform are shown in Figures 4.11 (basement signal) and 4.12 (10th-floor signal). As before, we see in Figure 4.11 that there is no significant difference in the three input signals, except maybe a harmonic high-frequency component which appears more clearly in the first run than in the others. We did not see such a component when we analyzed the Gabor transform because we did not consider large enough values of the frequency. The phenomenon suggested by the changes in the CGT in Figure 3.9, namely, the decrease of the fundamental frequency from run to run, can be observed as well. Note that the representation of the signal in the time-scale domain is quite different, in the sense that the wavelet transform is much more precise at small scales and gives a better description of the singular parts of the signal.

Of course, it is possible to construct a "wavelet transfer function" along the same lines as in Section 3.3.2 and Figure 3.10. The results (not displayed here) lead to the same conclusions.

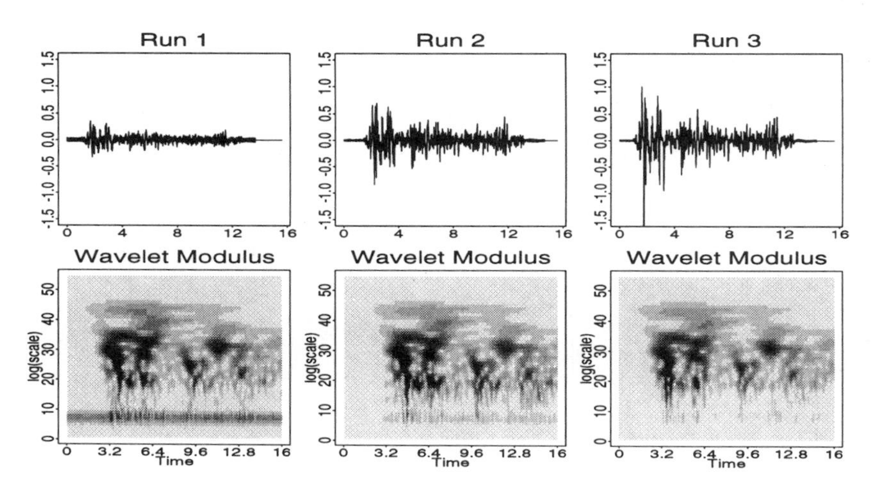

Figure 4.11. CWT of the three runs of the quake data for the accelerations measured on the basement. The original signal is shown on top and the modulus of the CWT is imaged on the bottom.

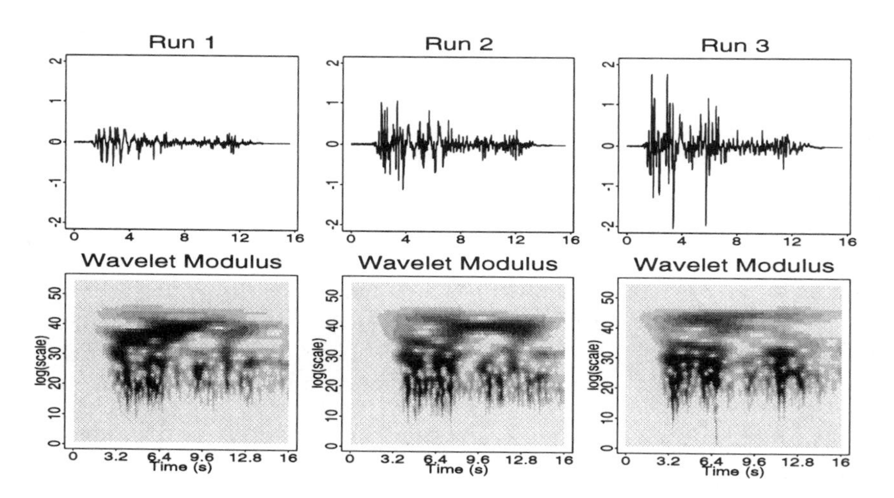

Figure 4.12. CWT of the three runs of the quake data for the accelerations measured on the 10th floor. The original signal is shown on top and the modulus of the CWT is imaged on the bottom.

4.6 Examples and S-Commands

4.6.1 Morlet Wavelets

All the examples presented in this book are treated using Morlet's wavelet. Examples of Morlet's wavelets (see Figure 4.3) may be generated by using the Splus command `morlet`, as follows:

Example 4.1 Morlet wavelet
Generate a Morlet wavelet at given scale (here 40*) and location (here* 256*).*

```
> mor <-morlet(512,256,40)
> par(mfrow=c(1,2))
> tsplot(Re(mor))
> title(''Real part'')
> tsplot(Im(mor))
> title(''Imaginary part'')
```

4.6.2 Wavelet Transforms

The basic continuous wavelet transform function is named `cwt`. It implements continuous wavelet transform with respect to the Morlet wavelet

$$\psi(x) = e^{-x^2/2} e^{i\omega_0 x}$$

for values of the scale variable of the form

$$a = a_{j,k} = 2^{j+k/n_v}$$

Here $j = 1, \cdots, n_o$ labels the *octaves* (i.e., the powers of 2), $k = 0, \cdots, n_v - 1$ labels the *voices*, and n_v (the number of "voices per octave") is an integral number.

`cwt` basically takes three arguments: the input signal, the number of octaves n_o, and the number of voices per octave n_v. The *central frequency* ω_0 is by default set to 2π, but may be modified as an argument of `cwt`.

The function `cwt` returns a complex two-dimensional array, the modulus and the argument of which may be visualized using the `image` or `persp` functions. By default, the function `cwt` "images" the wavelet transform modulus.

As alternatives to `cwt`, Swave contains two other continuous wavelet transform functions, implementing two other families of wavelets: the Cauchy wavelets (function `cwtTh`) and the (progressive) derivatives of Gaussians (function `DOG`). Both functions have essentially the same syntax as `cwt`. The only differences are the central frequency (automatically set to 2π), and an extra variable controlling the vanishing moments of the wavelets. We first consider simple examples of non-oscillating singularities.

Example 4.2 : A First Example of Singularity

Generate the sequence and plot a delta signal:

```
> sing <- numeric(256)
> sing[30:60] <- sqrt(1:30)/sqrt(30)
> sing[60:90] <- 1
> sing[140] <- 0.5
> sing[200] <- 0.5
> sing[201] <- -0.5
> par(mfrow=c(2,2))
> tsplot(sing)
> title(''Some singularities'')
```

Compute the wavelet transform, and plot. Notice that the lines of constant phase tend to "converge" towards the singularity:

```
> cwtsing <- cwt(sing,5,12)
> persp(Mod(cwtsing))
> image(Arg(cwtsing))
```

The case of a sine wave can be treated as another simple example.

Example 4.3 Sine Wave

Generate the sine wave of frequency 1/32 Hz, sampled with unit sampling frequency.

```
> x <- 1:512
> sinwave <- sin(2*pi*x/32)
> par(mfrow=c(1,3))
> tsplot(sinwave)
> title(''Sine wave'')
```

Compute the wavelet transform with $n_o = 5$ *and* $n_v = 12$ *(the choice* $n_v = 12$ *is very common for musical applications, since* 12 *voices per octave is the resolution of a piano, for example).*

```
> cwtsinwave <- cwt(sinwave,5,12)
> image(Arg(cwtsinwave))
> title(''Phase'')
```

Example 4.4 Sum of Three Sine Waves

Generate a signal with three harmonic components (with frequencies 1/16, 1/8, *and* 3/16 *Hz, respectively,) sampled with unit sampling frequency. For the sake of comparison we compute the Gabor and wavelet transforms with the appropriate parameters.*

```
> x <- 1:512
> triwave <- sin(2*pi*x/16)+sin(2*pi*x/8)sin(3*2*pi*x/16)
> par(mfrow=c(1,3))
> tsplot(triwave)
> title(''Signal with 3 harmonic components'')
> cgttri <- cgt(triwave,50,.01,25)
> cwttri <- cwt(triwave,5,12)
```

We now turn to the example of a chirp signal.

Example 4.5 Chirp
Generate the chirp and compute its wavelet transform:

```
> x<- 1:512
> chirp <- sin(2*pi*(x + 0.002*(x-256)*(x-256))/16)
> par(mfrow=c(1,3))
> tsplot(chirp)
> title(''Chirp signal'')
> cwtchirp <- cwt(chirp,5,12)
> image(Arg(cwtchirp))
> title(''Phase of wavelet transform'')
```

4.6.3 Real Signals

We return to the quake data: the results in Figure 4.12 were obtained from the following Swave commands.

Example 4.6 10th-Story Quake Signal
Plot the three quake signals (10th story) and compute their wavelet transform (nine octaves, six voices per octave).

```
> par(mfrow=c(3,2))
> tsplot(H1.10[2,])
> tsplot(H2.10[2,])
> tsplot(H3.10[2,])
> cwtH1.10 <- cwt(H1.10[2,], 9, 6)
> cwtH2.10 <- cwt(H2.10[2,], 9, 6)
> cwtH3.10 <- cwt(H3.10[2,], 9, 6)
```

Finally, we give the S-commands used to generate Figures 4.9 and 4.10.

Example 4.7 Simulation of a Noisy Gravitational Wave
Get the files from the disk and compute their wavelet transforms, with five octaves and ten voices per octave.

```
> purwave <- scan("signals/pure.dat")
> cwtpure <- cwt(purwave, 5, 8)
> purwave <- scan("signals/noisy.dat")
> cwtnoisy <- cwt(noisywave, 5, 8)
```

4.7 Notes and Complements

Wavelet transforms seem to have been discovered and rediscovered many times during the past 30 years. The *wavelet new age* started with a series of papers by Goupillaud, Grossmann, and Morlet [147] and Grossmann and Morlet [147, 151], where the continuous wavelet transform was identified as such, and its relevance for signal analysis was clearly identified. Continuous wavelet analysis actually existed already in the harmonic analysis context since the early work of Calderón [60], and in the group theoretical context thanks to works of Duflo and Moore [117], Carey [63], and Moscovici and Verona [229] (see also the paper by Godement [145] in the case of unimodular groups). In fact, the theory of square-integrable group representations developed in the latter publications ended up being the right theoretical framework for unifying wavelet and Gabor analyses [153, 154] (see also [124, 125]).

The group theoretical approach is closely related to quantum mechanics and the so-called coherent states formalism (see [193] for a review of the early articles and applications and [12] for a somewhat more modern presentation). Wavelets were shown by T. Paul to be especially well adapted to the radial wave function of the harmonic oscillator [238].

Systematic applications of continuous wavelet analysis to signal processing (and in particular to computer music) are due to A. Grossmann, R. Kronland-Martinet, and J. Morlet and co-workers (see, for example, [150, 199, 200]). Specific extensions of the standard continuous wavelet analysis to higher dimensions were considered among others by R. Murenzi [231] who was motivated by image processing applications.

The characterization of the regularity properties of functions and distributions using wavelet-type decompositions goes back to early works on the Littlewood-Paley theory (see [136] for a detailed account). The first results using the modern form of wavelets are due to Holschneider [167] and Jaffard [181] (see also [171]). A detailed account of more recent results can be found in the memoir published by Jaffard and Meyer [183]. These works eventually led to a set of wavelet-based methods for the characterization of fractal measures, signals and functions. These methods were introduced in [167, 21]. A review can be found in the book published by Arneodo and collaborators [Arneodo95].

Wavelet analysis has received considerable attention since the early days and early developments reported in this chapter. This extraordinary popularity came with the discovery of orthonormal bases and the concept of multiresolution analysis. They are discussed in the next chapter.

Chapter 5

Discrete Time-Frequency Transforms and Algorithms

We now turn to the case of the discrete forms of the wavelet and Gabor transforms. As we shall see, there are several possible ways of discretizing the transforms (especially in the wavelet case), and which one to choose really depends on the application. The redundancy of the continuous transforms is obviously reduced by the discretization, and we show how to take advantage of the flexibility in the choice of the discretization of the time-frequency representation to control the level of redundancy.

Obviously, if the application is close to signal coding or compression, the redundancy has to be reduced as much as possible. In such cases, the discretization grid becomes sparse, and in the wavelet case one turns to the so-called *dyadic grid* of discretization. On the other hand, if the purpose is signal analysis or the use of a model to fit a curve to the data, it is often convenient to keep some redundancy in the transform.

We begin this chapter with a brief description of the theory of frames which we apply both to the wavelet and Gabor cases. We then turn to several forms of the discrete wavelet transform, emphasizing the aspects which we find relevant for our purpose. Finally, we close this chapter with a description of possible algorithms for implementing the various transforms described in Chapters 3 and 4, and in the current chapter.

5.1 Frames

The notion of frame is a very useful generalization of the notion of basis in the context of Hilbert space. For the sake of clarity, and to avoid duplications due to the separate treatments of the Gabor and the wavelet transforms, we give the definition of frames in an abstract setting.

Definition 5.1 *A family $\{\psi_\lambda;\ \lambda \in \Lambda\}$ of elements of a Hilbert space $\mathcal{H}$ is said to be a frame if there exist finite positive constants A and B such that*

$$A\|f\|^2 \le \sum_{\lambda\in\Lambda} |\langle f, \psi_\lambda\rangle|^2 \le B\|f\|^2, \tag{5.1}$$

for all $f \in \mathcal{H}$. The constants A and B are called the frame bounds.

Here Λ is a countable set which serves as index for the elements of the frame. Continuous frames have also been studied [11, 12, 166], but we shall not consider such generalizations here. Be aware of the fact that the index set Λ will change from an application to another.

Frames provide the right theoretical framework (sorry, but this is not intended as a joke) for the analysis of possibly redundant decompositions of elements of a Hilbert space. In this sense, and this should be clear from the preceding definition, they are generalizations of the notion of basis. In fact, any orthonormal basis is a frame with frame bounds $A = 1$ and $B = 1$. Multiplying all the elements of a frame by the same scalar changes the bounds A and B, but it does not affect the ratio B/A. The latter is indeed the relevant constant, and a frame will be said to be tight whenever $A = B$. Notice that orthonormal bases are tight frames but there exist tight frames which are not orthonormal bases.

The notion of frame is close to (though different from) the notion of Riesz basis, with which it should not be confused.

Definition 5.2 *A countable family $\{\psi_\lambda;\ \lambda \in \Lambda\}$ in a Hilbert space $\mathcal{H}$ is said to be a Riesz basis if there exist positive finite constants C' and C'' for which:*

$$C' \sum_{\lambda\in\Lambda} |a_\lambda|^2 \le \|\sum_{\lambda\in\Lambda} a_\lambda\psi_\lambda\|^2 \le C'' \sum_{\lambda\in\Lambda} |a_\lambda|^2 , \tag{5.2}$$

for all sequences of (complex) numbers $\{a_\lambda\}_{\lambda\in\Lambda}$.

Obviously, a Riesz basis is a frame, but the converse is not true in general.

Given a frame $\{\psi_\lambda;\ \lambda \in \Lambda\}$, we introduce the associated *frame operator* F, mapping $\mathcal{H}$ into the Hilbert space $\ell^2(\Lambda)$ of square summable sequences indexed by Λ. It is defined by the formula

$$[Ff](\lambda) = \langle f, \psi_\lambda\rangle , \qquad f \in \mathcal{H}. \tag{5.3}$$

Notice that the definition property (5.1) of a frame can be rewritten in term of a property of the self adjoint operator $\mathcal{R} = F^*F$ on $\mathcal{H}$. Indeed, since the action of $\mathcal{R}$ is given by the formula

$$[\mathcal{R}f] = \sum_{\lambda\in\Lambda} \langle f, \psi_\lambda\rangle \psi_\lambda, \qquad f \in \mathcal{H},$$

the definition of a frame can be rewritten as

$$A \leq \mathcal{R} \leq B ,$$

where the inequalities are to be understood in the sense of quadratic forms on the Hilbert space $\mathcal{H}$. These inequalities imply that $\mathcal{R}$ is bounded (and that $\|\mathcal{R}\| \leq B$) and has a bounded inverse (satisfying $\|\mathcal{R}^{-1}\| \leq A^{-1}$). Consequently, we have the following decomposition formula:

$$f = \sum_{\lambda\in\Lambda} \langle f, \psi_\lambda\rangle \tilde{\psi}_\lambda , \tag{5.4}$$

provided we set

$$\tilde{\psi}_\lambda = \mathcal{R}^{-1}\psi_\lambda . \tag{5.5}$$

The family $\{\tilde{\psi}_\lambda;\ \lambda \in \Lambda\}$ is also a frame, called the *dual frame*. It is in general difficult to compute the dual frame explicitly. Clearly, if $B = A$ (i.e., if the frame is tight), $\mathcal{R}$ is the identity operator and $\tilde{\psi}_\lambda = \psi_\lambda$. The inversion formula is in such a case the simplest possible. It mimics the inversion formulas already encountered in the case of the continuous transforms

$$f = \frac{1}{A}\sum_{\lambda\in\Lambda} \langle f, \psi_\lambda\rangle \psi_\lambda . \tag{5.6}$$

A straightforward calculation yields

$$\mathcal{R}^{-1}f = \frac{2}{A+B}\sum_{n=0}^{\infty} T^n f , \tag{5.7}$$

where

$$T = 1 - \frac{2\mathcal{R}}{A+B} . \tag{5.8}$$

Of course, the closer A and B are, the faster the convergence of (5.7). Equation (5.7) provides an iterative algorithm for the inversion of the frame operator. This iterative algorithm may be written as follows: Start from decomposition (5.6), use (5.17), and set

$$f^{(k)} = \frac{2}{A+B} \sum_{\lambda \in \Lambda} \langle f, \psi_\lambda \rangle [T^k \psi_\lambda] .$$

Then we have

$$f^{(k)} = f^{(k-1)} - \frac{2}{A+B} \sum_{\lambda \in \Lambda} \langle f^{(k-1)}, \psi_\lambda \rangle T^k \psi_\lambda .$$

Therefore, the kth term is simply obtained using a simple *decomposition-reconstruction* scheme. Other inversion schemes have been investigated by several authors; see, e.g., [123, 98].

5.1.1 Gabor Frames

The Gabor frames are given by specializing the general theory just presented to the particular case of the Hilbert space $\mathcal{H} = L^2(\mathbb{R})$, the index set $\Lambda = \mathbb{Z} \times \mathbb{Z}$, and the elements $\psi_\lambda = g_{m,n}$ the Gabor functions defined by

$$g_{mn}(x) = e^{in\omega_0(x-mb_0)} g(x - mb_0) , \tag{5.9}$$

where b_0 and ω_0 are two nonzero constants.

The idea of introducing windows into Fourier analysis is a very natural (and old) idea. However, the study of discrete versions of the method goes back to Gabor in the mid-1940s [139]. Gabor's original idea was to construct a representation in which the time and frequency variables would appear simultaneously, and to suppress the redundancy present in the continuous representation. He then discretized the transform in the natural way. According to the definition of the continuous Gabor transform given in Chapter 3, if $g \in L^2(\mathbb{R})$ is the analyzing window, the Gabor coefficients of $f \in L^2(\mathbb{R})$ are given by

$$G_f(b, \omega) = \int f(x) \overline{g(x-b)} e^{-i\omega(x-b)} dx . \tag{5.10}$$

It follows from general arguments (namely, the time and frequency translation invariance of the continuous Gabor transform) that the natural discretization lattice should be of the form

$$(b_m, \omega_n) = (mb_0, n\omega_0), \qquad m, n \in \mathbb{Z} . \tag{5.11}$$

Indeed, the Gabor function being of constant size in the time (and thus the frequency) domain, a uniform sampling is the most natural (see Figures 5.1

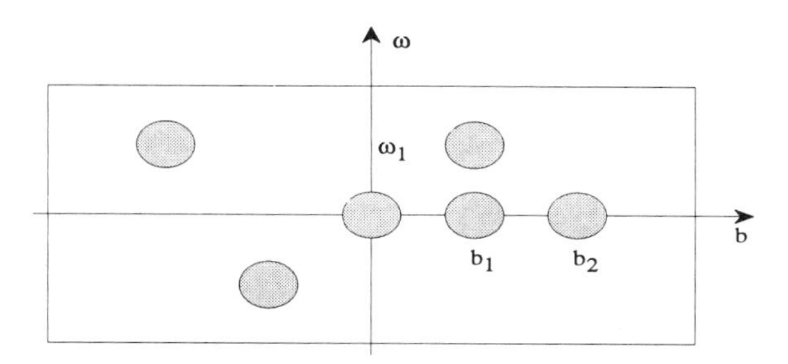

Figure 5.1. Schematics of essential supports of Gabor functions in the time-frequency space.

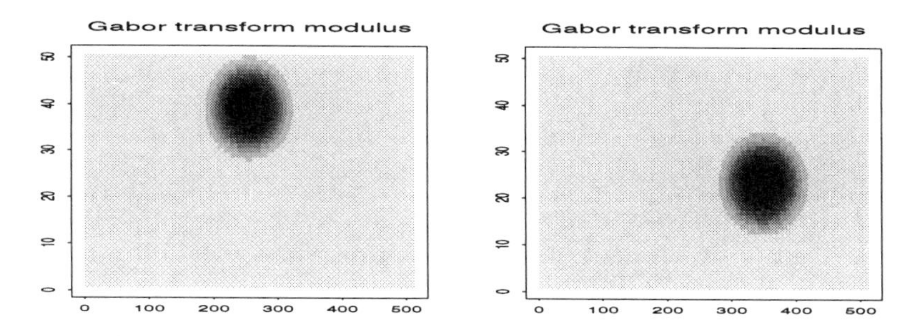

Figure 5.2. CGT of two Gabor functions, with different time and frequency parameters as generated in Example 5.1.

and 5.2). The discrete Gabor transform is defined as the restriction to this lattice of the continuous Gabor transform. In other words, given a window $g \in L^2(\mathbb{R})$, we define the Gabor functions as in (5.9) and the discrete Gabor transform of any $f \in L^2(\mathbb{R})$ by

$$[G_d f](m,n) = G_f(b_m, \omega_n) = \langle f, g_{m,n} \rangle, \qquad (m,n) \in \mathbb{Z} \times \mathbb{Z}. \quad (5.12)$$

It is natural to expect (and easy to prove) that the values of the (continuous) Gabor transform on the grid still determine the function f when the values of b_0 and ω_0 are small. The natural question is now the following: Do there exist limits on the values of the sampling periods b_0 and ω_0 beyond which the representation is not complete anymore? Such a problem has been analyzed carefully by I. Daubechies [93]; it turns out that the most important quantity is the product $b_0\omega_0$ and how it compares to the so-called *Nyquist density* 2π. We shall prove the following:

- If $b_0\omega_0 > 2\pi$, the discrete Gabor representation is *not* complete [256].

- If $b_0\omega_0 < 2\pi$, the discrete Gabor representation may be *over-complete.*

At the critical density, there is room for bases of Gabor functions, but this is also the point where the so-called *Balian-Low phenomenon* comes into play restricting the properties of these bases (see later discussion).

To illustrate the technicalities involved in the existing sufficient conditions for the existence of frames we quote without proof the following result:

Theorem 5.1 *Let $g \in L^2(\mathbb{R})$ be a window, let b_0 be a sampling period and let us set:*

$$\begin{array}{rcl} m(g,b_0) & = & \text{ess } \inf_{x\in[0,b_0]} \sum_n |g(x-nb_0)|^2 \\ M(g,b_0) & = & \text{ess } \sup_{x\in[0,b_0]} \sum_n |g(x-nb_0)|^2 \ , \end{array} \tag{5.13}$$

$$\beta(s) = \sup_{x\in[0,b_0]} \sum_n |g(x-nb_0)||g(x+s-nb_0)| \ ,$$

and

$$C_\epsilon = \sup_{s\in\mathbb{R}} \left[(1+s^2)^{(1+\epsilon)/2}\beta(s)\right] \ .$$

Then if $m(g,b_0) > 0$, $M(g,b_0) < \infty$ and $C_\epsilon < \infty$ for some positive ϵ, there exists a critical value ω_0^c such that for any $\omega_0 < \omega_0^c$, the corresponding family $\{g_{mn}\}$ is a frame of $L^2(\mathbb{R})$. In addition, setting

$$A = m(g,b_0) - \sum_{k\neq 0} \sqrt{\beta\left(\frac{2\pi k}{\omega_0}\right)\beta\left(\frac{-2\pi k}{\omega_0}\right)} \ , \tag{5.14}$$

$$B = M(g,b_0) + \sum_{k\neq 0} \sqrt{\beta\left(\frac{2\pi k}{\omega_0}\right)\beta\left(\frac{-2\pi k}{\omega_0}\right)} \ , \tag{5.15}$$

A and B are frame bounds for the frame $\{g_{mn}, m,n \in \mathbb{Z}\}$.

A proof of the theorem may be found in [Daubechies92a].

Frame bounds have been discussed in [93] for some special cases of windows g, and numerical estimates for selected values of b_0 and ω_0 are given there.

Given a Gabor frame $\{g_{mn}\}$, the corresponding frame operator F maps $L^2(\mathbb{R})$ onto $\ell^2(\mathbb{Z}\times\mathbb{Z})$, and it is given by

$$[Ff](m,n) = \langle f, g_{mn}\rangle.$$

We have the decomposition formula

$$f(x) = \sum_{mn,} \langle f, g_{mn}\rangle \, \tilde{g}_{mn}(x) \ . \tag{5.16}$$

A remarkable fact is that the dual frame $\{\tilde{g}_{mn};\ m,n \in \mathbb{Z}\}$ is also a Gabor frame because

$$\tilde{g}_{mn} = \mathcal{R}^{-1} g_{mn} = \left(\mathcal{R}^{-1} g\right)_{mn} . \tag{5.17}$$

This result does not hold in the case of the wavelet transform.

5.1.2 Critical Density: The Balian-Low Phenomenon

The two bulleted points given earlier emphasize the fact that there exists a critical value for the sampling rates of the time and frequency variables. We restate the first result in a proposition.

Proposition 5.1 *Let $\{g_{mn};\ m,n \in \mathbb{Z}\}$ be the Gabor functions defined in (5.9). Then, if $b_0\omega_0 > 2\pi$, there exists $f \in L^2(\mathbb{R})$ such that $f \neq 0$ and $\langle f, g_{mn}\rangle = 0$ for all $m,n \in \mathbb{Z}$.*

The proof of the following result can be found in [256]. The case $b_0\omega_0 = 2\pi$ of the critical density is interesting by itself. Indeed, in this case there is an insurmountable obstruction that prevents building tight frames of Gabor functions with arbitrary regularity and localization. This result is known as the *Balian-Low phenomenon.* It was discovered by Balian and Low independently.

Theorem 5.2 *Let $g \in L^2(\mathbb{R})$ be such that the associated family of Gabor functions g_{mn} constructed at the Nyquist density $b_0\omega_0 = 2\pi$ is a frame of $L^2(\mathbb{R})$. Then we have either*

$$\int x^2 |g(x)|^2 dx = \infty \qquad \text{or} \qquad \int \xi^2 |\hat{g}(\xi)|^2 d\xi = \infty . \tag{5.18}$$

This result, which may be seen as a consequence of Heisenberg's uncertainty principle (see, for example, the proofs given in [37]), has important theoretical and practical implications. It prevents building "nice" bases of Gabor functions, and it is indirectly one of the reasons for the success of wavelets. Some ways of avoiding the obstruction (the so-called Wilson and Malvar -or local trigonometric- bases; see below) have been proposed in [216, 97], for example, but they are not, strictly speaking, based on Gabor-type constructions.

The situation for Gabor frames may then be summarized as follows:

- $b_0\omega_0 < 2\pi$: Frames are possible, with arbitrarily high regularity.
- $b_0\omega_0 = 2\pi$: Frames are possible, but the corresponding functions are either poorly localized, or have poor regularity.
- $b_0\omega_0 > 2\pi$: No frame is possible!

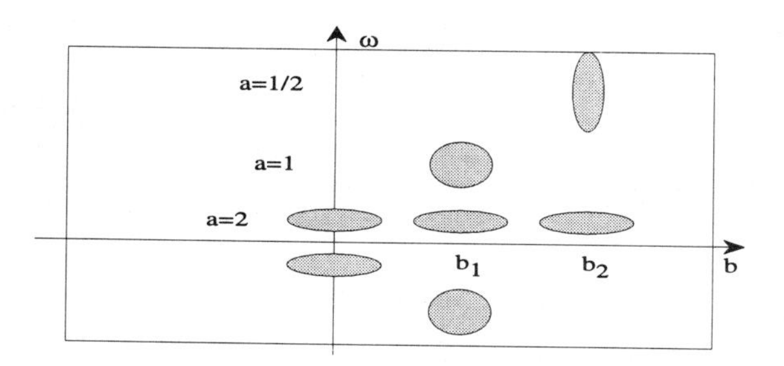

Figure 5.3. Schematic essential supports of wavelets in the time-frequency space.

Remark 5.1 It is important to stress that the density condition $b_0\omega_0 < 2\pi$, even though necessary, is not a sufficient condition for the family of Gabor functions to form a frame. Several examples have been discussed in [93].

5.1.3 Wavelet Frames

The same discussion (with the exception of the Balian Low theorem) applies to the case of the wavelet transform. The main difference lies in the way of discretizing the transform domain. Indeed, the wavelets are functions of constant shape, but variable size (see Figure 5.3). The natural discretization grid for wavelets is thus a lattice of the form

$$(b_{jk}, a_j) = (kb_0a_0^j, a_0^j) \qquad j, k \in \mathbb{Z}\ , \tag{5.19}$$

for some nonzero constants a_0 and b_0, and we are left with the following family of wavelets:[1]

$$\psi_{jk}(x) = a_0^{j/2}\psi\left(a_0^j x - k\right)\ . \tag{5.20}$$

The same questions as before have to be answered: Do there exist values of b_0 and a_0 such that the corresponding family of wavelets forms a frame of $L^2(\mathbb{R})$?

Theorem 5.3 *Let $\psi \in L^2(\mathbb{R})$ be an analyzing wavelet and a_0 be a scale sampling period, and let us set*

$$m(\psi, a_0) = \operatorname{ess}\inf_{\xi\in[1,a_0]}\sum_j |\hat\psi(a_0^j\xi)|^2\ ,\quad M(\psi, a_0) = \operatorname{ess}\sup_{\xi\in[1,a_0]}\sum_j |\hat\psi(a_0^j\xi)|^2\ , \tag{5.21}$$

[1]Notice that, unlike in our discussion of the CWT in which the wavelets $\psi_{(b,a)}$ are normalized in such a way that $||\psi_{(b,a)}||_1 = ||\psi||_1$, we now choose a normalization such that $||\psi_{jk}||_2 = ||\psi||_2$. Such a switch in our normalization convention is motivated by the standard practice reported in the literature.

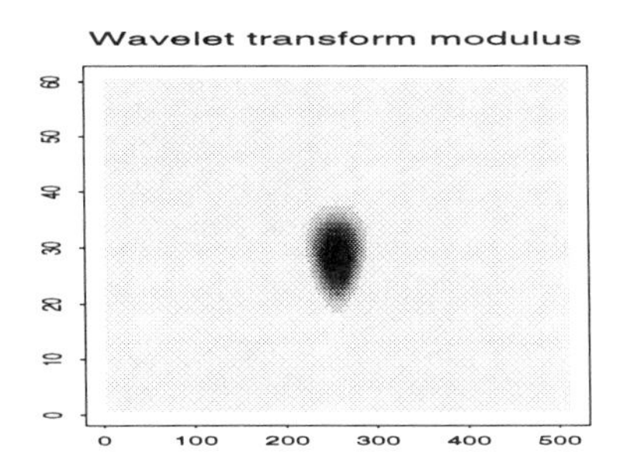

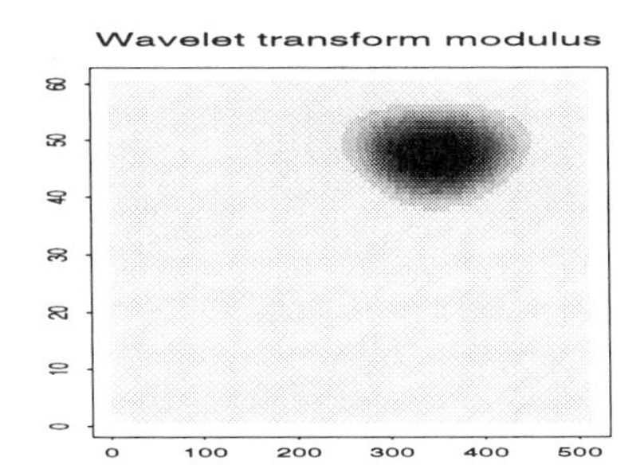

Figure 5.4. CWT of two wavelets, with different time and scale parameters as generated in Example 5.2.

$$\beta(s) = \sup_{\xi \in [1,a_0]} \sum_j |\hat{\psi}(a_0^j \xi)||\hat{\psi}(a_0^j \xi + s)| ,$$

and

$$C_\epsilon = \sup_{s \in \mathbb{R}} \left[(1+s^2)^{(1+\epsilon)/2} \beta(s) \right] .$$

Then if $m(\psi, a_0) > 0$, $M(\psi, a_0) < \infty$, and $C_\epsilon < \infty$ for some positive ϵ, there exists a critical value b_0^c such that for any $b_0 < b_0^c$, the corresponding family $\{\psi_{jk}\}$ is a frame of $L^2(\mathbb{R})$. In addition, the numbers A and B defined by

$$A = m(\psi, b_0) - \sum_{k \neq 0} \sqrt{\beta\left(\frac{2\pi k}{b_0}\right) \beta\left(\frac{-2\pi k}{b_0}\right)} , \tag{5.22}$$

$$B = M(\psi, b_0) + \sum_{k \neq 0} \sqrt{\beta\left(\frac{2\pi k}{b_0}\right) \beta\left(\frac{-2\pi k}{b_0}\right)} , \tag{5.23}$$

are frame bounds for the frame $\{\psi_{mn}, m, n \in \mathbb{Z}\}$.

Again, the same discussion as before applies, and the frame operator F defined by

$$[Ff](j,k) = \langle f, \psi_{jk} \rangle$$

may be inverted on its range. This yields a decomposition formula

$$f(x) = \sum_{j,k} \langle f, \psi_{jk} \rangle \widetilde{\psi_{jk}}(x) ,$$

where the functions $\widetilde{\psi_{jk}}$ of the dual frame are defined by

$$\widetilde{\psi_{jk}}(x) = (F^* F)^{-1} \psi_{jk}(x) ,$$

and where the convergence is as before in the L^2 sense. Notice, however, that contrary to the Gabor case, the dual frame is not a frame of wavelets any more, i.e., it is not obtained by scaling and shifting a single function. Again, when $B = A$ or when b_0 and a_0 are close enough to 0 and 1, respectively, so that $B \approx A$, a simple inversion formula may be used:

$$f(x) = \frac{1}{A}\sum_{j,k} \langle f, \psi_{jk}\rangle \psi_{jk}(x) \ .$$

In the general case, the same iterative algorithm as before may be used to compute the dual frame functions:

$$\widetilde{\psi_{jk}}(x) = \mathcal{R}^{-1}\psi_{jk}(x) = \frac{2}{A+B}\sum_{n=0}^{\infty} T^n \psi_{jk}(x) \ ,$$

where T has the same form as before, given in equation (5.8).

5.2 Intermediate Discretization: The Dyadic Wavelet Transform

In some situations, it is desirable to retain some of the invariance properties of the continuous wavelet transform which could be destroyed by brute force discretization procedures. This is especially true for the translation invariance. Indeed, if the purpose is signal analysis, it is natural to represent the signal in such a way that the transform of a shifted copy of a reference signal coincides with the shifted (by the same amount) copy of the transform of the reference signal. The price to pay is an extra level of redundancy. The so-called *dyadic wavelet transform* is a type of wavelet transform based on such requirements. With the dyadic wavelet transform, the scale variable is sampled on a dyadic grid while the time parameter is left untouched. This transform can also be thought of as a discrete version of Littlewood-Paley decompositions.

Let $\psi \in L^2(\mathbb{R})$ be a wavelet such that

$$0 < C \leq \sum_{j=-\infty}^{\infty} |\hat{\psi}(2^j\xi)|^2 \leq C' < \infty \quad a.e. \tag{5.24}$$

for some finite constants C and C', and let us set for all $j \in \mathbb{Z}$ and $b \in \mathbb{R}$

$$\psi_b^j(x) = 2^j \psi\left(2^j(x-b)\right) \ , \quad x \in \mathbb{R} \ . \tag{5.25}$$

Notice that since we are not aiming at constructing orthonormal bases and merely discretizing the continuous transform, we are going back to the L^1 normalization. Then for any $f \in L^2(\mathbb{R})$, we may compute the coefficients

$$T_f^j(b) = \langle f, \psi_b^j \rangle \ . \tag{5.26}$$

Definition 5.3 *Given a wavelet ψ such that (5.24) holds, the* dyadic wavelet transform *of any signal with finite energy $f \in L^2(\mathbb{R})$ is defined as the set $\{T_f^j(\cdot);\ j \in \mathbb{Z}\}$ of functions T_f^j defined in (5.26).*

The dyadic wavelet transform may be inverted as follows. Let $\tilde{\psi}$ (the *dual wavelet*) be defined in the Fourier domain by

$$\hat{\tilde{\psi}}(\xi) = \frac{\hat{\psi}(\xi)}{\sum_j |\hat{\psi}(2^j\xi)|^2} \ . \tag{5.27}$$

If we use the notation ψ^j for the scaled wavelet,

$$\psi^j(x) = 2^j \psi(2^j x),$$

then $\psi_b^j(x) = \psi^j(x-b)$ is merely the translated form of ψ^j, and also $\widehat{\psi^j}(\xi) = \hat{\psi}(2^{-j}\xi)$. For any $f \in L^2(\mathbb{R})$, elementary algebra (based on the above remarks and the definition of the dual wavelet) gives, at least formally,

$$f(x) = \sum_j \int T_f^j(b)\, \tilde{\psi}_b^j(x) db \ . \tag{5.28}$$

The only delicate step in the proof is the justification of the interchange of several integration signs. Such a Fubini-type result can be justified in the sense of integrals converging in the Hilbert space $L^2(\mathbb{R})$. In any case, this reconstruction formula is the discrete counterpart of the classical inversion formula. Note that in general, $\tilde{\psi} \neq \psi$.

Remark 5.2 Notice that, despite the fact that the dyadic wavelet transform is a sampling of the CWT, it is still redundant. In fact, the dyadic wavelet transform shares with the CWT a number of properties. In particular, there exists a corresponding reproducing kernel operator $\mathcal{K}_\psi$. Its action can be read off from the inversion formula (5.28): If $F \in L^2(\mathbb{Z} \times \mathbb{R})$, the action of $\mathcal{K}_\psi$ on $F(j,b)$ reads

$$[\mathcal{K}_\psi \cdot F](j,b) = \sum_{j'} \int \langle \tilde{\psi}_{b'}^{j'}, \psi_b^j \rangle F(j',b') db' \ . \tag{5.29}$$

Again, $\mathcal{K}_\psi$ may be thought of as an inverse dyadic wavelet transform, followed by a dyadic wavelet transform. Such reproducing kernels were extensively used by Mallat and Zhong [214] for projecting on the range of the dyadic wavelet transform.

Remark 5.3 There is a very simple way of generating a dyadic wavelet transform from a CWT. Let $g \in L^1(\mathbb{R})$ be a wavelet such that

$$c_g = \int_0^\infty |\hat{g}(\xi)|^2 \frac{d\xi}{\xi} = \int_0^\infty |\hat{g}(-\xi)|^2 \frac{d\xi}{\xi} = 1 ,$$

and let ψ be such that

$$\left|\hat{\psi}(\xi)\right|^2 = \int_1^2 |\hat{g}(a\xi)|^2 \frac{da}{a} .$$

Then it is clear that ψ satisfies (5.24) with $C = C' = 1$.

Let us consider the simple example of Gaussian wavelets. Let θ be the derivative of Gaussian window. In the Fourier domain it reads

$$\hat{g}(\xi) = \sqrt{2}\xi e^{-\xi^2/2} .$$

In this case, the wavelet ψ is given by

$$\left|\hat{\psi}(\xi)\right|^2 = e^{-\xi^2} - e^{-4\xi^2} .$$

5.2.1 Taking Large Scales into Account

For practical purposes, it does not make sense to consider the whole set of wavelet coefficients of a function $f \in L^2(\mathbb{R})$. Indeed, there are natural cutoffs for the values of the scale variable, namely the one provided by the sampling frequency of the signal, and the other one given by the overall length of the signal. It is necessary to take these facts into account. The first step is the introduction of a new function that takes care of the large-scale behavior of the analyzed signal.

Let $\psi \in L^2(\mathbb{R})$ be an analyzing wavelet such that equation (5.24) holds. Let χ be such that its Fourier transform satisfies

$$|\hat{\chi}(\xi)|^2 = \sum_{j=1}^{\infty} \overline{\hat{\psi}(2^{-j}\xi)}\hat{\tilde{\psi}}(2^{-j}\xi) . \qquad (5.30)$$

Then it is an immediate consequence of the inversion formula (5.28) that if we set

$$S_f^j(b) = \langle f, \chi_b^j \rangle , \qquad (5.31)$$

then for each integer $j_0 \in \mathbb{Z}$ we have

$$f(x) = \int \langle f, \chi_b^{j_0} \rangle \chi_b^{j_0}(x) db + \sum_{j=j_0}^{\infty} \int \langle f, \psi_b^j \rangle \tilde{\psi}_b^j(x) db, \tag{5.32}$$

where, as usual, the convergence of the integrals and the summations have to be understood in the L^2 sense. The first term in the sum represents an approximation of f at scale j_0, and the second term accounts for details at coarser scales.

Remark 5.4 Note that equation (5.30) does not define the function χ uniquely, and there is still a large freedom in the choice of phases, for example.

Remark 5.5 In the particular case of the wavelet defined in Remark 5.3, possible choices for χ are given by functions satisfying

$$|\hat{\chi}(\xi)|^2 = \int_0^1 |\hat{g}(a\xi)|^2 \frac{da}{a} .$$

If we come back to the example of Gaussian wavelets discussed at the end of the previous subsection, we see that the Gaussian function itself may be used as function χ. More generally, we have

$$|\hat{\chi}(\xi)|^2 = e^{-\xi^2} .$$

5.2.2 The Discrete Dyadic Wavelet Transform

Obviously, only discrete versions of the dyadic wavelet transform are available in practice, so the next step concerns discretization of time variable. To stay closer to practical applications, we have to focus on the case of *discrete signals* and make the connection with the preceding discussion. In this context, the function χ introduced earlier plays a prominent role, as we shall see. The discrete wavelet transform (DWT) relies on the existence of an underlying *scaling function.* Let us first specify more precisely what we mean by scaling function.

Definition 5.4 *A function $\chi \in L^2(\mathbb{R})$ is called a scaling function if the collection of its integer translates $\chi_k(x) = \chi(x-k)$, $k \in \mathbb{Z}$, is a Riesz basis of the closed linear subspace they span in $L^2(\mathbb{R})$.*

Notice that this closed linear span, say V_0, is by definition given by

$$V_0 = \overline{\text{span}\left\{ \sum_k \alpha_k \chi(x-k);\ \{\alpha_k\}\ \text{finite sequence} \right\}} \tag{5.33}$$

See Definition 5.2 for the definition of a Riesz basis. With this in mind it should be clear that, when the integer translates $\chi(x-k)$ form a Riesz basis of V_0, then the latter can in fact be written as

$$V_0 = \overline{\left\{ \sum_k \alpha_k \chi(x-k), \ \{\alpha_k\} \in \ell^2(\mathbb{Z}) \right\}}. \tag{5.34}$$

It is readily verified that for families of integer translates $\{\chi_k(x) = \chi(x-k),\ k \in \mathbb{Z}\}$ of a single function $\chi \in L^2(\mathbb{R})$, condition (5.2) is equivalent to the existence of two finite positive constants $A \leq B$ such that

$$A \leq \sum_k |\hat{\chi}(\xi + 2\pi k)|^2 \leq B \ . \tag{5.35}$$

Let us now consider $f \in V_0$. Then we may write

$$f(x) = \sum_k \alpha_k \chi(x-k) \tag{5.36}$$

for some square-integrable sequence $\{\alpha_k,\ k \in \mathbb{Z}\}$, given by $\alpha_k = \langle f, \tilde{\chi}_k \rangle$ with $\tilde{\chi}_k(x) = \tilde{\chi}(x-k)$ and

$$\hat{\tilde{\chi}}(\xi) = \frac{\hat{\chi}(\xi)}{\sum_k |\hat{\chi}(\xi + 2\pi k)|^2} \ . \tag{5.37}$$

Similarly, we may write $f \in V_0$ as

$$f(x) = \sum_k \langle f, \chi_k \rangle \tilde{\chi}_k(x) \ , \tag{5.38}$$

which shows how to recover $f(x)$ from the discrete set of coefficients $\langle f, \chi_k \rangle$. When such a decomposition holds we say that $\{(\chi_k, \tilde{\chi}_k);\ k \in \mathbb{Z}\}$ is a *biorthogonal basis* of V_0.

It is a well known fact that to each scaling function χ, one can associate another scaling function whose integer translates form an orthonormal basis of V_0, such a standard procedure is known as Gram's orthonormalization procedure. Applying Gram's procedure to the basis $\{\chi(x-k);\ k \in \mathbb{Z}\}$ yields a new scaling function ϕ whose Fourier transform satisfies

$$\hat{\phi}(\xi) = \frac{\hat{\chi}(\xi)}{\sqrt{\sum_k |\hat{\chi}(\xi + 2\pi k)|^2}} \ . \tag{5.39}$$

It is easy to check that the collection of integer translates of such a ϕ form an orthonormal basis of V_0 (see [Meyer89a] for example).

Let us now turn to the problem of evaluating the dyadic wavelet transform of sequences, i.e., discrete signals. The starting point is the following:

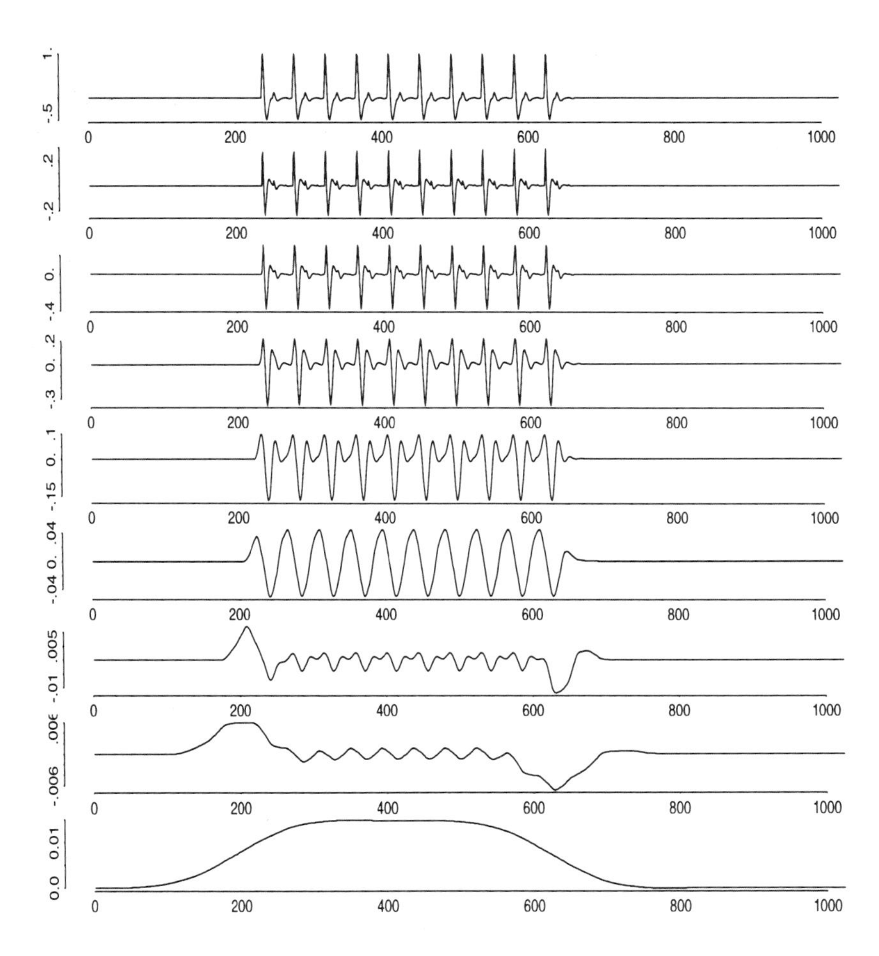

Figure 5.5. Discrete dyadic wavelet transform of the transient signal studied in Example 5.3.

Lemma 5.1 *For each finite energy sequence $s = \{s_k;\ k \in \mathbb{Z}\}$ there exists $f \in V_0$ such that $s_k = \langle f, \phi_k \rangle$ for all $k \in \mathbb{Z}$.*

The proof is immediate. Such a function f is simply given by $f(x) = \sum s_k \phi_k(x)$. It is now possible to introduce the following:

Definition 5.5 *The discrete dyadic wavelet transform of the finite energy sequence $s = \{s_k;\ k \in \mathbb{Z}\} \in \ell^2(\mathbb{Z})$ is by definition the family of sequences $\{T_f^j(k);\ k \in \mathbb{Z}\}$, where f is defined from s as before.*

Recall that the functions T_f^j were defined in equation (5.26). We shall see in the subsequent chapters that the discrete dyadic wavelet transform may be particularly efficient, especially from an algorithmic point of view.

Remark 5.6 Again, a lot of variations around such a definition are possible. Suppose for instance that the scaling function ϕ is such that the function

$$\gamma(\xi) = \sum_k |\hat{\phi}(\xi + 2\pi k)|^2$$

is real, bounded and bounded away from 0. Then for any sequence $\{s_k;\ k \in \mathbb{Z}\} \in \ell^2(\mathbb{Z})$, there exists a function $f \in V_0$ such that $f(k) = s_k$. More precisely, writing $f(x) = \sum_k \alpha_k \phi(x - k)$, we have

$$\alpha_k = \sum_\ell s_\ell \varrho_{k-\ell}\ ,$$

where the coefficients ϱ_k are the Fourier coefficients of $1/\gamma(\xi)$. Then an alternative to Definition 5.5 consists in defining the discrete dyadic wavelet transform of a sequence $\{s_k;\ k \in \mathbb{Z}\} \in \ell^2(\mathbb{Z})$ as the family of sequences $\{T_f^j(k);\ k \in \mathbb{Z}\}$, where T_f^j is defined in equation (5.26) with $f(x) = \sum_k \alpha_k \phi(x - k)$.

5.2.3 Local Extrema and Zero Crossings Representations

The dyadic wavelet transform is still redundant. To reduce the redundancy without going to sparser regular discretization, Mallat and Zhong [214] proposed an alternative representation, namely the local extrema representation. The main idea behind local extrema representations is borrowed from image analysis. Image processors claim that most of the information which is contained in an image (or at least the information which is analyzed by the human eye) lies in the edges, i.e., the singularities of the image. The first attempts to characterize signals by their singularities led to the celebrated Canny's edge detector (see also [175, 301]), which goes as follows.

We describe it in the 1D situation, for the sake of simplicity. The first step is a smoothing of the signal,

$$S(x, a) = f * \theta_a(x) ,$$

where θ is a low-pass filter, a is a reference scale, and $\theta_a(x) = \theta(x/a)$. Then the second step consists in differentiating $S(x, a)$ with respect to x,

$$T(x, a) = \partial_x S(x, a) .$$

In the presence of an edge at point $x = x_0$, i.e. a "steplike singularity," $S(x, a)$ has an inflection point at $x = x_0$, so that $T(x, a)$ has a local extremum at this point. This shows the correspondence between singularities and local extrema of $T(x, a)$. Actually, it was shown later (see [218, 214]) that in order to characterize the sharp variations in a signal, local extrema information had to be taken into account for several values of the scale variable a. We are then led to consider $T(x, a)$ as a function of x and a simultaneously, and to identify it with a wavelet transform (with a time-reversed copy of θ' as wavelet, i.e., a wavelet which is the derivative of a smoothing function and hence with only one vanishing moment).

Mallat and Zhong have shown (at least numerically) in [214] that the locations of local extrema of a dyadic wavelet transform, together with their values, provide sufficient information to characterize signals (within a certain accuracy). This problem has been studied independently by other groups (see, e.g., [260, 42]), who have proposed different methods to reconstruct signals from DWT local extrema. Meyer gave several counterexamples to the completeness of DWT local extrema representations, i.e., families of functions whose DWT share the same local extrema [Meyer93b]. However, such counterexamples seem to be too close to be distinguished numerically, and their value is more theoretical than practical. We shall describe in Chapter 8 an algorithm for signal reconstruction from DWT local extrema data, as well as some examples.

Remark 5.7 It is interesting to notice the similarity between the DWT local extrema data, and the information which is carried by the phase of complex continuous wavelet transform. Indeed, if one tries to track the local extrema across scales (which was done by Mallat and Hwang [212] for singularity analysis), this is equivalent to follow lines of constant phase for the complexified wavelet transform. This remark is due to A. Grossmann (see [150] and references therein).

5.3 Matching Pursuit

The purpose of this section is to present an interesting variant of the redundant representation of a signal as an additive superposition of elementary

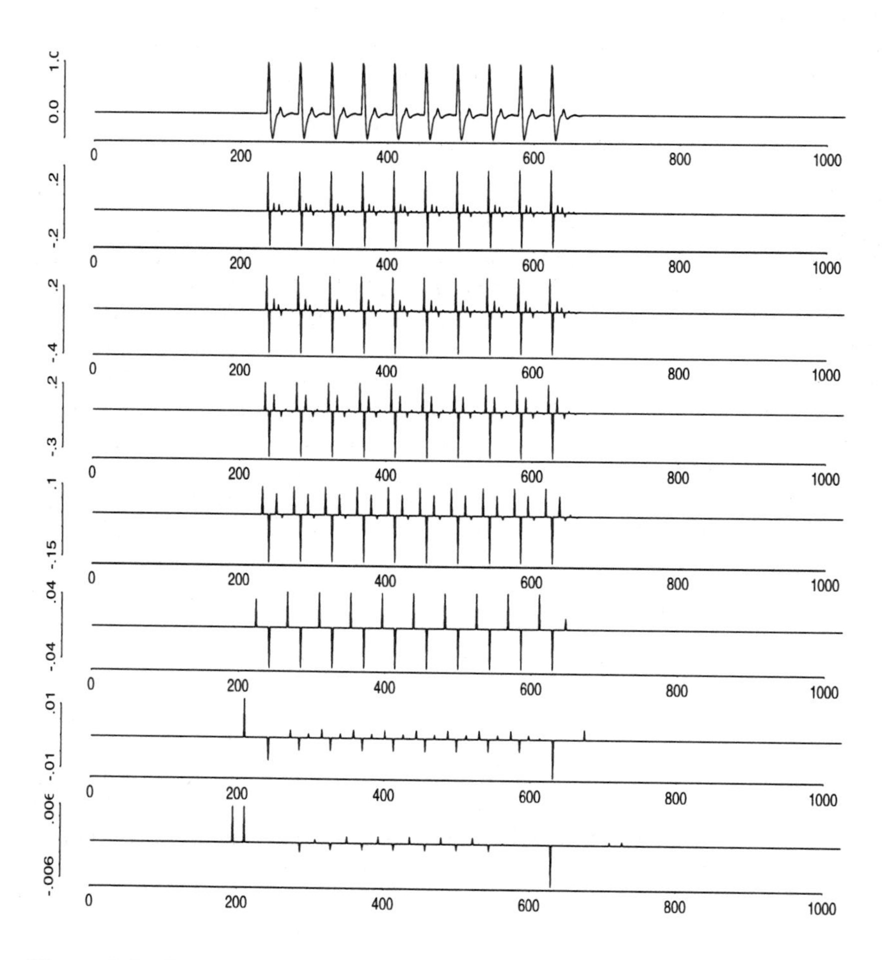

Figure 5.6. Local extrema of the wavelet transform of the transient signal of Figure 5.5 as studied in Example 5.3.

functions with a sharp localization in the time-frequency plane. The matching pursuit algorithm presented here is due to Mallat and Zhang (see, for example, [215]). The idea of the pursuit to represent a function has been known to statisticians for a long time. It can be found, for example, in a 1981 publication [138] by Friedman and Stuetzle, where the authors consider the problem of nonparametric regression, but it also appeared in the context of density estimation and discrimination analysis.

5.3.1 The Regression Pursuit Method

Let us start with some generalities on the matching pursuit method. As usual we consider the (possibly infinite-dimensional) Hilbert space $\mathcal{H} = L^2(\mathbb{R}, dx)$ of (possibly complex) signals with finite energy. If $\{g_\gamma; \gamma \in \Gamma\}$ is an unconditional basis of $\mathcal{H}$, then any signal with finite energy can be decomposed in a unique fashion as a linear combination of basis elements and finite linear combinations of elements of this basis can be used to provide approximations of the signal with arbitrary accuracy. The fact that we are using a basis forces on us the form of the function expansion.

On the other hand using a larger collection may provide more freedom in the representations of the signals, as in the case of the continuous decompositions described in Chapters 3 and 4 or the frames we described at the beginning of the present chapter. Unfortunately, the decomposition is in general not unique, the coefficients in the decompositions represent redundant information, and one has to compare several decompositions of the same signal and look for optimal representations.

Let us now assume that the system $\{g_\gamma; \gamma \in \Gamma\}$ is *total* in $\mathcal{H}$, in the sense that finite linear combinations of the g_γ's are dense in $\mathcal{H}$. Such a system will be called a *dictionary* and we shall denote it by the letter $\mathcal{D}$. We shall elaborate on the possible dictionaries in the next section.

Since for each signal f with finite energy there are many (in fact, infinitely many) decompositions on this dictionary, it is natural to look for an algorithm to decompose any finite energy signal. The matching pursuit is a typical greedy algorithm of this type. We call it greedy because it uses a gradient method to minimize the energy norm (i.e., L^2-norm for us) of the difference between the original signal and the approximation. Successive approximations are constructed by projections onto elements of the dictionary $\mathcal{D}$. In order to present the algorithm we choose a number $\alpha \in (0, 1]$. We shall later give a reason for needing such an extra parameter. Let us set $R^0 f = f$ for the residual of order 0. Let us now assume that the nth-order residual $R^n f$ has been constructed. We choose an element $g_{\gamma_n} \in \mathcal{D}$ which closely matches the residual $R^n f$ in the sense that:

$$|\langle R^n f, g_{\gamma_n} \rangle| \geq \alpha \sup_{\gamma \in \Gamma} |\langle R^n f, g_\gamma \rangle|. \tag{5.40}$$

Choosing $\alpha = 1$ would be best for the purpose of the matching of the residual, but such a choice for α may make it impossible (this is the case in infinite dimensions, for example) to find a solution to the inequality (5.40). The role of the parameter α is to guarantee the existence of an algorithm to pick a function $g_{\gamma_n} \in \mathcal{D}$ solving (5.40). Next we decompose the nth-order residual $R^n f$ into

$$R^n f = \langle R^n f, g_{\gamma_n} \rangle g_{\gamma_n} + R^{n+1} f,$$

defining this way the residual of order $n+1$. The residual $R^{n+1} f$ is orthogonal to g_{γ_n}, and one has

$$||R^n f||^2 = |\langle R^n f, g_{\gamma_n} \rangle|^2 + ||R^{n+1} f||^2.$$

In this way, the original signal f is decomposed into an orthogonal sum,

$$f = \sum_{n=0}^{m-1} \langle R^n f, g_{\gamma_n} \rangle g_{\gamma_n} + R^m f, \tag{5.41}$$

of dictionary elements which best match the residuals and a remainder term whose L^2-norm can be shown to converge to 0.

5.3.2 Time-Frequency Atoms

Many dictionary choices are possible. Most of them actually rely on the same idea, namely, generating a dictionary of functions with sharp localization in the time-frequency domain while keeping different localization properties. Let us illustrate these localization properties with simple examples.

First, consider the case of Gabor functions, whose time-frequency localization is displayed in Figures 5.2 and 5.1. Gabor functions are functions of constant size (which means that their Fourier transform is also of constant size). In other words, Gabor functions have a *constant bandwidth.*

On the contrary, wavelets, whose time-frequency localization is displayed in Figures 5.4 and 5.3, are functions of varying size; but their characteristic property is that they have a *constant relative bandwidth*, i.e., a bandwidth proportional to their frequency.

It has been known for a long time that Gabor analysis performs better in some cases and wavelets do better in other cases. It is then natural to look for "intermediate decompositions," involving simultaneously wavelets and Gabor functions. A possible approach amounts to generate such functions from a single one by simultaneously using translations, dilations, and modulations. In this case the analysis functions are labeled by $\gamma = (a, b, \omega) \in \Gamma = \mathbb{R}_+^* \times \mathbb{R}^2$, with a having the interpretation of a scale, b

of a time location, and ω of a frequency center. These atoms are defined as

$$g_\gamma(x) = \frac{1}{a} g\left(\frac{x-b}{a}\right) e^{i\omega(x-b)}, \quad x \in \mathbb{R}, \tag{5.42}$$

where the time window function g is a fixed unit vector of $\mathcal{H}$. We can suppose without loss of generality that g is real and centered at $x = 0$ (in most of the practical applications g is nonnegative and decaying fast at infinity, for example, a Gaussian), if not with compact support. In the Fourier domain, the definition (5.42) can be rewritten in the form

$$\hat{g}_\gamma(\xi) = \hat{g}(a(\xi - \omega))e^{-i\xi b}, \tag{5.43}$$

which shows that, since $\hat{g}_\gamma(\xi)$ is even, that $|\hat{g}_\gamma(\xi)|$ is concentrated in a neighborhood of $\xi = \omega$. In the same way (5.42) shows that $|g_\gamma(x)|$ is concentrated in a neighborhood of $x = b$.

Note that this (continuously labeled) time-frequency dictionary $\mathcal{D}$ is extremely redundant, and if one were to look for analysis-reconstruction formulas such as the wavelet and Gabor ones (see, for example, [277, 278]), one would have to introduce a dependency between the variables, for example, the scale and the modulation parameter. For instance, writing $a = \beta(\omega)$ yields functions which may be written in the Fourier domain as

$$\hat{g}_\gamma(\xi) = \hat{g}(\beta(\omega)(\xi\omega))e^{-i\xi b},$$

thus producing functions whose bandwidth varies as a function of the modulation parameter ω like $\beta(\omega)^{-1}$.

Even in the case of a discrete dictionary of functions g_γ of the form (5.42), the dictionary is still extremely redundant and is not a frame in general, since it contains both wavelet and Gabor frames [166].

The Swave package does not contain functions for the projection pursuit synthesis algorithm discussed in this section. The original proposal of Mallat and Zhang is implemented in the `matlab` package `wavelab`, but we do not know of a public domain package of S-functions with an implementation of the projection pursuit. Such an implementation exists in the S+wavelet commercial package of Splus for the wavelet packets and cosine packets dictionaries of time-frequency atoms (see later discussion), but not for the Gabor dictionary originally proposed by Mallat and Zhang.

5.4 Wavelet Orthonormal Bases

We have already discussed possible ways of reducing the redundancy of wavelet and Gabor transforms via discretizations. The limiting case is provided by the functions and discretization grids such that the corresponding functions form a Hilbert basis of $L^2(\mathbb{R})$. It is more convenient to avoid pathologies and to assume that the functions form a Riesz basis (see Definition 5.2 in Section 5.1).

As a consequence of the Balian-Low phenomenon, there does not exist a convenient (i.e., localized and regular) basis of Gabor functions. Alternatives are nevertheless provided by Wilson bases or local trigonometric bases; see [97, 216]. But these do not, strictly speaking, consist in families of Gabor functions. On the other hand, in the wavelet case, there now exists a large choice of wavelet bases of arbitrary regularity. In addition, they may all be constructed via a general algorithm, called multiresolution analysis (MRA), which also provides automatically fast decomposition and reconstruction algorithms. We describe here the main features of the constructions, and give a few examples. For more developments, we refer to textbooks such as [Meyer89a, Daubechies92a, Chui92a]. We also address the algorithmic aspects of MRAs.

5.4.1 Multiresolution Analysis and Orthonormal Bases

All of the known constructions of unconditional bases of wavelets rely on the concept of multiresolution analysis, which we describe next.

Definition 5.6 *A multiresolution analysis (MRA) of $L^2(\mathbb{R})$ is a collection of nested closed subspaces $V_j \subset L^2(\mathbb{R})$*

$$\cdots \subset V_{-2} \subset V_{-1} \subset V_0 \subset V_1 \subset V_2 \ldots \tag{5.44}$$

such that the following properties hold:

1. *$\overline{\cup V_j} = L^2(\mathbb{R})$ and $\cap V_j = \{0\}$.*
2. *If $f \in V_0$, then $f(\cdot - k) \in V_0$ for all $k \in \mathbb{Z}$; $f \in V_j$ if and only if $f(\cdot/2) \in V_{j-1}$.*
3. *There exists a function $\chi \in V_0$ such that the collection of the integer translates $\chi(\cdot - k)$ for $k \in \mathbb{Z}$ is a Riesz basis of V_0.*

χ is then a scaling function in the sense of Definition 5.2 and the Riesz basis may be orthonormalized as before by setting

$$\hat{\phi}(\xi) = \frac{\hat{\chi}(\xi)}{\sqrt{\sum_k |\hat{\chi}(\xi + 2\pi k)|^2}} \, . \tag{5.45}$$

Corollary 1 *With the same notation as before, the collection $\{\phi(\cdot - k);\ k \in \mathbb{Z}\}$ is an orthonormal basis of V_0.*

An immediate consequence of the definition of ϕ in (5.45) is that the Fourier transform satisfies

$$\sum_k |\hat{\phi}(\xi + 2\pi k)|^2 = 1 \; . \tag{5.46}$$

From now on, we shall essentially work with the scaling function ϕ associated with orthonormal basis. We shall come back to χ and other functions later in this chapter.

It follows directly from the inclusion of the V_j spaces that $\phi(x)$ (and also $\chi(x)$) may be expressed as a linear combination of the $\sqrt{2}\phi(2x - k)$ (which form an orthonormal basis of V_1). This yields the so-called *two-scale difference equation* (or *refinement equation*)

$$\phi(x) = \sqrt{2} \sum_k h_k \phi(2x + k) \; . \tag{5.47}$$

Let us now introduce the 2π-periodic function $m_0(\xi)$ whose Fourier coefficients are (up to a proportionality factor) the coefficients h_k. We set

$$m_0(\xi) = \frac{1}{\sqrt{2}} \sum_k h_k e^{ik\xi} \; , \tag{5.48}$$

and let us denote by W_j the orthogonal complement of V_j in V_{j+1}. Then one may prove that there exists a function $\psi \in W_0$ such that the collection $\{\psi(\cdot \, - k), k \in \mathbb{Z}\}$ is an orthonormal basis of W_0. More precisely, if we define the 2π-periodic function $m_1(\xi)$ by

$$m_1(\xi) = e^{i\xi} \overline{m_0(\xi + \pi)}, \tag{5.49}$$

and if we denote by $2^{-1/2} g_k$ its Fourier coefficients, i.e., if we set

$$m_1(\xi) = \frac{1}{\sqrt{2}} \sum_k g_k e^{ik\xi}, \tag{5.50}$$

then the coefficients g_k are related to the coefficients h_k by

$$g_k = -(-1)^k \overline{h}_{1-k} \; , \quad k \in \mathbb{Z} \; . \tag{5.51}$$

Then we have

Theorem 5.4 *If $\psi(x)$ is defined by*

$$\psi(x) = \sqrt{2} \sum_k g_k \phi(2x + k), \qquad x \in \mathbb{R}, \tag{5.52}$$

then the collection $\{\psi(\cdot \, - k); \; k \in \mathbb{Z}\}$ is an orthonormal basis of W_0.

See, e.g., [Meyer89a] for a proof. Notice that ψ constructed in this way is in fact an analyzing wavelet in the sense of the continuous wavelet transform introduced in Chapter 4. Formula (5.52) is a second example of refinement equation. Let us now set

$$\left\{ \begin{array}{lcl} \psi_{jk}(x) & = & 2^{j/2}\psi\left(2^j x - k\right) , \\ \phi_{jk}(x) & = & 2^{j/2}\phi\left(2^j x - k\right) . \end{array} \right. \tag{5.53}$$

Notice the normalization is similar to the one we used in the case of frames, but different from that of dyadic wavelet transform. By definition of the multiresolution analyses, it follows that for a given $j \in \mathbb{Z}$, the collection $\{\phi_{jk};\, k \in \mathbb{Z}\}$ (resp. $\{\psi_{jk};\, k \in \mathbb{Z}\}$) is an orthonormal basis of V_j (resp. W_j), and we have

Corollary 2 *With the same notation as before, the collection of wavelets $\{\psi_{jk};\, j, k \in \mathbb{Z}\}$ is an orthonormal basis of $L^2(\mathbb{R})$.*

The remarkable property of $m_0(\xi)$ and $m_1(\xi)$ is that they satisfy the so-called *Quadrature Mirror Filter* (QMF) condition, which is a direct consequence of equations (5.46) and (5.47):

Lemma 5.2 *The periodic functions m_0 and m_1 defined in* (5.48) *and* (5.49) *satisfy*

$$|m_0(\xi)|^2 + |m_1(\xi)|^2 = 1, \qquad \xi \in \mathbb{R}. \tag{5.54}$$

5.4.2 Simple Examples

Shannon Wavelet

One of the simplest multiresolution analyses is provided by band-limited functions. Let V_0 be the space of square-integrable functions whose Fourier transforms vanish outside the interval $[-\pi, \pi]$, and let $\hat{\phi}(\xi) = \mathbf{1}_{[-\pi,\pi]}(\xi)$ be the characteristic function (also called indicator function) of the interval $[-\pi, \pi]$. Then ϕ is the cardinal sine function:

$$\phi(x) = \frac{\sin(\pi x)}{\pi x} , \tag{5.55}$$

and the collection of its integer translates is an orthonormal basis of V_0 (this is the keystone of the classical sampling theory). The corresponding QMFs and wavelet are easily obtained. Clearly,

$$m_0(\xi) = \sum_{\ell} \mathbf{1}_{[-\pi/2,\pi/2]}(\xi + 2\pi\ell)$$

and

$$\hat{\psi}(\xi) = e^{-i\xi}\left(\mathbf{1}_{[-\pi,-\pi/2]}(\xi) + \mathbf{1}_{[\pi/2,\pi]}(\xi)\right) .$$

In the time domain the wavelet reads

$$\psi(x) = \frac{1}{\pi(x-1)}\left[\sin(\pi(x-1)) - \sin\left(\frac{\pi}{2}(x-1)\right)\right] .$$

We have sharp localization (i.e., compact support) in the Fourier domain, but poor localization in the time domain, since both $\phi(x)$ and $\psi(x)$ decay as $1/|x|$ at infinity. To overcome this shortcoming, Meyer proposed a family of wavelets (associated with the so-called *Littlewood-Paley MRAs*) with compact support in the Fourier domain and any prescribed algebraic decay in the x domain (see [Meyer89a] for a review of the properties of this class of wavelets). Such wavelets are generally regarded as being of poor interest for applications, since the corresponding quadrature mirror filters have a large number of nonnegligible coefficients. However, they become particularly efficient when frequency localization becomes an important issue, or as soon as the implementation is done in the Fourier domain.

Spline Wavelets

There exist a number of very simple and well-known wavelet bases. Among them, the simplest is probably the Haar basis constructed from dilates and translates of the Haar wavelet,

$$\psi(x) = \mathbf{1}_{[0,\frac{1}{2}]}(x) - \mathbf{1}_{[\frac{1}{2},1]}(x) ,$$

and associated with the scaling function

$$\phi(x) = \mathbf{1}_{[0,1]}(x) .$$

The computation of the filter coefficients yields $h_0 = h_{-1} = 1/\sqrt{2}$, $g_0 = -g_{-1} = -1/\sqrt{2}$, and $h_k = g_k = 0$ for $k \neq 0, -1$. The corresponding space V_0 is the space of square integrable functions which are constant over integer intervals of the form $[k, k+1)$ for $k \in \mathbb{Z}$.

A direct generalization of the MRA associated with the Haar basis is provided by the spline MRAs (see, e.g., [Chui92a] for an extensive treatment of splines and their connections to wavelets). Let $r = 1, 2, \ldots,$ be an integer, and let us set

$$V_0^r = \{f \in C^{r-1}(\mathbb{R});\ f \text{ polynomial of degree } r \text{ on } (k, k+1),\ k \in \mathbb{Z}\}. \tag{5.56}$$

Let χ_r be the B-spline of degree r with knots on the integer lattice $\mathbb{Z}$. In the Fourier domain, we have

$$\hat{\chi}_r(\xi) = \left(\frac{1 - e^{-i\xi}}{i\xi}\right)^{r+1} , \tag{5.57}$$

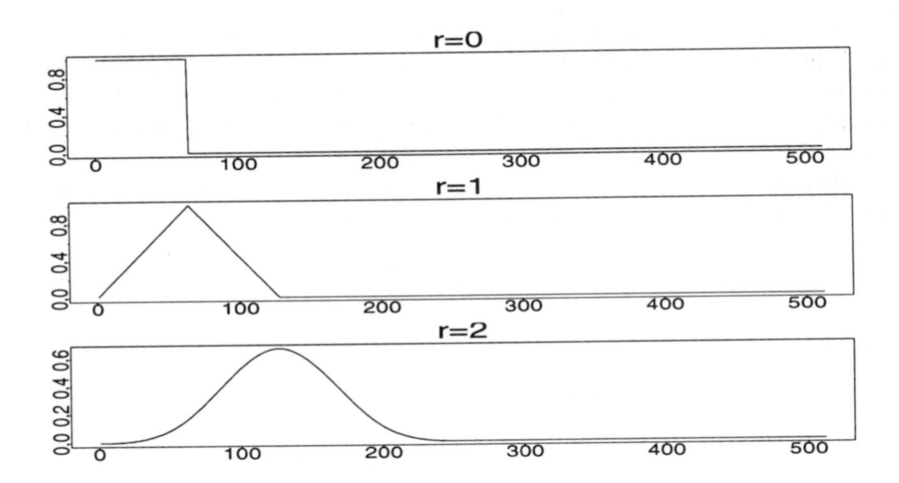

Figure 5.7. Basic spline functions of degrees 1 to 3.

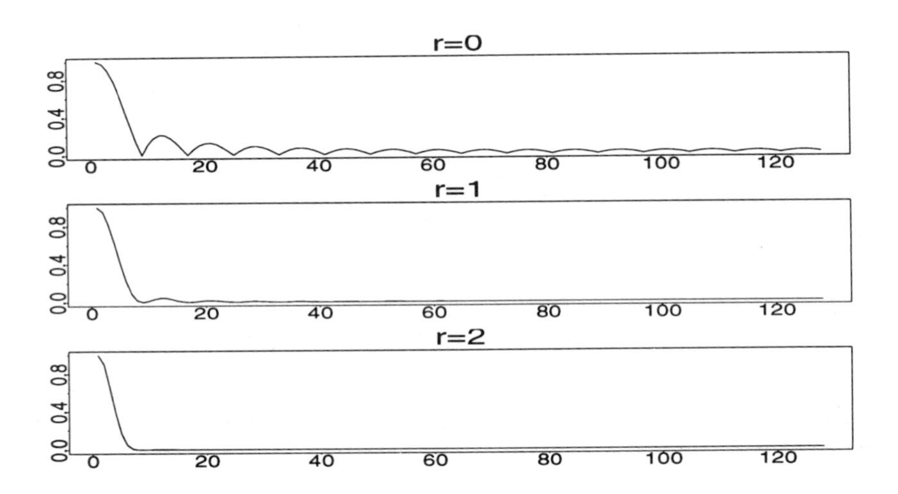

Figure 5.8. Fourier transforms of basic spline functions of degrees 1 to 3.

and the integer translates of χ_r form a basis of V_0^r. The basic splines of degrees 1 to 3 are displayed in Figure 5.7, and their Fourier transforms are in Figure 5.8. Denoting by V_j^r the corresponding scale subspaces, condition 2 of Definition 5.6 implies that all the other axioms are fulfilled, so that we have an MRA. If one associates with χ_r the function ϕ_r defined by equation (5.45), the collection of integer translates of ϕ_r is an orthonormal basis of V_0.

The denominator in Equation (5.45) may be calculated explicitly (see, e.g., [Meyer89a, Chui92a] for a derivation) by using the equality

$$\frac{1}{\tan z} = \sum_k \frac{1}{z + \pi k} ,$$

and we obtain

$$\begin{aligned} \Omega_r(\xi) &= \sum_k |\hat{\chi}_r(\xi + 2\pi k)|^2 \\ &= -\left(\sin\frac{\xi}{2}\right)^{2r+2} \frac{1}{(2r+1)!} \left(\frac{d^{2r+1}}{dz^{2r+1}} \frac{1}{\tan z}\right)_{z=\xi/2} . \end{aligned} \tag{5.58}$$

The corresponding filters and wavelet are finally obtained using the classical formulas (5.49) and (5.52).

As an example, let us consider the cases $r = 1, 2, 3$. We obtain

$$\left\{ \begin{array}{lcl} \Omega_1(\xi) &=& \frac{2+\cos(\xi)}{3} , \\ \hat{\phi}_1(\xi) &=& \left(\frac{1-e^{-i\xi}}{i\xi}\right)^2 \sqrt{\frac{3}{2+2\cos(\xi)}} , \end{array} \right. \tag{5.59}$$

$$\left\{ \begin{array}{lcl} \Omega_2(\xi) &=& \frac{33+26\cos(\xi)+\cos(2\xi)}{60} , \\ \hat{\phi}_2(\xi) &=& \left(\frac{1-e^{-i\xi}}{i\xi}\right)^4 \sqrt{\frac{60}{33+26\cos(\xi)+\cos(2\xi)}} , \end{array} \right. \tag{5.60}$$

and

$$\left\{ \begin{array}{lcl} \Omega_3(\xi) &=& \frac{1208+1191\cos(\xi)+120\cos(2\xi)+\cos(3\xi)}{2520} , \\ \hat{\phi}_3(\xi) &=& \left(\frac{1-e^{-i\xi}}{i\xi}\right)^6 \sqrt{\frac{2520}{1208+1191\cos(\xi)+120\cos(2\xi)+\cos(3\xi)}} . \end{array} \right. \tag{5.61}$$

It is easily seen that starting from B-spline of degree r, one ends up with a wavelet with $r+1$ vanishing moments. Spline wavelets have been extensively studied in [Chui92a].

A remarkable property of spline multiresolution analyses and corresponding wavelets is their asymptotic behavior as the degree r goes to infinity

$$\lim_{r\to\infty} \phi_r(x) = \frac{\sin(\pi x)}{\pi x} .$$

In other words, the spline orthogonal wavelets approach Shannon's wavelet as the number of vanishing moments tends to infinity.

Compactly Supported Wavelets

The most popular family of wavelets is provided by the compactly supported wavelets which were introduced by Ingrid Daubechies. The construction goes essentially as follows. Let us start by writing

$$m_0(\xi) = \left(\frac{1+e^{-i\xi}}{2}\right)^N Q\left(e^{-i\xi}\right) , \tag{5.62}$$

where $Q(z)$ is a polynomial with real coefficients. Then the following result was proved in [94]:

Theorem 5.5 *Any trigonometric polynomial solution of the QMF relation is given by formula* (5.62) *where Q is a polynomial satisfying*

$$\left|Q\left(e^{-i\xi}\right)\right|^2 = \sum_{j=0}^{N-1} \binom{N-1+j}{j} \sin^{2j}\left(\frac{\xi}{2}\right) + \sin^{2N}\left(\frac{\xi}{2}\right) R\left(\frac{1}{2}\cos\xi\right) , \tag{5.63}$$

and R is an odd polynomial chosen in such a way that $|Q(z)|^2 \geq 0$.

The classical Daubechies wavelets correspond to the choice $R = 0$. The polynomial $Q(z)$ is obtained from $|Q(z)|^2$ in (5.63) using a factorization result due to Riesz. In the present case, it may be formulated as follows (we mainly follow the lines of [Daubechies92a]). Let us write

$$\left|Q\left(e^{-i\xi}\right)\right|^2 = q(\cos\xi)$$

and denote by c_k , $k = 1, \ldots, N$, the zeroes of $q(z)$. Because of the form of $q(z)$, there are $N-1$ of them: real singletons and complex pairs $(c_k, \overline{c_k})$. We may then write

$$\begin{aligned} |Q(z)|^2 &= q\left(\frac{1}{2}(z+z^{-1})\right) \\ &= Kz^{1-N} \prod_{k=1}^{N-1} \left(\frac{1}{2} - c_k z + \frac{1}{2}z^2\right) \end{aligned}$$

for some constant K. If c_k is real, the zeroes of $\frac{1}{2} - c_k z + \frac{1}{2}z^2$ are r_k and r_k^{-1}, where

$$r_k = c_k + \sqrt{c_k^2 - 1} \, .$$

When $|c_k| \geq 1$, r_k is real. When $|c_k| < 1$, r_k is a complex number of modulus 1 and we write $r_k = e^{i\alpha_k}$. If c_k is complex, the zeroes of $(\frac{1}{2} - c_k z + \frac{1}{2}z^2)(\frac{1}{2} - \overline{c_k} z + \frac{1}{2}z^2)$ are $z_k, z_k^{-1}, \overline{z_k}$, and $\overline{z_k}^{-1}$, where

$$z_k = c_k + \sqrt{c_k^2 - 1} \ .$$

Now, observing that for $|z| = 1$,

$$\left|(z - z_0)(z - \overline{z_0}^{-1})\right| = |z_0|^{-1}|z - z_0|^2,$$

we conclude that

$$\begin{aligned} \left|Q\left(e^{-i\xi}\right)\right|^2 &= C' \prod_{j=1}^J |z_j|^{-2} \prod_{k=1}^K |r_k|^{-1} \left|\prod_{j=1}^J \left(e^{-i\xi} - z_j\right)\left(e^{-i\xi} - \overline{z_j}\right)\right|^2 \\ &\quad \times \left|\prod_{k=1}^K \left(e^{-i\xi} - e^{i\alpha_k}\right)\left(e^{-i\xi} - e^{-i\alpha_k}\right)\right|^2 \\ &\quad \times \left|\prod_{\ell=1}^L \left(e^{-i\xi} - r_\ell\right)\right|^2 . \end{aligned} \tag{5.64}$$

From this last expression we easily obtain a square root, and thus the function $m_0(\xi)$.

However, the square root is far from unique, for one has to choose the zeroes in (5.64) to make up $Q(z)$ from $|Q(z)|^2$. Several variations are possible and have been investigated [95, 79]. The simplest solution, proposed in [94], amounts to taking all the zeroes within the unit circle in the complex plane. The corresponding function $m_0(\xi)$ may be written as

$$m_0(\xi) = \frac{1}{\sqrt{2}} \sum_{k=0}^{2N-1} h_k e^{ik\xi} \ , \tag{5.65}$$

and the coefficients h_k, $k = 0, \ldots, 2N-1$, are given in Table 6.1 on page 195 of [Daubechies92a]. The corresponding wavelets are far from symmetric[2] and are called *minimal phase wavelets*. Other choices for the zeroes of $Q(z)$ have been examined (see, e.g., [95]).

There is an extra degree of freedom in (5.63), namely, the polynomial $R(z)$, which is set to zero in the classical Daubechies bases. This freedom may be used for another purpose. We shall come back to that later when we discuss Coiflets.

These compactly supported wavelet bases have been thoroughly discussed and used in the literature. We refer to [Daubechies92a] for a more complete discussion, together with a description of several variations on the construction which we outlined.

[2] In fact, it may be shown that except for the Haar case, real compactly supported wavelets and scaling functions obtained from a multiresolution analysis cannot have a symmetry or an antisymmetry axis.

Remark 5.8 The goal of compactly supported wavelets is to achieve "optimal" localization in the x domain. A consequence of the good x-localization is poor localization in the Fourier domain, i.e., poor regularity. This aspect was discussed by I. Daubechies and various authors (see, for example, [79]). Again, we refer to [Daubechies92a] for a summary of these results.

5.4.3 Computations of the Wavelet Coefficients

Let us now assume that we have a wavelet orthonormal basis of the form $\{\psi_{jk};\, j,k \in \mathbb{Z}\}$. Any function $f \in L^2(\mathbb{R})$ is completely characterized by the coefficients of its decomposition in the basis and may be decomposed as

$$f(x) = \sum_{k=-\infty}^{\infty} s_k^{j_0} \phi_{j_0 k}(x) + \sum_{j=j_0}^{\infty} \sum_{k=-\infty}^{\infty} t_k^j \psi_{jk}(x), \tag{5.66}$$

where we use the following notation[3]:

$$\begin{cases} t_k^j = \langle f, \psi_{jk} \rangle \,, \\ s_k^j = \langle f, \phi_{jk} \rangle \,. \end{cases} \tag{5.67}$$

The coefficients t_k^j and s_k^j carry information about the content of the signal at various scales. They are often referred to as the approximation (s) and detail (t) coefficients, or sums and differences (in reference to the particular case of the Haar wavelet). As a direct consequence of the refinement equations (5.47) and (5.52), we have the following relations between the approximation and detail coefficients across scales:

$$\begin{cases} s_k^{j-1} &= \sum_\ell \overline{h_\ell} s_{2k-\ell}^j \,, \\ t_k^{j-1} &= \sum_\ell \overline{g_\ell} s_{2k-\ell}^j \,. \end{cases}$$

We shall see in Section 5.6 that these equations yield fast algorithms (the so-called pyramidal algorithm) for computing wavelet coefficients, as soon as a sequence of approximation coefficients $\{s_k^{j_0}, k \in \mathbb{Z}\}$ is known, i.e., as soon as a projection of the function onto a space V_{j_0} is known.

Taking for granted the existence of such pyramidal algorithms, the remaining problem lies in the initial data, i.e., in the computation of coefficients $s_k^{j_0}$, j_0 being the finest scale taken into account. This problem was already partly discussed in the section devoted to the discrete dyadic wavelet transform. If an explicit expression for $f(x)$ is available, the coefficients $s_k^{j_0}$ may be evaluated directly, if necessary by means of an appropriate

[3] The details coefficients are usually denoted by d_k^j. We chose to use the notation t_k^j to be consistent with the notation T, which we use for the wavelet transform.

quadrature formula. But this is not the general situation since in general only discrete values are available. In such a case, the simplest solution (which is widely used in the electrical engineering community[4]) consists in directly using the samples $f(2^{-j_0}k)$ in place of the coefficients $s_k^{j_0}$, and start the pyramid algorithm. This solution is sometimes called the "second wavelet crime." In any case this is a safe operation, since the pyramid algorithm is yields perfect reconstruction. The crime may be justified by arguing that the coefficient $s_k^{j_0}$ essentially represents a local average of the signal, provided by the scaling function, and that "it should be close to the sample." More precisely, let us assume that $f \in V_j$. Then we have

$$2^{-j/2}f(\ell 2^{-j}) = \sum_k s_k^j \phi(\ell - k) , \tag{5.68}$$

and the problem amounts to recovering the coefficients s_k^j from the samples $f(\ell 2^{-j})$, i.e., a deconvolution problem. Of course, if the scaling function has the interpolation property, i.e., if $\phi(n) = \delta_{n,0}$, the identification of the coefficient s_k^j with a multiple of the sample $f(k2^{-j})$ is exact. But even in more general cases, the result of the approximation turn out to be of acceptable accuracy.

Coiflets

We now reconsider the property $s_k^j \approx f(\ell 2^{-j})$, and we examine such a claim in more detail with the goal of making it rigorous for some specific choices of scaling functions. Indeed, it is possible to construct multiresolution analyses whose scaling function has vanishing moments, i.e., such that

$$\int x^m \phi(x) dx = 0 , \quad m = 1, 2 \ldots M' - 1 \tag{5.69}$$

(of course, the first moment cannot vanish, and we still have $\int \phi(x)dx = 1$). Such multiresolution analyses were introduced in [95] (see also [47]), and the corresponding wavelets are named *Coiflets*. The corresponding filters are designed as follows. The polynomial $R(z)$ in (5.63) is chosen in such a way that condition (5.69) is satisfied. The price for this extra generalization is that the new filters are longer than Daubechies'.

The following result is an easy consequence of Taylor's formula.

Lemma 5.3 *Let ϕ be a scaling function such that equation* (5.69) *holds. Then for any $f \in C^n(\mathbb{R})$, $n > M' - 1$,*

$$s_k^j = 2^{-j/2}\left(f(k2^{-j}) + O(2^{-jM'})\right) . \tag{5.70}$$

[4]In fact, in most of the signal processing problems, signals are already discrete, and Lemma 5.1 is implicitly used.

This result simply means that for such scaling functions, as soon as the analyzed function is smooth enough and the samples are given at sufficiently fine scale, they are hardly distinguishable from the coefficients s_k^j (up to a multiplicative constant only depending on the scale). This provides a justification to the "crime."

Table 8.1 on page 261 of [Daubechies92a] gives the filter coefficients for Coiflets with maximal number of vanishing moments (for the scaling function).

Spline MRAs Revisited

We have seen that spline MRAs provide some of the simplest examples of MRAs and wavelet bases. The spline MRAs have another advantage. They are intrinsically associated with interpolation schemes; this fact makes the connection between the coefficients s_k^j and the samples $f(k2^{-j})$ explicit. This can be seen from the following lemma, whose proof is easily obtained by inspection.

Lemma 5.4 *Let ϕ_r be the scaling function associated with a spline MRA, and let us set*

$$\Omega'(\xi) = \sum \hat{\phi}_r(\xi + 2\pi k) \ . \tag{5.71}$$

Then there exist two positive and finite constants A' and B' such that

$$A' \leq |\Omega'(\xi)| \leq B' \ . \tag{5.72}$$

Now, observing that the 2π-periodic function $\Omega'(\xi)$ is nothing but the Fourier transform of the sequence $\{\phi_r(n), n \in \mathbb{Z}\}$, it is a direct consequence of the preceding lemma that the convolution in equation (5.68) may be inverted, and then the coefficients s_k^j obtained from the samples. More precisely, by taking the Fourier transform of equation (5.68), we obtain

$$\sum_k f(k2^{-j})e^{ik\xi} = \Omega_j'(\xi) \sum_k s_k^j e^{ik\xi} \ ,$$

where

$$\Omega_j'(\xi) = 2^{j/2} \sum_k \phi(k) e^{ik\xi} = 2^{j/2}\Omega'(-\xi)$$

because of Poisson's summation formula. Now, because of Lemma (5.72), we know that $1/\Omega'(\xi)$ is a bounded 2π-periodic function. If we set

$$\frac{1}{\Omega'(\xi)} = \sum_k \beta_k e^{ik\xi} \ , \tag{5.73}$$

we obtain the following relationship between samples and approximation coefficients.

Lemma 5.5 *With the same notation as above,*

$$s_\ell^j = \sum_k \beta_k 2^{-j/2} f((\ell + k)2^{-j}) \ . \tag{5.74}$$

Therefore, given a function f, and assuming that $f \in V_j$ for some integer j, equation (5.74) allows one to compute the coefficients s_k^j for $k \in \mathbb{Z}$ from the corresponding samples $f(k2^{-j})$ of the function.

The General Case

A thorough analysis of the interpolation aspect of the wavelet coefficients has been carried out in [103], leading to similar conclusions in the general case. We describe here the main results obtained in this direction (see also [170]). They can be expressed as follows. Suppose that we are given samples $f(k2^{-j_0})$ of a function f at a given scale j_0. Then the following algorithm can be used to initialize the pyramidal computation of the wavelet coefficients:

1. Find a sequence $\{\beta_k\}$ such that

$$\int x^\ell \phi(x) dx = \sum_k \beta_k k^\ell \,, \quad \ell = 0, \dots M, \tag{5.75}$$

   for some integer M.

2. Set

$$\widetilde{s_\ell^{j_0}} = \sum_k \beta_k 2^{-j_0/2} f((\ell + k)2^{-j_0}) \ . \tag{5.76}$$

3. Run the pyramidal algorithm with $s_\ell^{j_0} = \widehat{s_\ell^{j_0}}$ as initialization.

Setting $\widetilde{P_{j_0} f}(x) = \sum_k \widetilde{s_\ell^{j_0}} \phi_{j_0 k}(x)$, one can estimate the precision of the corresponding approximation (for example, in terms of some Besov norms of $\widetilde{P_{j_0} f}(x) - P_{j_0} f(x)$) and derive optimality results. We refer to [103] for a precise description as well as formulations in more general contexts.

Let us emphasize that this algorithmic procedure covers the two cases considered earlier: in the case of Coiflets, thanks to vanishing moments, we have $\beta_k = \delta_{0,k}$, and in the case of splines (as well as for many other cases in which the function $\Omega'(\xi)$ is bounded from above and from below), a possible solution amounts to taking for β_k the Fourier coefficients of $1/\Omega'(\xi)$.

5.5 Playing with Time-Frequency Localization: Wavelet Packets, Local Cosines, and the Best Basis Method

Since the beginning of this chapter, we have only considered bases made from shifting and dilating a single function. As stressed in Section 5.1, this has some consequences for the localization properties of the basis elements in the time-frequency domain. However (as we also saw when discussing the matching pursuit method), it may be desirable to modify the time-frequency localization of the functions in order to "match" the features of specific signals or problems. This is what the wavelet packet and local cosines bases are about. See [Wickerhauser94] for a detailed account of the theory.

5.5.1 Wavelet Packets

The main idea is the following. It is a well understood fact that the role of the filters $m_0(\xi)$ and $m_1(\xi)$ in equations (5.47) and (5.52) is to split the frequency band $[0, \pi]$ into low and high frequencies, respectively. This is how the wavelet and the scaling function are generated. The wavelet packets construction essentially amounts to keep splitting the frequency band, still keeping the same pair of filters. More precisely, the construction goes as follows.

Fixed-Scale Wavelet Packets

The main feature of the construction of wavelet bases is the existence of the two-scale relations. This point was generalized by Coifman, Meyer, and Wickerhauser as follows. Start with a fixed MRA with wavelet ψ and associated scaling function ϕ, which we rename w_0 for convenience. In the standard multiresolution paradigm, the first refinement equation yields different scaled copies of the scaling function, and the second refinement equation yields the wavelet ψ:

$$\begin{cases} \phi(x) &= \sqrt{2}\sum_k h_k \phi(2x+k) \ , \\ \psi(x) &= \sqrt{2}\sum_k g_k \phi(2x+k) \ . \end{cases}$$

The main idea of the wavelet packet construction is to make systematic use of iterated refinement equations. Given a function w_n, set

$$\begin{cases} w_{2n}(x) &= \sqrt{2}\sum_k h_k w_n(2x+k) \ , \\ w_{2n+1}(x) &= \sqrt{2}\sum_k g_k w_n(2x+k) \ . \end{cases} \tag{5.77}$$

In the Fourier domain, we have the expressions

$$\begin{cases} \widehat{w_{2n}}(\xi) &= m_0\left(\frac{\xi}{2}\right)\widehat{w_n}\left(\frac{\xi}{2}\right) , \\ \widehat{w_{2n+1}}(\xi) &= m_1\left(\frac{\xi}{2}\right)\widehat{w_n}\left(\frac{\xi}{2}\right) , \end{cases} \tag{5.78}$$

or, more generally,

$$\widehat{w_n}(\xi) = \prod_{j=1}^{\infty} m_{\epsilon_{j-1}}(2^{-j}\xi) , \tag{5.79}$$

where

$$n = \sum_j \epsilon_j 2^j \tag{5.80}$$

is the dyadic expansion of the integer n. It may be shown that the index n is actually a frequency index; more precisely, introduce the inverse Gray code $z : \mathbb{Z} \to \mathbb{Z}$, defined recursively as

$$z(2n) = \begin{cases} 2z(n) & \text{if } n \text{ is even,} \\ 2z(n)+1 & \text{if } n \text{ is odd,} \end{cases} \tag{5.81}$$

$$z(2n+1) = \begin{cases} 2z(n)+1 & \text{if } n \text{ is even,} \\ 2z(n) & \text{if } n \text{ is odd.} \end{cases} \tag{5.82}$$

Then the wavelet packet w_n is localized near the frequencies $\pm\pi z(n)$ in the frequency domain. Such a property may be checked directly in the cases of the Haar and Shannon MRAs. However, the frequency localization is generally regarded as the weak point of wavelet packet bases, in the sense that for large values of the frequency index n, the Fourier transforms of the wavelet packets do not have good decay properties.

Denote by Ω_n the closed linear span of the functions $w_n(\cdot - k)$ when $k \in \mathbb{Z}$. It follows from the properties of quadrature mirror filters that the family $\{w_n(\cdot - k);\, k \in \mathbb{Z}\}$ is an orthonormal basis of Ω_n. Putting the spaces Ω_n together, Coifman and Meyer proved the following:

Theorem 5.6 *The family $\{w_n(\cdot - k),\, n, k \in \mathbb{Z}\}$ is an orthonormal basis of $L^2(\mathbb{R})$.*

This basis is called the *fixed scale wavelet packet basis* of $L^2(\mathbb{R})$.

Multiscale Wavelet Packets

So far, we did not use the possibility of dilating the wavelet packets. Let us denote by D the operator of contraction by 2, defined by

$$Df(x) = \sqrt{2} f(2x)$$

(we chose an L^2-normalization for D to be a unitary operator in L^2), and let us introduce the new functions

$$w_{j,n,k}(x) = 2^{j/2} w_n \left(2^j x - k\right) . \tag{5.83}$$

The function $w_{j,n,k}$ is called the wavelet packet at position k, scale j, and frequency n. The first remark is that the family of wavelet packets $\{w_{j,n,k};\ k \in \mathbb{Z}\}$ is an orthonormal basis of the dilated copy $D^j \Omega_n$ of the space Ω_n. As a consequence of the perfect reconstruction condition (5.54), we have that

$$\Omega_0 = D\Omega_0 \oplus D\Omega_1$$

and, more generally,

$$\Omega_n = D\Omega_{2n} \oplus D\Omega_{2n+1} .$$

Iterating this equation, we obtain the following lemma.

Lemma 5.6 *For any positive integer j, we have*

$$D^{-j}\Omega_n = \Omega_{2^j n} \oplus \Omega_{2^j n+1} \oplus \Omega_{2^j n+2} \oplus \cdots \oplus \Omega_{2^j (n+1)-1} . \tag{5.84}$$

At this point, it is useful to recall the definition of a dyadic interval. The dyadic interval I_{jn} is the semi-open interval:

$$I_{jn} = [2^{-j}n, 2^{-j}(n+1)) .$$

There is a one-to-one correspondence between dyadic intervals and dilated copies $D^{-j}\Omega_n$ of the space Ω_n. In particular,

$$\langle w_{j,n,k}, w_{j',n',k'} \rangle = 0 \qquad \text{if } I_{jn} \cap I_{j'n'} = \emptyset,$$

and

$$\langle w_{j,n,k}, w_{j',n',k'} \rangle = \delta_{kk'} \qquad \text{if } I_{jn} = I_{j'n'} .$$

We proceed with an attempt to precise this correspondence between properties of the dyadic intervals and the orthogonality properties of the wavelet packets. We first need to introduce the notion of bounded dyadic cover.

Definition 5.7 *A dyadic cover of the half-line $\mathbb{R}^+$ is a collection $\mathcal{I}$ of disjoint dyadic intervals I_{jn} such that*

$$\bigcup I_{jn} = \mathbb{R}_+ .$$

The dyadic cover $\mathcal{I}$ is said to be bounded if there exists J such that $j \le J$ for all $I_{jn} \in \mathcal{I}$.

With this definition in mind we can state

Theorem 5.7 *Let $\mathcal{I}$ be a bounded dyadic cover of the half-line. Then the collection of wavelet packets $\{w_{j,n,k};\ I_{jn} \in \mathcal{I}, k \in \mathbb{Z}\}$ is an orthonormal basis of $L^2(\mathbb{R})$.*

A large number of wavelet packet decompositions may be obtained by considering all the possible bounded dyadic covers of the half-line.

Remark 5.9 The boundedness assumption on the dyadic cover may actually be relaxed, but this does not correspond to practical situations (in practice, there is always a limit on the scales to be considered). We shall not dwell on this point here.

5.5.2 Local Trigonometric Bases

We already discussed Gabor frames and the Balian-Low phenomenon. One of the main aspects of the theory of Gabor frames is the existence of an obstruction which prevents from obtaining orthonormal bases of smooth and localized Gabor functions. However, the obstruction can be circumvented by modifying a little bit the construction rule. Several approaches deserve to be mentioned in this respect (see, e.g., [86, 97, 216]), but we shall limit ourselves to the local cosine bases.

The starting point is a partition of the real line into a series of intervals (*segments*) of the form

$$I_k = [a_k, a_{k+1})\ ,$$

where $\{a_k;\ k \in \mathbb{Z}\}$ is a family of real numbers such that

$$\begin{cases} \cdots < a_{-1} < a_0 < a_1 < a_2 < \dots \\ \lim_{k\to\pm\infty} a_k = \pm\infty\ . \end{cases} \tag{5.85}$$

It is obvious that any $f \in L^2(\mathbb{R})$ may be expanded as $f = \sum_k f_k$ where $f_k(x) = f(x)\mathbf{1}_{I_k}(x)$ is the restriction of f to the interval I_k. One can then expand each f_k into a cosine (or sine) series,

$$f_k(x) = \alpha_{k0}\sqrt{\frac{1}{\ell_k}} + \sum_{\nu=1}^{\infty} \alpha_{k\nu}\sqrt{\frac{2}{\ell_k}}\cos\left(\frac{\pi\nu}{\ell_k}(x-a_k)\right),$$

where

$$\ell_k = a_{k+1} - a_k\ ,\quad k \in \mathbb{Z}\ , \tag{5.86}$$

to obtain a trigonometric expansion of the original function f. Several variations on this idea are described in Chapter 3 of [Wickerhauser94].

The purpose of local cosine bases is to replace the characteristic functions $\mathbf{1}_k$ with smoother windows v_k, so as to enforce the decay at infinity of the coefficients α_k when f is a smooth function. Coifman and Meyer proposed to introduce smooth windows in the following way. First, consider a sequence of numbers $\{\eta_k;\ k \in \mathbb{Z}\}$, such that

$$\eta_k + \eta_{k+1} < \ell_k\ , \quad k \in \mathbb{Z}. \tag{5.87}$$

Then, let $\{v_k;\ k \in \mathbb{Z}\}$ be a collection of real-valued functions such that

$$\begin{cases} v_k(x) = 0 & \text{if } x \notin [a_k - \eta_k, a_{k+1} + \eta_{k+1}]\ , \\ v_k(x) = 1 & \text{if } x \in [a_k + \eta_k, a_{k+1} - \eta_{k+1}]\ , \\ v_k(a_k + \tau)^2 + v_{k-1}(a_k + \tau)^2 = 1 & \forall \tau < \eta_k\ , \\ v_k(a_k + \tau) = v_{k-1}(a_k - \tau) & \forall \tau < \eta_k\ . \end{cases} \tag{5.88}$$

Then one proves the following [86] (see also [Wickerhauser94]):

Theorem 5.8 *Let $\{a_k;\ k \in \mathbb{Z}\}$ be a family of real numbers satisfying* (5.85)*, and let $\{v_k;\ k \in \mathbb{Z}\}$ be a collection of real-valued functions such that* (5.88) *holds. Then the collection of functions*

$$u_{k\nu}(x) = \sqrt{\frac{2}{\ell_k}}\, v_k(x) \cos\left(\frac{\pi(\nu + 1/2)}{\ell_k}(x - a_k)\right) \tag{5.89}$$

is an orthonormal basis of $L^2(\mathbb{R})$.

Definition 5.8 *The collection $\{u_{k\nu};\ k \in \mathbb{Z}, \nu = 1, 2, \dots\}$ is called the local cosine basis associated with the family of windows $\{v_k;\ k \in \mathbb{Z}\}$.*

Remark 5.10 The same result may be proved if the cosines in (5.89) are replaced with sines, or if one alternates sines and cosines and replaces $\nu + 1/2$ with ν in (5.89). These aspects have been carefully analyzed in [25, 26].

As a result, we obtain an infinite number of orthonormal bases of $L^2(\mathbb{R})$, associated with different "pavings" of the real line. Searching for an "optimal" basis (whatever that means) in such a family is practically impossible, and it is therefore necessary to reduce the search to subfamilies. The solutions developed in [Wickerhauser94] are particularly elegant. They amount to generating windows from each other using a "split and merge" strategy, which we review now.

For the sake of simplicity, suppose that we start with a family of windows v_k of the form

$$v_k(x) = v(x - a_k)\ , \quad k \in \mathbb{Z},$$

on a regular tiling of the real axis:

$$a_{k+1} - a_k = \ell \ ,$$

with a constant $\eta_k = \eta$ such that equations (5.88) are satisfied. Given such a window v_k, the corresponding interval I_k may be split into two subintervals of equal lengths $[a_k, a_{k'}]$ and $[a_{k'}, a_{k+1}]$, with $a_{k'} = a_k + \ell/2$. If the new intervals fulfill (5.87), i.e., if $\ell > 4\eta$, then the window v_k may be replaced with a pair of windows v_k^1 and v_k^2, defined by

$$v_k^1(x) = \begin{cases} v_k(x - a_k) & \text{if } x \in [a_k - \eta, a_k + \eta] \\ 1 & \text{if } x \in [a_k + \eta, a_{k'} - \eta] \\ v_k(x - a_{k'} + \ell) & \text{if } x \in [a_{k'} - \eta, a_{k'} + \eta] \ , \end{cases}$$

and

$$v_k^2(x) = \begin{cases} v_k(x - a_{k'}) & \text{if } x \in [a_{k'} - \eta, a_{k'} + \eta] \\ 1 & \text{if } x \in [a_{k'} + \eta, a_{k+1} - \eta] \\ v_k(x - a_k) & \text{if } x \in [a_{k+1} - \eta, a_{k+1} + \eta] \ . \end{cases}$$

This procedure may be iterated as long as condition (5.87) is fulfilled. In that way, one generates a collection of bases associated with a binary tree, as in the case of the wavelet packets construction. The next step is to search for an "optimal" basis within the binary tree.

5.5.3 The "Best Basis" Strategy

Suppose now that a family of bases associated with a binary tree has been chosen, and consider all the corresponding decompositions of a given function $f \in L^2(\mathbb{R})$. The best basis strategy amounts to searching among all such decompositions the one which represents f best, the latter being understood in the sense of some criterion fixed in advance.

To this end, one generally introduces an *information cost functional*, which one tries to minimize. Information-based distances and cost functionals have been used by statisticians for a long time. See, for example, [284] or [92]. But the present implementation of these ideas is specific to the need for comparing decompositions of a given function f in a Hilbert space $\mathcal{H}$ onto the various complete orthonormal systems B of the set $\mathcal{B}$ of all the orthonormal bases of $\mathcal{H}$.

Definition 5.9 *Let M be an additive functional on $\ell^2(\mathbb{Z})$, i.e., a mapping $M : \ell^2(\mathbb{Z}) \to \mathbb{C}$ such that*

$$M(s) = \sum_n \mu(|s_n|), \qquad s = (s_n)_n \in \ell^2(\mathbb{Z}),$$

for some real-valued nonnegative function μ defined on the real line and such that $\mu(0) = 0$. If $f \in L^2(\mathbb{R})$, the M-information cost of f in a given orthonormal basis $B = \{b_\ell;\ \ell \in \mathbb{Z}\} \in \mathcal{B}$ is the number

$$\mathcal{M}_f(B) = M\left(\{\langle f, b_\ell\rangle;\ \ell \in \mathbb{Z}\}\right) . \tag{5.90}$$

Information cost functionals are generally chosen to represent the amount of data needed to describe the information present in the signal. Classical choices include the following:

- *Number of coefficients after thresholding:* Let $\epsilon > 0$ be a fixed threshold, and define

$$\mu_\epsilon(x) = \begin{cases} 1 & \text{if } |x| \geq \epsilon , \\ 0 & \text{otherwise} . \end{cases} \tag{5.91}$$

- *Concentration in ℓ^p norm:* Let $0 < p < 2$, and set

$$\mu_p(x) = |x|^p . \tag{5.92}$$

- *Entropy:* The entropy functional possesses a number of interesting statistical properties. Given any sequence $\{s_n\} \in \ell^2(\mathbb{Z})$, consider $||s||^2 = \sum_\ell |s_\ell|^2$, and

$$p_\ell = \frac{|s_\ell|^2}{||s||^2} . \tag{5.93}$$

  The p_ℓ form a probability distribution on $\mathbb{Z}$ (or, more generally, the countable index set of the basis), and the corresponding entropy is defined by

$$S(s) = -\sum_\ell p_\ell \log p_\ell . \tag{5.94}$$

  Note that the entropy is not, strictly speaking, an information cost functional. However, it is easily checked that if one sets

$$\mu(x) = -|x|^2 \log |x|^2 , \tag{5.95}$$

  the corresponding information cost functional is related to the entropy by

$$M(s) = ||s||^2 \left(S(s) - \log ||s||^2\right) , \tag{5.96}$$

  so that minimizing $S(s)$ and minimizing $M(s)$ are two equivalent problems.

Once an M-information cost functional has been chosen, the best basis strategy may be summarized as follows:

1. Compute all the decompositions of the signal with respect to the bases associated with a given binary tree (i.e., a given pair of filters in the wavelet packet case, or a given initial window in the local cosine case).
2. Select the basis which minimizes the chosen information cost functional.

Obviously, such a strategy is only possible when the decompositions with respect to the different bases may be computed efficiently. We discuss the algorithmic aspects below.

5.6 Algorithms and Implementation

There are several different ways to evaluate Gabor and wavelet transforms numerically. These different approaches are adapted to different forms of the transform we have studied thus far (for example, redundant or non-redundant transform), but also to the accuracy requirements (some algorithms are exact, others are approximate). We describe some of them in this chapter.

5.6.1 Direct Quadratures

The simplest (but presumably not the best) way of computing numerically a wavelet transform (or a Gabor transform) consists in discretizing by brute force the integral defining the transform, using classical quadrature (such as a trapezoidal rule). Although it may look attractive because of its simplicity, this approach has several drawbacks. We merely stress the two main ones.

- First, such an approach is far from precise. Although numerical accuracy is often not a major concern in many signal analysis applications, it is well known that such an approach tends to be quite inaccurate in the case of oscillatory signals (such as audio signals) or signals containing singularities. Unfortunately, these are the two kinds of signals we are mostly interested in.
- Second, direct discretizations are also inappropriate from the numerical point of view because they are not efficient. Let us consider the following example. Let us assume that the signal f is sampled with sampling frequency ν_s, $f_n = f(n/\nu_s)$, $n = 1 \cdots N = 2^J$, and let us suppose that we want to compute a discrete wavelet transform, with

J different scales $a_j = 2^j$, and N points per scale. Suppose that we simply approximate the integral using the trapezoidal rule (the precision of such an approach has been discussed in [170]):

$$\frac{1}{a_j} \int f(x) \overline{\psi\left(\frac{x - n/\nu_s}{a_j}\right)} dx \approx \frac{1}{a_j \nu_s} \sum_k f(k) \overline{\psi\left(\frac{k-n}{a_j \nu_s}\right)} .$$

If we choose to take into account only complex multiplications, we have for each coefficient at scale a_j to perform $O(La_j)$ operations, where L is the number of samples of the considered wavelet at unit scale. The total cost of the algorithm will then be of the order of

$$NL + 2NL + 4NL + \cdots + 2^J NL = O(LN^2),$$

which is *not* efficient (recall that we are computing $N \log_2 N$ coefficients only).

A systematic analysis of this problem is presented in [170] (see also [263]), where the problem of evaluating a wavelet transform from samples is studied.

5.6.2 FFT-Based Algorithms

Since we are dealing with time-frequency representations, it may seem natural to use fast Fourier transform algorithms to evaluate the transforms.

Gabor Transform

Let us consider first the case of the Gabor transform. Given a window $g \in L^2(\mathbb{R})$, we want to evaluate the Gabor transform of the function $f \in L^2(\mathbb{R})$. Recall that the latter is given by the integral

$$G_f(b, \omega) = \int f(x) \overline{g(x-b)} e^{-i\omega(x-b)} dx.$$

Such an expression takes the form of a Fourier transform. Once discretized, it may be evaluated with a standard FFT routine. The complexity of such an approach may be estimated as follows. Let us assume that we start from N samples of a signal and that we want to compute the Gabor transform for M values of the frequency variable. We have to compute exactly $MN \log N$ multiplications.

An alternative is to use the expression of $G_f(b, \omega)$ given by Plancherel's formula,

$$G_f(b, \omega) = \frac{1}{2\pi} \int \hat{f}(\xi) \overline{\hat{g}(\xi - \omega)} e^{i\xi b} d\xi,$$

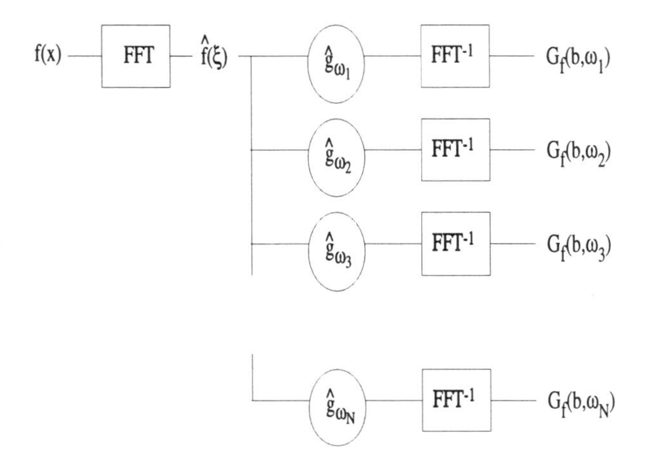

Figure 5.9. Organization of the FFT-based algorithm for continuous Gabor transform.

and to use FFT again to evaluate the integral. The resulting algorithm is also an $O(N \log N)$ one. The structure of the algorithm is represented graphically in Figure 5.9.

Wavelet Transform

If we now consider the case of the continuous wavelet transform, we first remark that it may be written in the form

$$T_f(b,a) = \frac{1}{a}\int f(x)\overline{\psi\left(\frac{x-b}{a}\right)}dx = \frac{1}{2\pi}\int \hat{f}(\xi)\overline{\hat{\psi}(a\xi)}e^{i\xi b}d\xi \ , \tag{5.97}$$

i.e., in the form of the inverse Fourier transform of the product $\hat{f}(\xi)\overline{\hat{\psi}(a\xi)}$. Again, FFT methods may be used to evaluate such an integral.

We can then estimate the number of operations as before. We have to compute the first FFT once for all, and then for each one of the $\log N$ values of the scale variable, the product $\hat{f}(\xi)\overline{\hat{\psi}(a\xi)}$ (N multiplications) and the inverse FFT ($N \log N$ operations). The overall complexity of the algorithm is then $O(N \log^2 N)$. The general organization of FFT-based wavelet algorithm is the same as that described in Figure 5.9.

Remark 5.11 Another element which makes these methods very convenient is that unlike the dedicated algorithms described later, they allow us to use arbitrary values for the scale variable (and not only powers of 2), without any extra effort.[5]

5.6.3 Filter-Bank Approaches to the Wavelet Transform

A major reason for the success of wavelets in the scientific community lies in the existence of dedicated fast algorithms. The existence of such algorithms relies on a certain number of assumptions. One of these is that the wavelet transform is evaluated on a lattice which is invariant under a specific group of dilations. To this end we shall restrict ourselves to dilations by powers of 2 only (the generalization to powers of arbitrary prime numbers is obvious). The other assumption is the existence of a scaling function, and of an associated pair of *quadrature mirror filters.* So let us assume that we have at hand a pair (ϕ, ψ) consisting of a scaling function and a wavelet, related by

$$\left\{ \begin{array}{rcl} \hat{\phi}(2\xi) & = & m_0(\xi)\hat{\phi}(\xi) , \\ \hat{\psi}(2\xi) & = & m_1(\xi)\hat{\phi}(\xi) , \end{array} \right. \tag{5.98}$$

with two 2π-periodic functions m_0 and m_1. As before we denote by $\{h_k;\ k \in \mathbb{Z}\}$ and $\{g_k;\ k \in \mathbb{Z}\}$ the Fourier coefficients of $m_0(\xi)$ and $m_1(\xi)$, normalized so that

$$\left\{ \begin{array}{rcl} m_0(\xi) & = & \frac{1}{\sqrt{2}} \sum_\ell h_\ell e^{i\ell\xi} , \\ m_1(\xi) & = & \frac{1}{\sqrt{2}} \sum_\ell g_\ell e^{i\ell\xi} . \end{array} \right. \tag{5.99}$$

Those filters may be used within a pyramidal algorithms to compute the wavelet coefficients of any finite-energy signal f.

Redundant Wavelet Decompositions

Let us start with the case in which the continuous wavelet transform is sampled on a fine grid, i.e., a lattice of the form $\{(k2^j, 2^j),\ j,k \in \mathbb{Z}\}$. This corresponds to the dyadic wavelet transform. We then consider the following set of functions:

$$\left\{ \begin{array}{rcl} \psi_k^j(x) & = & 2^j \psi\left(2^j(x-k)\right) , \\ \phi_k^j(x) & = & 2^j \phi\left(2^j(x-k)\right) , \end{array} \right. \tag{5.100}$$

[5]It is worth mentioning that filter bank approaches may be adapted to accommodate arbitrary scales, at least if one is willing to use approximate filters.

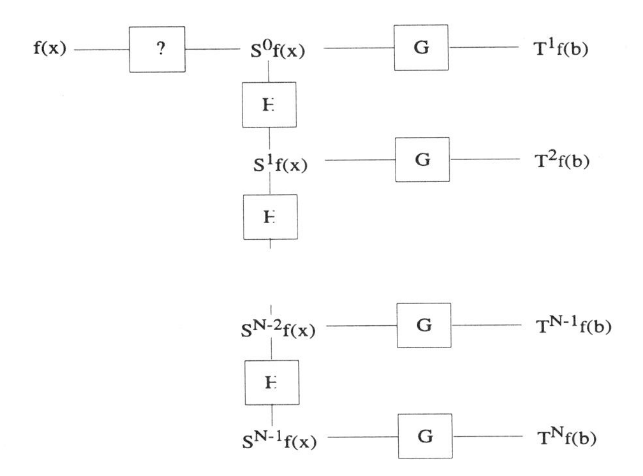

Figure 5.10. Organization of a filter-bank algorithm for redundant wavelet transform.

We are interested in the following (dyadic wavelet) coefficients of a signal $f \in L^2(\mathbb{R})$:

$$\begin{cases} T^j(k) &= \langle f, \psi_k^j \rangle \ , \\ S^j(k) &= \langle f, \phi_k^j \rangle \ . \end{cases} \tag{5.101}$$

It follows directly from equations (5.98) and (5.99) that these coefficients may be computed as follows.

Proposition 5.2 *The coefficients $T^j(k)$ and $S^j(k)$ satisfy*

$$\begin{cases} T^j(k) &= \frac{1}{\sqrt{2}} \sum_\ell \overline{g_\ell} S^{j-1}(k - 2^{j-1}\ell) \ , \\ S^j(k) &= \frac{1}{\sqrt{2}} \sum_\ell \overline{h_\ell} S^{j-1}(k - 2^{j-1}\ell) \ . \end{cases} \tag{5.102}$$

The general organization of a filter-bank wavelet algorithm is described in Figure 5.10. The boxes represent convolution with the filters $H = \{h_k;\ k \in \mathbb{Z}\}$ and $G = \{g_k;\ k \in \mathbb{Z}\}$. The first box has been labeled with a question mark, because of the various possible ways of dealing with the first step of the algorithm (see the discussion of this point in Subsection 5.4.3).

It is interesting to notice that the same filters (or, more precisely, dilated copies of these filters) are used throughout the calculations. However, the discrete dilation used here is a trivial one in the sense that intermediate coefficients for the dilated filter are set to zero. One sometimes speaks of

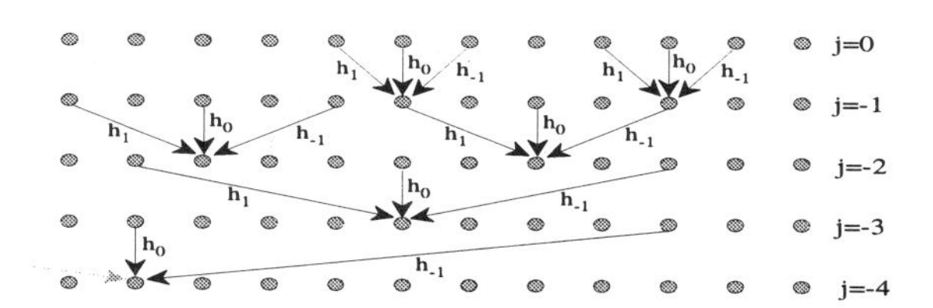

Figure 5.11. Pyramidal algorithm for a (dyadic) redundant wavelet transform, with a filter of length 3.

dilation with holes, and this algorithmic structure is referred to as the *à trous algorithm.* This dilation structure has been studied in [170, 135, 263].

The complexity of the algorithm is readily evaluated. In the same situation as before, the number of multiplications is the same for each value of the scale variable j and equals $2MN$, where M is the length of the H and G filters (assumed to be the same for simplicity), and $N = 2^J$ is the length of the (discrete) signal, or more precisely the number of the coefficients at scale $j = 0$ taken into account (see Remark 5.12, later, for a discussion of this point). Then if we assume that the wavelet transform is computed for $J = \log_2(N)$ values of the scale variable, the overall complexity of the algorithm is

$$O(N \log_2(N)) .$$

To illustrate the algorithm, consider, for example, the case of an $H = \{h_\ell\}$ filter with only three nonzero coefficients $\{h_{-1}, h_0, h_1\}$. Then the resulting algorithm is displayed in Figure 5.11.

Non-redundant Wavelet Decompositions

The same kind of analysis may be carried out if we turn to less redundant versions of the wavelet transform, namely, wavelet transforms on dyadic grids. We then have to consider the functions

$$\left\{ \begin{array}{lcl} \psi_{jk}(x) & = & 2^{j/2}\psi\left(2^j x - k\right) , \\ \phi_{jk}(x) & = & 2^{j/2}\phi\left(2^j x - k\right) , \end{array} \right. \tag{5.103}$$

and we are interested in the following wavelet coefficients of a signal $f \in L^2(\mathbb{R})$:

$$\left\{ \begin{array}{lcl} t_k^j & = & \langle f, \psi_{jk} \rangle , \\ s_k^j & = & \langle f, \phi_{jk} \rangle . \end{array} \right. \tag{5.104}$$

As before, it is easy to see that these coefficients may be computed from the following property:

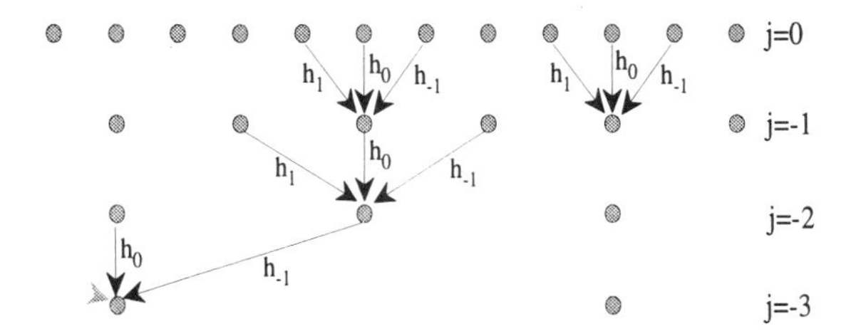

Figure 5.12. Pyramidal algorithm for a wavelet transform on a dyadic grid, with a filter of length 3.

Proposition 5.3 *The coefficients t_k^j and s_k^j satisfy*

$$\left\{ \begin{array}{lcl} t_k^j & = & \sum_\ell \overline{g_\ell} s_{2k-\ell}^{j-1} , \\ s_k^j & = & \sum_\ell \overline{h_\ell} s_{2k-\ell}^{j-1} . \end{array} \right. \tag{5.105}$$

The corresponding algorithm is illustrated in Figure 5.12. Notice that it is simply obtained by throwing $\log_2 N$ nodes (and the corresponding arrows) away from Figure 5.11. Using the same arguments as before, it is easily shown that the algorithm has complexity $O(N \log N)$.

Remark 5.12 *Computation of the coefficients at scale $j = 0$:* In general, the user is provided with a discrete signal, and it is not obvious to make the connection between the samples and the S_k^j coefficients. The usual answer to that question amounts to saying that in general the samples are not true samples of a continuous-time signal, but rather the result of a smoothing + sampling procedure. The sampling function being unknown, it is a common practice known as the "second wavelet crime" to identify the values of the discrete signal with the S_k^j coefficients, and to use them directly to initialize the algorithm. Alternatives are described in [103, 109] and have been already discussed in Section 5.4.3.

Wavelet Packet Algorithms

The construction of wavelet packets reviewed in Section 5.5.1 is based upon recursive use of the two-scales difference equations associated with a pair of filters (H, G). Therefore, it is not difficult to extend the previous discussion to obtain the following iterative scheme for evaluating wavelet packet coefficients:

$$\left\{\begin{array}{lcl} \langle f, w_{j\,2n\,k}\rangle & = & \sum_\ell \overline{h_\ell}\langle f, w_{j+1\,n\,2k-\ell}\rangle\ ; \\ \langle f, w_{j\,2n+1\,k}\rangle & = & \sum_\ell \overline{g_\ell}\langle f, w_{j+1\,n\,2k-\ell}\rangle . \end{array}\right. \tag{5.106}$$

Therefore, the coefficients may be computed using a scheme similar to the previous one. The complexity of the resulting algorithm is easily seen to be $O(N\log(N))$ if all wavelet packet decompositions are computed. Implementation details are described in [Wickerhauser94].

Thanks to such a pyramidal organization, the best basis search may be implemented in a very simple way. The first step is a decomposition into all possible wavelet packet bases. The second one is a "bottom-to-top" search for the optimal basis. We refer to [Wickerhauser94] for a detailed treatment of wavelet packets and best basis methods, with several applications.

5.6.4 Approximate Algorithms

As we have seen in the last section, the existence of pyramidal algorithms relies on the existence of a scaling function and a pair of 2π-periodic functions m_0 and m_1 satisfying specific relations. However, for a given wavelet, say the Laplacian of a Gaussian for the sake of the present discussion, it is not difficult to construct a corresponding scaling function, but in general there does not exist any pair of 2π-periodic (notice that periodicity is the main restriction) functions such that equations (5.98) hold. If one insists on pyramidal algorithms for such wavelets, one has to make some approximations. Several approaches to this problem have been proposed. They all basically rely on a simple recipe: either find a wavelet (and a scaling function) with associated pyramidal algorithms that approximates the desired wavelet, or keep the wavelet and find *approximate filters* that would yield approximate fast algorithms.

The first approach is often obtained by projecting the considered wavelet onto an approximation space associated with a given (your favorite) multiresolution analysis. Spline multiresolution analyses turn out to be extremely convenient for this purpose. One then obtains an approximate wavelet and a corresponding scaling function which satisfy a two-scale difference equation. This is enough to yield pyramidal algorithm. Such an approach has been described in several places, in particular in [3, 208, 283], and is extensively used. Interestingly enough, it may also be extended to wavelet transforms with arbitrary scale factors (not only powers of 2) [208, 283] See also the discussion in [170]. This approach has the advantage of leading to a pair of filters which satisfy the perfect reconstruction property (5.54), and such that the high-pass filter $m_1(\xi)$ is such that $m_1(0) = 0$, which ensures that the wavelet used in the algorithm has zero integral, as it should.

The second approach [232] is somewhat more intrinsic in the sense that it amounts to find the best approximate filters for a given wavelet. Such an approach may be summarized as follows. Let $\psi \in L^1(\mathbb{R}) \cap L^2(\mathbb{R})$, and let $\phi \in L^1(\mathbb{R}) \cap L^2(\mathbb{R})$ be a scaling function (for example, a scaling function from the family constructed in our discussion of continuous multiresolutions). The assumption is that the collection of the integer translates of ϕ form a Riesz basis of the space V_0 they span (see Definition 5.2). Then we have the following:

Definition 5.10 *Assume that equation* (5.2) *holds. Then the best filters associated with ϕ and ψ are defined by*

$$\begin{cases} \tilde{m}_0(\xi) &= \dfrac{\sum_\ell \hat{\phi}(2\xi + 4\pi\ell)\overline{\hat{\phi}(\xi + 2\pi\ell)}}{\sum_\ell |\hat{\phi}(\xi + 2\pi\ell)|^2}, \\ \tilde{m}_1(\xi) &= \dfrac{\sum_\ell \hat{\psi}(2\xi + 4\pi\ell)\overline{\hat{\phi}(\xi + 2\pi\ell)}}{\sum_\ell |\hat{\phi}(\xi + 2\pi\ell)|^2}. \end{cases} \tag{5.107}$$

The interpretation of these formulas is quite simple. In the standard multiresolution case when there exist 2π-periodic filters, the function $\phi(x/2)/2$ may be expanded into the basis of integer translates of ϕ (and the same for ψ as well), and the Fourier coefficients of the filters are nothing but the coefficients in this expansion.

If 2π-periodic filters do not exist, $\phi(x/2)/2$ cannot be expanded anymore into the basis of integer translates of ϕ. However, it is in general true that the distance (with respect to the $L^2(\mathbb{R})$ norm, for example) between $\phi(x/2)$ and the space V_0 is expected to be small, simply because of frequency localization arguments. If this distance is indeed small enough to be numerically negligible, then it makes sense to consider the element of V_0 closest to $\phi(x/2)/2$, namely, the orthogonal projection of $\phi(x/2)/2$ onto V_0. And we are back to the same situation as before.

As we can see from the result of the next proposition, the natural control parameters are the following:

$$\begin{cases} \mu_0 &= \left[\displaystyle\int \left|\hat{\phi}(2\xi) - \tilde{m}_0(\xi)\hat{\phi}(\xi)\right|^2 d\xi\right]^{1/2}, \\ \mu_1 &= \left[\displaystyle\int \left|\hat{\psi}(2\xi) - \tilde{m}_1(\xi)\hat{\phi}(\xi)\right|^2 d\xi\right]^{1/2}. \end{cases} \tag{5.108}$$

If in addition we set

$$C_i = \operatorname{ess\,sup} |\tilde{m}_i|, \qquad i = 0, 1 ,$$

the following bounds can be proven (see [232] for proofs).

Proposition 5.4 *Let $f \in L^2(\mathbb{R})$, and let us denote by $T^j(k)$ and $S^j(k)$ the corresponding wavelet and scaling function coefficients defined in (5.101). Let $\tilde{T}^j(k)$ and $\tilde{S}^j(k)$ be the coefficients obtained by using the pyramidal algorithm (5.102) with the best filters defined in (5.107). Then we have*

$$||\tilde{S}^j - S^j||_\infty \le \mu_0 2^{(1-j)/2} \frac{1-(C_0\sqrt{2})^j}{1-C_0\sqrt{2}} ||f|| \,, \tag{5.109}$$

and

$$||\tilde{T}^j - T^j||_\infty \le 2^{(1-j)/2} \left[\mu_1 + C_1 \mu_0 \sqrt{2} \frac{1-(C_0\sqrt{2})^{j-1}}{1-C_0\sqrt{2}} \right] ||f|| \,. \tag{5.110}$$

Similar estimates may be obtained for the case of the s_k^j and t_k^j coefficients.

This approach is interesting also because it suggests a way of generalizing the pyramidal algorithms to other situations, for instance, the more redundant cases in which more than one scale per octave is considered.

Remark 5.13 The functions $\tilde{m}_0$ and $\tilde{m}_1$ are particularly well adapted to the case where the wavelet and scaling functions are related by

$$\hat{\psi}(2\xi) = \hat{\phi}(2\xi) - \hat{\phi}(\xi) \,.$$

In such a case we immediately see that

$$\tilde{m}_0(\xi) + \tilde{m}_1(\xi) = 1 \,.$$

This implies the simple reconstruction formula

$$S^0(k) = S^J(k) + \sum_{j=J}^{-1} T^j(k) \,,$$

which is the discrete counterpart of Morlet's simple reconstruction formula. However, one does not have $m_1(0) = 0$ in general.

In any case, once a pair of 2π-periodic approximate filters has been chosen, the decomposition algorithm follows exactly the same steps as before.

5.7 Examples and S-Commands

The package S+wavelets implements various tools we described in this chapter. We do not reproduce the discussion here, and we refer to [Bruce94] for details and illustrations. Notice, nevertheless, that it has a projection pursuit function for the wavelet packet dictionaries, but not for the dictionary comprising the time frequency atoms. We focus here on the tools provided by Swave.

5.7.1 Localization of the Wavelets and Gabor Functions

We start by illustrating the time-frequency localization properties of wavelets and Gabor functions, using the S-functions `cwt` and `cgt` described in the previous chapters. Consider, for example, the case of the Gabor transform. We can get an idea of the time-frequency localization properties of the Gabor transform simply by considering the CGT of various Gabor functions. For example, the Gabor transform of the Gabor function $g_{(b_0,\omega_0)}$ is given by $\langle g_{(b_0,\omega_0)}, g_{(b,\omega)}\rangle$, which is essentially the reproducing kernel for the CGT, for fixed values of the first arguments.

Example 5.1 : CGT and time-frequency localization
Generate a Gabor function and compute its Gabor transform:

```
> x <- 1:512
> par(mfrow=c(1,2))
> time <- 256
> x <- x - time
> frequency <- 50
> gab <- exp(- x*x/500)*cos(2*pi*frequency*x/512)
> cgtgab <- cgt(gab,50,.005,25)
```

Generate a second Gabor function and compute its Gabor transform:

```
> time <- 350
> frequency <- 30
> gab <- exp(- x*x/500)*cos(2*pi*frequency*x/512)
> cgtgab <- cgt(gab,50,.005,25)
```

Considering this example with various values for the time and frequency variables shows that Gabor functions localize in the time-frequency plane as in Figure 5.1. Corresponding examples for wavelet transform are obtained similarly:

Example 5.2 : CWT and time-frequency localization
Generate a wavelet and compute its wavelet transform:

```
> par(mfrow=c(1,2))
> time <- 256
> scale <- 10
> wave <- morlet(512,time,scale)
> cwtwave <- cwt(wave,5,12)
```

Same thing with different values for the time and scale parameters:

```
> time <- 350
> scale <- 30
> wave <- morlet(512,time,scale)
> cwtwave <- cwt(wave,5,12)
```

The output is displayed in Figure 5.4.

5.7.2 Dyadic Wavelet Transform and Local Extrema

The Splus function `mw` implements the discrete dyadic wavelet transform with respect to the wavelet which was used by S. Mallat and his collaborators [212, 214] (`mw` stands for Mallat's wavelet), namely,

$$\psi(x) = \frac{d}{dx}\chi_4(x), \tag{5.111}$$

where $\chi_4(x)$ is the cubic B-spline function

$$\chi_4(x) = \begin{cases} x^3/6 & \text{if } 0 \le x \le 1 \\ -x^3/2 + 2x^2 - 2x + 2/3 & \text{if } 1 \le x \le 2 \\ x^3/2 - 4x^2 + 10x - 22/3 & \text{if } 2 \le x \le 3 \\ -x^3/6 + 2x^2 - 8x + 32/3 & \text{if } 3 \le x \le 4 \\ 0 & \text{elsewhere.} \end{cases} \tag{5.112}$$

The implementation is based on the sub-band coding schemes (described in detail in Section 5.6), and the two-scales relation

$$\frac{1}{2}\psi\left(\frac{x}{2}\right) = \sum_k g_k \chi_3(x+k) \ , \tag{5.113}$$

and

$$\frac{1}{2}\chi_3\left(\frac{x}{2}\right) = \sum_k h_k \chi_3(x+k) \ , \tag{5.114}$$

where h_k and g_k are the Fourier coefficients of the functions

$$m_0(\xi) = \sum_k h_k e^{ik\xi} \ , \tag{5.115}$$

and

$$m_1(\xi) = \sum_k g_k e^{ik\xi} \ . \tag{5.116}$$

`mw` essentially needs two arguments: an input signal and the number of scales for which the dyadic wavelet transform is to be computed. It returns a structure containing the original signal, the wavelet coefficients, the scaling function coefficients, the number of scales and the size of the signal (padded with zeroes to be a power of 2).

The Splus function `ext` computes the local extrema of the output of function `mw`. It returns a structure of the same format as `mw`, with the local extrema in place of the dyadic wavelet transform.

In addition, the Splus utilities `wpl` and `epl` display the structures (outputs of `mw` and `ext`, respectively) on the graphic device.

We now give an example of local extrema analysis for a transient signal.

Example 5.3 *Given a signal (here a model of transient signal named* `A0`*), compute its dyadic wavelet transform with the cubic spline derivative wavelet on seven consecutive scales:*

```
> A0 <- scan(''signals/A0'')
> dwA0 <- mw(A0)
```

Compute the local extrema of the dyadic wavelet transform:

```
> extA0 <- ext(dwA0)
```

The result is displayed in Figure 5.5.

We shall come back to this example in the last chapter devoted to statistical reconstructions of signals from dyadic wavelet transform local extrema.

5.8 Notes and Complements

The theory of frames (initiated by Duffin and Schaeffer [116]) was shown by Daubechies, Grossmann, and Meyer [96] and later on in [93] in a context more closely related to signal processing to provide the right framework for the discretization of the continuous Gabor transform. The theory of frames of wavelets and Gabor functions was first developed in [93] and has enjoyed considerable developments since then (see, for example, the contributions of Feichtinger and Gröchenig [124, 125], and the review articles by Heil and Walnut [161] and Benedetto and Walnut [38]). In particular, the Balian-Low phenomenon for Gabor frames has received great attention (see, for example, [37]).

The reader interested in a more detailed treatment of the theory of frames can consult [Daubechies92a], for example, or [Chui92a]. More details on the discrete transforms may be found in the various tutorials now available. See, for example, [Chui92a, Hernández97, Kahane96, Meyer89a] for mathematics-oriented books, and [A.Cohen95, Daubechies92a, Mallat97, Meyer93b, Strang96, Vetterli96, Wickerhauser94] for more application-oriented texts.

Dyadic wavelet transforms were introduced by Mallat in an image processing context (see [214], for example). They were used to provide multiscale representations of images, in order to simplify other tasks such as edge detection, stereo matching, ... The local maxima representation has been extensively used for multifractal analysis by Arneodo and co-workers (see [Arneodo95] for a review).

The matching pursuit algorithm was introduced in [215]. Its shortcomings were documented in [184]. Details on the basis pursuit algorithm and its implementations can be found in [71]. Basis pursuit can be viewed as a global optimization form of the greedy form of the matching pursuit algorithm. It is formulated as a linear programming algorithm which can be solved either by the standard simplex algorithm or by interior point methods (see [285] for details), but because of the size of the problems, we believe that interior point methods are superior to the implementations based on the simplex method.

The theory of wavelets was significantly rejuvenated by the introduction by S. Mallat and Y. Meyer of the concept of multiresolution analysis [210]. Numerous examples of MRAs have been proposed. We mention, for example, the Littlewood-Paley-type MRAs by Y. Meyer, the spline MRAs (by G. Battle and P. G. Lemarié, who introduced the same MRA independently), and the MRAs associated to compactly supported wavelets by I. Daubechies. We refer to [Meyer89a, Daubechies92a] for a review of several of these constructions, and to [268] for an elementary introduction to the connection between dilation equations and the construction of wavelet orthonormal bases. The connection between wavelets and QMFs was pointed out by Mallat [210] and clarified later on by Cohen [78] and Lawton [202]. The interested reader can find a detailed analysis in [Daubechies92a].

There are many possible variations around the concept of wavelet basis and multiresolution analysis. We quote some of them here for completeness. One of the simplest modifications of the notion of MRA consists in replacing the scale factor 2 by another number. In the case of integer or rational scale factors, the machinery of multiresolution analysis goes essentially as before, except that one generally needs several wavelets instead of a unique one to describe the spaces W_j. Such constructions were discussed in detail by several authors in the electrical engineering community (see, e.g., Vaidyanathan [Vaidyanathan92], or the book by Vetterli and Kovacevic [Vetterli96]). For irrational scale factors, Auscher proved that there does not exist any corresponding multiresolution analysis or wavelet basis.

Other variations around wavelet bases are possible. Let us revisit the example of spline wavelets, and let us consider the function $\chi(x)$ defined by its Fourier transform in equation (5.57). The integer translates of $\chi(x)$ form a Riesz basis of V_0, which is very convenient for many reasons and in particular because the basis functions are compactly supported. Also, as a consequence of the central limit theorem, $\chi_r(x)$ converges to a Gaussian function as $r \to \infty$, and Gaussian functions are known to optimize the time-frequency concentration (i.e., minimize Heisenberg's inequality). The basis is not orthonormal. However, it is easy to find a function $\tilde{\chi}(x)$ such that its integer translates form a basis of V_0, biorthogonal to the basis $\{\chi_k, k \in \mathbb{Z}\}$. This *dual scaling function* is simply given by $\hat{\tilde{\chi}}(\xi) = \hat{\chi}(\xi)/\sum_k |\hat{\chi}(\xi+2\pi k)|^2$ (note that the denominator is in this case a trigonometric polynomial). The dual functions are not compactly supported, and in fact retain most of the nice properties usually associated with orthonormal wavelet bases. In particular, $\chi(x)$ also satisfies a refinement equation (with filters that are not compactly supported, but may have fast decay.)

The general theory of biorthogonal wavelet has been developed by A. Cohen, I. Daubechies, and J.C. Feauveau [80]. Other generalizations involve the construction of wavelet bases on bounded domains, such as the circle or intervals in the one-dimensional case, or more general domains in higher dimensions (see, [81], for example). These generalizations go beyond the scope of the present book.

The pyramidal algorithms for evaluating wavelet and scaling function coefficients come from older techniques developed in image processing, namely the Laplacian pyramid of Burt and Adelson [59], and sub-band coding ideas from Esteban and Galand [121] and Smith and Barnwell [266]. The relation with wavelet bases was emphasized by Mallat [211] and used later on by Cohen [78] and Lawton [202]. Sub-band coding-type algorithms for dyadic and continuous wavelet transforms were proposed in [170], and later on in [263]. A review of several algorithms for discrete and continuous wavelet transforms may be found in [258]. More recent treatments of these algorithmic aspects appeared in the books by Vetterli and Kovacevic [Vetterli96] and Wickerhauser [Wickerhauser94].

The wavelet packet algorithm described in the text is designed to identify the best of the bases given by a tree, best being understood as a criterion which depends on a given signal. The way we present it shows that it is easy to modify slightly the criterion to find the basis best adapted to a finite set of signal. This idea was exploited by Wickerhauser, who coined the term approximate KL basis. Here KL stands for Karhunen-Loeve, as we shall see in the next chapter. The KL basis plays an important role in statistical data analysis, and Wickerhauser's thesis is that, if we cannot determine and use this optimal basis, why not shoot for the best basis among those for which we have fast computational algorithms! See [Wickerhauser94] for details. This idea of best basis for a finite family of signals was extended by Coifman and Saito to two (and more generally to any finite number of) finite families of signals. The goal is not only to express the signal efficiently in a basis, but also to make sure that the coefficients in the developments make it easier to decide to which family a given signal belongs to. This problem is of crucial importance in statistical pattern recognition and in discriminant analysis. The interested reader is referred to [89, 90].

Beyond the applications to signal analysis, wavelet bases have found many applications in several areas of applied mathematics. Let us mention for the record applications in statistics (see [Antoniadis94] for a review) and to the numerical resolution of ordinary and partial differential equations (and their corresponding numerical analyses). We refer to [46, 47] for an introduction to the latter topic.

Part III

Signal Processing Applications

Chapter 6

Time-Frequency Analysis of Stochastic Processes

We now turn to the discussion of time-frequency representations of stochastic processes. As we shall see, the case of stationary processes is relatively easy to understand. We give a complete analysis of the continuous Gabor transform and the continuous wavelet transform of these processes. We relate these transforms to the spectral characteristics of the processes and we augment the spectral analysis toolkit with the tools provided by these transforms. This analysis is also interesting as it suggests some possible extensions to non-stationary situations. We limit ourselves here to simple arguments, since the spectral theory of non-stationary processes is still in its infancy and despite a very active research, the notation, the definitions, and the results of this theory are presumably not in their final form yet. We illustrate some of the techniques with the examples of processes with stationary increments and especially the process of fractional Brownian motion.

6.1 Second-Order Processes

Our presentation of the classical spectral theory of stationary processes given in Chapter 2 was based on a bottom-to-top approach: we introduced the notation and the concepts for finite sequences first, then we worked our way up in generality to doubly infinite time series and finally to continuous processes parameterized by the whole real line $\mathbb{R}$. Using this preparatory work (i.e., assuming that the reader went through Chapter 2 or is familiar with the classical spectral theory), we deal with continuous processes from the get-go.

6.1.1 Introduction

Our theoretical discussion will be restricted to (possibly complex valued) stochastic processes $\{f(x);\ x \in \mathbb{R}\}$ which have moments of order 2 in the sense that $\mathbb{E}\{|f(x)|^2\} < \infty$ for all $x \in \mathbb{R}$. Moreover, except when dealing with actual data, we shall assume implicitly that the mean is zero in the sense that $\mathbb{E}\{f(x)\} = 0$ for all $x \in \mathbb{R}$. In other words, we assume that the mean has been computed and subtracted from the signal. For processes of order 2, the main statistic is the covariance function $C_f(x,y)$ defined by

$$C_f(x,y) = \mathbb{E}\{f(x)\overline{f(y)}\}, \qquad x, y \in \mathbb{R}.$$

Even though the covariance function C_f is usually regular, it happens quite often that one would like to analyze processes for which this covariance is not necessarily a function of the two variables x and y, but instead a generalized function (or a Schwartz distribution). The famous example of a white noise for which one would like to have $C(x,y) = \delta(x-y)$ is presumably the simplest instance of this situation. In order to include these mathematically pathological cases with those of regular processes, it is convenient to define and analyze the covariance structure of the process by means of the integral operator $\mathcal{C}_f$ whose kernel is the function C_f, i.e., the operator defined formally by

$$[\mathcal{C}_f\varphi](x) = \int C_f(x,y)\varphi(y)\,dy. \tag{6.1}$$

Notice that this operator is the identity in the case of a white noise: no doubt, this is more *regular* than the *delta* distribution. So considering the covariance operator instead of the covariance function makes it possible to handle the covariance structure of more processes. But there is no free lunch and what has been gained in generality will be lost in regularity. Indeed, the *sample paths* of these *generalized* processes will in general be extremely irregular: the sample paths of the white noise cannot be defined as functions but only as generalized functions (in the sense of the theory of Schwartz distributions). One way to introduce this white noise process is to define it as the derivative of the process of Brownian motion. The latter has more reasonable sample paths (they are continuous functions which are Hölder continuous of order α for any $\alpha < 1/2$), but this process is not stationary, its increments are. We shall come back later to a similar situation when we discuss the fractional white noise (fWN for short) and the fractional Brownian motion (fBM for short.)

Let us revisit for a moment the notion of wide-sense stationary processes (or even of wide-sense stationary generalized processes). Such processes are defined as those processes for which the operator $\mathcal{C}_f$ is shift invariant, i.e., commutes with translations. In other words, using the notation T_b introduced in Chapter 3 for the time translation, we have

$$\mathcal{C}_f = T_b \mathcal{C}_f T_{-b}$$

for all $b \in \mathbb{R}$. The autocovariance function of such a process is necessarily of the form $C_f(x,y) = C_f(x-y)$, where we used the same notation C_f for the function of the difference $x-y$. In this situation the covariance operator is a convolution operator,

$$[\mathcal{C}_f \varphi](x) = \int C_f(x-y)\varphi(y)dy \ ,$$

and (recalling the derivation of formula (1.30) in Chapter 1), since C_f is nonnegative definite, according to Bochner's theorem it is the Fourier transform of a nonnegative finite measure ν_f.

Under a mild decay condition at infinity (integrability of the function C_f is enough) the spectral measure ν_f is absolutely continuous, i.e., $\nu_f = \nu_{f,ac}$ with the notation of the Lebesgue decomposition (1.31) and its density $\mathcal{E}_f(k)$ is called the spectral density. $\mathcal{E}_f$ happens to be the Fourier transform of the covariance function in the sense that

$$C_f(x) = \frac{1}{2\pi}\int \mathcal{E}_f(k)e^{ikx}dk \ ,$$

and the Fourier representation of the covariance function can be used to obtain

$$[\mathcal{C}_f \varphi](x) = \frac{1}{2\pi}\int\int \mathcal{E}_f(k)e^{ik(x-y)}\varphi(y)\,dy\,dk \ .$$

This shows that this operator is *diagonalized* by the complex exponential functions e^{ikx}. As we saw in Chapter 2, the latter form the essential building block of the spectral theory of stationary processes, which can be understood as the spectral decomposition of the (nonnegative) covariance operator $\mathcal{C}_f$. There are excellent textbooks on the spectral theory of operators on Hilbert spaces. We shall refer to [254, 255] for details on the results alluded to in this chapter.

In the general case of non-stationary processes, the lack of shift invariance makes it impossible to have a common basis diagonalizing all the covariance operators. It is nevertheless possible in some cases to take advantage of a good spectral decomposition of the covariance operator when the latter exists. This is the purpose of the following section.

6.1.2 The Karhunen-Loeve Transform

Throughout this section we restrict our analysis to the class of *Hilbert-Schmidt* covariance operators. Recall that an operator $\mathcal{C}$ on $L^2(D,dx)$

with kernel $C(x,y)$ is said to be a Hilbert-Schmidt operator (HS operator for short) if

$$||\mathcal{C}||_{HS}^2 = \int_D \int_D |C(x,y)|^2 dx dy < \infty. \tag{6.2}$$

Here D can be the whole real line $D = \mathbb{R}$ as well as a bounded interval. The space $\mathcal{H}_{HS}$ of all the Hilbert-Schmidt operators is a Hilbert space for the inner product,

$$\langle \mathcal{C}, \mathcal{C}' \rangle_{HS} = \int_D \int_D C(x,y)\overline{C'(x,y)} dx dy, \tag{6.3}$$

obtained by polarization of the definition of the norm defined earlier. It is a remarkable property of Hilbert-Schmidt operators that they are arbitrarily well approximated by finite-rank operators. More precisely, assuming that $\mathcal{C}$ is a Hilbert-Schmidt operator, for any given precision $\epsilon > 0$, one can always find a finite-rank operator $\mathcal{C}'$ such that $||\mathcal{C} - \mathcal{C}'||_{HS}^2 \leq \epsilon$. This point is of considerable importance for numerical applications. Its proof is a consequence of the eigenfunction expansion which we are now presenting.

First we note that Hilbert-Schmidt operators may not necessarily yield a complete system of eigenfunctions when they are not symmetric. In such situations, one turns to the operators $\mathcal{C}\mathcal{C}^*$ and $\mathcal{C}^*\mathcal{C}$, which are nonnegative (and in particular self-adjoint.) As such, they admit a common set of eigenvalues $\{\lambda_k^2\}_{k \geq 0}$ and a complete orthonormal system of eigenvectors, denoted by $\{v_k(x)\}_{k \geq 0}$ and $\{u_k(x))\}_{k \geq 0}$, respectively. The nonnegative numbers λ_k are called the *singular values* of $\mathcal{C}$. They are usually ordered

$$\lambda_0^2 \geq \lambda_1^2 \geq \lambda_2^2 \geq \cdots\cdots \geq 0$$

and repeated according to their multiplicity. The representation

$$C(x,y) = \sum_k \lambda_k u_k(x) \overline{v_k(y)} \ ,$$

where the convergence holds in the sense of $L^2(D \times D, dx\, dy)$, is known as the *singular value decomposition* or *svd* of the operator $\mathcal{C}$. If in addition the operator $\mathcal{C}$ is *normal*, i.e., such that

$$\mathcal{C}^*\mathcal{C} = \mathcal{C}\mathcal{C}^*, \tag{6.4}$$

then necessarily $v_k = u_k$ and the kernel $C(x,y)$ has the representation

$$C(x,y) = \sum_k \lambda_k u_k(x) \overline{u_k(y)} \tag{6.5}$$

in terms of the eigenvalues and the eigenfunctions of the operator $\mathcal{C}\mathcal{C}^* = \mathcal{C}^*\mathcal{C}$.

Let us now concentrate on the operators appearing as covariance operators $\mathcal{C}_f$ of a process $f = \{f(x);\ x \in \mathbb{R}\}$. Let us assume (at least temporarily) that the covariance operator $\mathcal{C}_f$ is a Hilbert-Schmidt operator (notice that this will not be possible for stationary processes unless we are restricting them to finite intervals) and let us use the notation introduced earlier. The orthonormal basis $\{u_k;\ k \geq 0\}$ of eigenvectors is usually called the Karhunen-Loeve basis (KL-basis) associated to the covariance operator $\mathcal{C}_f$ and/or the covariance kernel C_f, or even to the process $f = \{f(x);\ x \in D\}$. We denote by $f_k = \langle f, u_k \rangle$ the components of each realization of the process $f = \{f(x);\ x \in D\}$ in the KL-basis. Then we have the representation

$$f(x) = \sum_k f_k u_k(x) \ , \tag{6.6}$$

where the random variables f_k's are uncorrelated in the sense that they satisfy

$$\mathbb{E}\{f_k \overline{f_\ell}\} = \langle \mathcal{C}_f u_k, u_\ell \rangle = \lambda_k \delta_{k\ell} \ . \tag{6.7}$$

The transform which takes the random function $f = \{f(x);\ x \in D\}$ into the sequence $\{f_k;\ k \geq 0\}$ of random variables is called the *Karhunen-Loeve transform* of the process. This transform is in fact a form of the spectral representation of a stochastic process. As in the case of the spectral representation of stationary processes, the process is written as the superposition of basis functions computed from the spectral decomposition of the covariance operator: the Karhunen-Loeve basis plays the role of the complex exponential "basis" and the singular values play the role of the (values of the) spectral density. Note that we talk about "complex exponential basis" even though the functions $x \hookrightarrow e^{i\xi x}$ are sometime in uncountable number and that they are not even in the Hilbert space in question when $D = \mathbb{R}$. Nevertheless, this terminology is so suggestive that we shall keep on using it.

It is important to stress that even though the Karhunen-Loeve transform provides a representation of nonstationary processes which is theoretically "optimal" in a specific sense, it is far from being optimal when it comes to many practical issues. Let us list some of the main reasons:

1. The Karhunen-Loeve transform is based on the assumption that the covariance is known perfectly; in practice this is generally not the case, and the covariance has to be estimated from one (or in some very rare cases several) realization(s) of the process.

2. Even in the cases where the covariance is completely known, the problem of the singular value decomposition remains. Even though Hilbert-Schmidt operators are well approximated by finite rank operators, the rank needed for the approximation to be reasonable is generally very large and this makes the diagonalization cumbersome, if not impossible.

Therefore, the Karhunen-Loeve transform is certainly not the final answer and there is room for alternatives, or at least approximate versions (as long as they can be computationally efficient).

6.1.3 Approximation of Processes on an Interval

In order to put Karhunen-Loeve expansions into perspective and compare them with the other types of expansions we are considering, we revisit the problem of the approximation of random functions by expansions in a basis. This problem was rapidly touched upon in Chapter 5. We shall assume that $D = [0, 1]$ (any other bounded interval would do) and that $f = \{f(x);\, x \in [0, 1]\}$ is a stochastic process of order 2 with a continuous autocovariance kernel $(x, y) \hookrightarrow C_f(x, y)$. In this case, the corresponding covariance operator $\mathcal{C}_f$ is a Hilbert-Schmidt operator and the results above apply. The notation λ_k, u_k, and f_k will have the meaning given in the previous subsection.

Given a complete orthonormal system $\{e_n;\, n \geq 0\}$ in $L^2([0, 1], dx)$, an approximation scheme for a random function f in $L^2([0, 1], dx)$ is an increasing sequence of subsets $E_N(f)$ such that $E_N(f)$ has exactly N elements and the actual Nth-order approximation f_N of the function f is defined as

$$f_N(x) = \sum_{n \in E_N(f)} \langle f, e_n \rangle e_n(x) \ . \tag{6.8}$$

The corresponding error is defined by

$$\epsilon_N(f) = \mathbb{E}\{||f - f_N||^2\} = \mathbb{E}\{||f - \sum_{n \in E_N(f)} \langle f, e_n \rangle e_n||^2\} \ , \tag{6.9}$$

where the norm $||\cdot||$ denotes the norm in the Hilbert space $L^2([0, 1], dx)$. For the sake of simplicity we shall restrict ourselves to two simple examples of approximations. We review the results of [10] and we refer the interested reader to the original article for the proofs. The first example corresponds to $E_N(f) \equiv \{0, 1, \cdots, N - 1\}$. In this case the approximation is obtained as the (linear) projection on the span of the first N basis vector. This approach is discussed first. For the second one, $E_N(f)$ is chosen to be the set of indices of the N largest coefficients. This approximation is no longer a linear function of the function to approximate since the set of indices $E_N(f)$ over which the summation occurs changes with the realization of f. We shall discuss this case as an example of nonlinear approximation.

Linear Approximations

In this case the approximation is given by

$$f_N = \sum_{j=0}^{N-1} \langle f, e_j \rangle e_j(x) \ . \tag{6.10}$$

A simple computation gives

$$\begin{aligned} \mathbb{E}\{|\langle f, e_n \rangle|^2\} &= \mathbb{E}\{|\int_0^1 f(x)\overline{e_n(x)}\, dx|^2\} \\ &= \mathbb{E}\{\int_0^1 \int_0^1 f(x)\overline{f(y) e_n(x)} e_n(y)\, dxdy\} \\ &= \int_0^1 \int_0^1 C(x,y)\overline{e_n(x)} e_n(y)\, dxdy\} \\ &= \langle \mathcal{C}_f e_n, e_n \rangle, \end{aligned} \tag{6.11}$$

and consequently the error is given by

$$\epsilon(N) = \sum_{n \geq N} \mathbb{E}\{|\langle f, e_n \rangle|^2\} = \sum_{n \geq N} \langle \mathcal{C}_f e_n, e_n \rangle \ . \tag{6.12}$$

The orthonormal basis which minimizes this quantity for every N is the K-L basis, and in this case the error is given by the tail of the sum of the eigenvalues in the sense that

$$\epsilon(N) = \sum_{n \geq N} \lambda_n^2 \, . \tag{6.13}$$

Example 6.1 *If the autocovariance function is given by*

$$C_f(x, y) = e^{-|x-y|}, \tag{6.14}$$

then it is a plain exercise to prove that the K-L basis is given by

$$u_n(x) = \pi(n+1)\cos(\pi(n+1)x) + \sin(\pi(n+1)x), \qquad x \in \mathbb{R}, \quad n \geq 0. \tag{6.15}$$

It is interesting to notice that, whether one uses the complex exponential basis, the K-L basis, or a wavelet basis one always gets

$$\epsilon_N(f) \approx \text{cst} N^{-1}, \qquad \text{as} \quad N \to \infty. \tag{6.16}$$

Notice that we quoted a result for an orthonormal basis of wavelets on a bounded interval. This is a nontrivial generalization of the notion of wavelet basis presented in Chapter 5. See the section Notes and Complements of this Chapter 5 for details. In fact, it is possible to prove that for such a wavelet basis,

$$\epsilon_N(f) \leq \text{cst} N^{-\alpha}, \qquad \text{as} \quad N \to \infty, \tag{6.17}$$

provided $C_f(x,x)$ is C^α on the diagonal and all the wavelets have sufficiently many vanishing moments.

Nonlinear Approximations

We now consider the case of the approximation given by

$$f_N(x) = \sum_{n \in E_N(f)} \langle f, e_n \rangle e_n(x) \ ,$$

where $E_N(f)$ denotes the set of indices giving the largest coefficients. This set depends upon the realization of the process, and consequently it is impossible to interchange the expectation and the summation in the computation of the error $\epsilon_N(f)$. In the present situation, we have

$$\epsilon_N(f) = \mathbb{E}\{ \sum_{n \notin E_N(f)} |\langle f, e_n \rangle|^2 \}. \tag{6.18}$$

It is possible to construct a specific class of stationary stochastic processes on the unit interval such that the K-L basis and the complex exponential basis do not perform any better than in the case of the linear approximation in the sense that

$$\epsilon(N) \geq \text{cst} N^{-1}$$

for some positive constant cst and N is large enough provided $C_f(0) > C_f(1)$. On the contrary, the nonlinear approximations can potentially do significantly better in the sense that the bound (6.17) holds under the same conditions on the same assumptions on the regularity of the autocovariance function C_f and the vanishing of the moments of the wavelet. Such processes can be constructed in the following way. One first chooses points at random on the unit interval according to a (stationary) Poisson point process. Stationary point processes are the simplest point processes.

They are very popular because of their ergodic properties and the independence built into them. Typically, the values of a Poisson point process are given by a random ordered finite set of points in such a way that the numbers of points in two disjoint Borel sets are independent and the expected number of points in any given bounded Borel set is proportional to the Lebesgue measure of the set (the coefficient of proportionality being called the intensity of the point process). Say that we have picked points $0 = t_0 < t_1 < t_2 < \cdots < t_k < t_{k+1} = 1$ in this way. Then, we pick k independent copies, say, $f_0, f_1, \cdots, f_k$ of a stationary process, and we choose our process $f = \{f(x);\ x \in [0,1]\}$ in such a way that it coincides with $f_j(x)$ on the interval $[t_j, t_{j+1})$ for $j = 0, 1, \cdots, k$. Details are given in [10].

Remark 6.1 The processes in the class described above are not Gaussian. In fact, we believe that for Gaussian processes, the rates achieved by the linear and the nonlinear approximations are the same and that one needs to search for non-Gaussian processes to force the rates to be different.

6.1.4 Time-Varying Spectra

The previous subsection took us on an excursion in the world of (random) function approximation. We now return to spectral analysis. Clearly, the notion of spectral density is not well defined in the non-stationary situation. However, since this notion has proven to be extremely useful because of its strong physical and intuitive interpretation and it is sensible to attempt to define an adapted notion of spectral density which could bear the same relationship to nonstationary processes. Several approaches have been proposed in the literature. We review some of them here, referring to [196] for a more complete discussion.

Weighted Spectra

The solution which comes to mind immediately would amount to make use of a time-frequency representation to build a local spectrum. Let us consider, for example, a window function $g \in L^2(\mathbb{R})$, and let us use the square modulus of the corresponding Gabor transform as a local spectrum (the *Gabor spectrum*). In other words, if $f = \{f(x);\ x \in \mathbb{R}\}$ is a second-order mean zero process, we set

$$\mathcal{E}_{f,CGT}(b,\omega) = \frac{1}{||g||^2}\mathbb{E}\{|G_f(b,\omega)|^2\} . \tag{6.19}$$

This expression may be given a simple interpretation when the covariance operator $\mathcal{C}_f$ is Hilbert-Schmidt. Indeed, let us denote by Π^g the (rank one) orthogonal projection onto the linear subspace of $L^2(\mathbb{R})$ spanned by the window function g, i.e.,

$$[\Pi^g \varphi](x) = \frac{\langle \varphi, g\rangle}{||g||^2} g(x) , \qquad \varphi \in L^2(\mathbb{R}), \quad x \in \mathbb{R} ,$$

and for each (b,ω) by $\Pi^g_{(b,\omega)}$ the orthogonal projection onto the subspace spanned by the Gabor function $g_{(b,\omega)}$. Using the translation and modulation operators already introduced in Chapter 3, we may write

$$\Pi^g_{(b,\omega)} = T_b E_\omega \Pi^g \left(T_b E_\omega\right)^{-1}.$$

Remark 6.2 Note that for any pair (b,ω), $\Pi^g_{(b,\omega)}$ defines an action of (b,ω) on the operator Π^g. We denote this action by $ad(b,\omega)$ (it is called the *adjoint action*) . In other words:

$$ad(b,\omega) \cdot \Pi^g := \Pi^g_{(b,\omega)} = T_b E_\omega \Pi^g \left(T_b E_\omega\right)^{-1} .$$

A simple calculation shows that the Gabor spectrum is given by the inner product in the space of Hilbert-Schmidt operators of the covariance operator $\mathcal{C}_f$ with the family of rank one projectors, i.e., with the orbit of Π^g by the group of translations and modulations (the Weyl-Heisenberg group):

$$\mathcal{E}_{f,CGT}(b,\omega) = \langle \mathcal{C}_f, \Pi^g_{(b,\omega)}\rangle_{HS} = \langle \mathcal{C}_f, ad(b,\omega)\cdot \Pi^g\rangle_{HS}. \tag{6.20}$$

This formula is easily derived from the definitions of the various objects entering it. Formula (6.20) bears a close similarity with the construction of the Gabor transform of signals: test the signal against copies of a fixed window, acted on by translations and modulations. Then, the Gabor spectrum may be interpreted in some sense as some generalized Gabor transform of the covariance operator.

A similar approach is based on wavelet transform. If $\psi \in L^1(\mathbb{R}) \cap L^2(\mathbb{R})$ is an admissible wavelet, one defines the *wavelet spectrum* of f by

$$\mathcal{E}_{f,CWT}(b,a) = \frac{1}{||\psi||^2} \mathbb{E}\{ |T_f(b,a)|^2 \} = \langle \mathcal{C}_f, \Pi^\psi_{(b,a)}\rangle_{HS}, \tag{6.21}$$

with Π^ψ and $\Pi^\psi_{(b,a)}$ denoting the orthogonal projector on the one-dimensional linear spaces generated by ψ and $\psi_{(b,a)}$, respectively. Formula (6.21) is an easy consequence of the definitions. It can also be read of formula (6.41) below.

Remark 6.3 Like the Gabor spectrum, the wavelet spectrum may be given a geometric interpretation in terms of the action of the affine group on Π^ψ:

$$\mathcal{E}_{f,CWT}(b,a) = \langle \mathcal{C}_f, ad(b,a)\cdot \Pi^\psi\rangle_{HS} ,$$

with $ad(b,a)$ denoting the adjoint action of the affine group, i.e.,

$$\Pi^{\psi}_{(b,a)} = ad(b,a) \cdot \Pi^{\psi} = T_b D_a \Pi^{\psi} (T_b D_a)^{-1} \ .$$

The Gabor and wavelet spectra are included in a wider class of spectra generically called ***weighted spectra***. Such weighted spectra (sometimes called ***physical spectra***) are usually not considered suitable as definitions of local spectra, since they depend on an analyzing function. However, we shall see in a subsequent section that they can nevertheless be of interest.

Wigner Spectrum

A natural alternative to the approach just presented is provided by the Wigner's function. We shall give more details on that when we discuss locally stationary processes. To introduce Wigner's functions, it is convenient to write the covariance function in the following way:

$$C_f(x,y) = C_0\left(\frac{x+y}{2}, x-y\right) . \tag{6.22}$$

This formula is to be understood as the definition of the function C_0. The latter is intended to suggest that when we compute $C(x,y)$, we are in fact trying to see a process which is stationary, at least in a neighborhood of the mid-point $(x+y)/2$. Hence, it is natural to study its covariance as a function of $x-y$ and do its spectral analysis in this form. At this stage, the notation introduced by the definition of C_0 is only suggestive of this interpretation, but we shall find it extremely convenient when we consider locally stationary processes. Notice that, with this notation, the covariance operator takes the form

$$\begin{aligned}
[\mathcal{C}_f\varphi](x) &= \int C_f(x,y)\varphi(y)dy \\
&= \int C_0\left(\frac{x+y}{2}, x-y\right)\varphi(y)dy \\
&= \frac{1}{2\pi}\int \mathcal{E}_0\left(\frac{x+y}{2}, k\right) e^{ik(x-y)} f(y)dkdy,
\end{aligned}$$

provided the spectral function $\mathcal{E}_0$ is defined by

$$\mathcal{E}_0(b,\omega) = \int C_f\left(b+\frac{x}{2}, b-\frac{x}{2}\right) e^{-i\omega x} dx \ . \tag{6.23}$$

This function is called the Wigner-Ville spectrum. It is the so-called *Weyl symbol* of the covariance operator $\mathcal{C}_f$.

Remark 6.4 The Wigner-Ville spectrum is nothing but the expectation of the Wigner-Ville transform we introduced in Chapter 1. Indeed, according to formula (1.48), we have

$$\mathcal{E}_0(b,\omega) = \mathbb{E}\{W_f(b,\omega)\} .$$

This formula is one of the reasons for the popularity of the Wigner-Ville transform and related tools in the signal processing community.

Remark 6.5 Remarkably enough, there exists a close relationship between the Wigner-Ville spectrum and the Gabor spectrum we discussed earlier. Indeed, a simple calculation shows that the Gabor spectrum is simply a (two-dimensional) smoothed version of the Wigner-Ville spectrum:

$$\mathcal{E}_{f,CGT}(b,\omega) = \frac{1}{2\pi||g||^2}\int \mathcal{E}_0(u,\xi)\overline{W_g(u-b,\xi-\omega)}dud\xi \ , \tag{6.24}$$

where $W_g(b,\omega)$ is the Wigner-Ville transform of the window function g introduced in (1.48). This remark goes back to early works of Flandrin and collaborators (see, e.g., [Flandrin93] and [196]). It suggests that one should use weighted spectra to estimate Wigner-Ville spectra, as they are generally easier to implement and interpret. Note that the wavelet spectrum is also related to Wigner-Ville spectrum, but the relation is not as natural. A natural generalization of (6.24) requires using the so-called *affine Wigner representation* (see, e.g., [Flandrin93].) We shall not dwell on such details here, the study of families of Wigner distributions being beyond the scope of the present text.

The Wigner-Ville spectrum $\mathcal{E}_0(b,\omega)$ is a potential candidate for the generalized spectral density, but as we have seen it is not the only one. In addition, contrary to the functions e^{ikx} in the stationary case, the generalized spectral density does not diagonalize $\mathcal{C}_f$. However, it may perform such a diagonalization in an approximate sense in some situations. We shall come back to this point when discussing locally stationary processes.

But before going any further in the controversial domain of the spectral theory of non-stationary processes, we find it instructive to revisit the spectral theory of stationary processes in the light of the time-frequency and time-scale transforms introduced in the previous chapters.

6.2 Time-Frequency Analysis of Stationary Processes

Before looking at examples of non-stationary processes, we revisit the spectral theory of stationary processes presented in Chapter 2 to investigate in more detail the role which the time-frequency transforms introduced in Chapters 3 and 4 can play. In the case of stationary processes, the Wigner-Ville spectrum $\mathcal{E}_0(b,\omega)$ defined in formula (6.23) does not depend upon b and it reduces to the usual spectral density $\mathcal{E}_f(\omega)$. But it is interesting to look more closely at the behavior of weighted spectra and of the corresponding estimators. We first consider the case of the Gabor transform.

6.2.1 Gabor Analysis

Let us consider a mean zero wide-sense stationary stochastic process $f = \{f(x);\ x \in \mathbb{R}\}$ of order 2; let us assume that it has an absolutely continuous spectrum; and let us denote by $\mathcal{E}$ its spectral density. For simplicity we assume that $f(x)$ is real-valued. Its spectral representation has the form

$$f(x) = \frac{1}{\sqrt{2\pi}} \int e^{ikx} \sqrt{\mathcal{E}(k)}\, W(dk)\ , \tag{6.25}$$

for some orthogonal white noise L^2 measure W. Let $g \in L^2(\mathbb{R})$ be a window function. We assume that $\|g\| = 1$ for the sake of simplicity. If we use the representation (6.25) to compute the Gabor transform of f, we get

$$\begin{aligned} G_f(b,\omega) &= \frac{1}{\sqrt{2\pi}} \int \left(\int e^{ikx} \sqrt{\mathcal{E}(k)} W(dk) \right) \overline{g(x-b) e^{-i\omega(x-b)}} dx \\ &= \frac{1}{\sqrt{2\pi}} \int e^{ikb} \sqrt{\mathcal{E}(k)} \left(\int e^{i(k-\omega)x} \overline{g(x)} dx \right) W(dk) \\ &= \frac{1}{\sqrt{2\pi}} \int_{-\infty}^{\infty} e^{ikb}\ \overline{\hat{g}(k-\omega)}\ \sqrt{\mathcal{E}(k)} W(dk). \end{aligned} \tag{6.26}$$

Notice that we interchanged the stochastic integration with respect to the white noise measure $W(dk)$ and the integral defining the Gabor transform (see Figure 6.1). It is not difficult to justify this interchange (proving a Fubini's type result for stochastic integrals) simply by using the definition of the stochastic integral reviewed in Chapter 2. This computation gives directly the spectral representation of the process $\{G_f(b,\omega);\ b \in \mathbb{R}\}$ for each fixed frequency ω. This representation shows that this process is stationary and that its power spectrum $\mathcal{E}_\omega(k)$ can be obtained by multiplying the power spectrum of the original process $f(x)$ by the quantity $|\hat{g}(k-\omega)|^2$:

$$\mathcal{E}_\omega(k) = \mathcal{E}(k) |\hat{g}(k-\omega)|^2. \tag{6.27}$$

Then, it is clear that as long as the Fourier transform of the window is still localized near the origin, the Gabor transform provides, for each fixed ω, information on the part of the original signal which comes from the frequency contributions localized near ω. From there one can recover the time

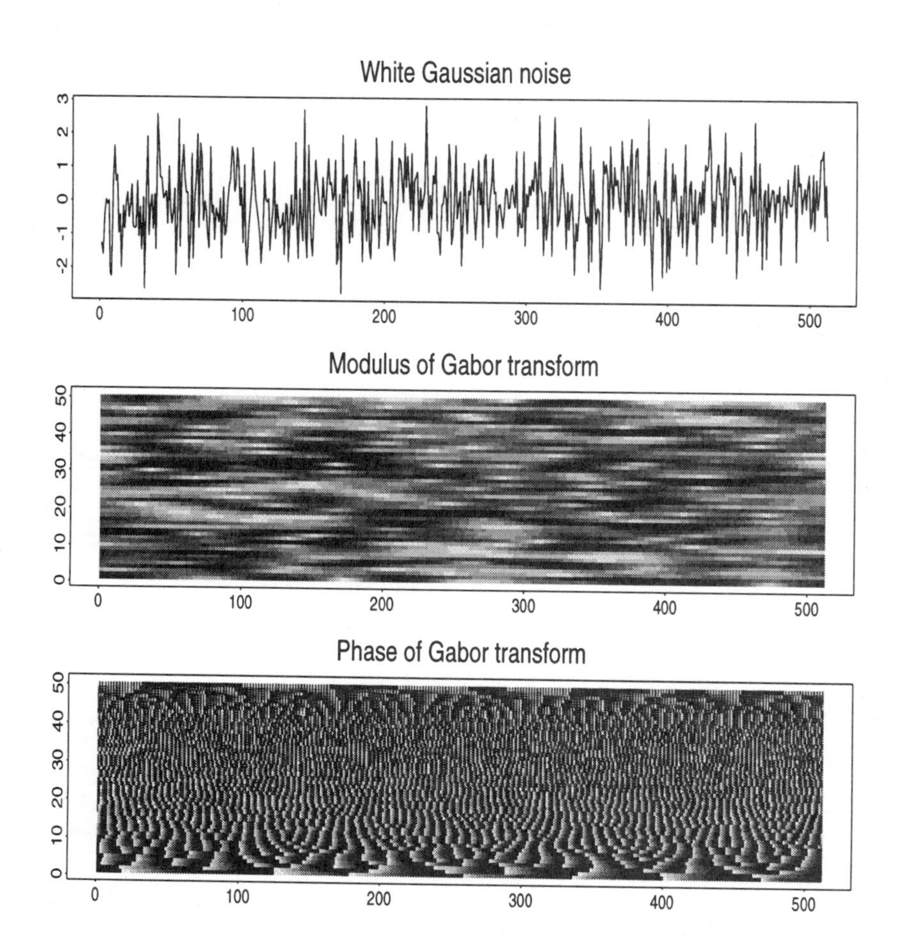

Figure 6.1. Plots of the modulus and the phase of the CGT of a Gaussian white noise signal (with unit variance), as generated in Example 6.2.

domain statistics of the process $G_f(\cdot, \omega)$. In particular, one gets its autocovariance function $C_\omega(b)$ by inverse Fourier transform of the spectrum, i.e.,

$$C_\omega(b) = \frac{1}{2\pi}\int_{-\infty}^{\infty} e^{-ikb}|\hat{g}(k-\omega)|^2\mathcal{E}(k)dk\ , \tag{6.28}$$

and the common variance is nothing but the Gabor spectrum we already introduced (recall that we suppose throughout this section that $||g|| = 1$), which in this case depends only upon the frequency variable:

$$\mathbb{E}\{|G_f(b,\omega)|^2\} = C_\omega(0) = \frac{1}{2\pi}\int_{-\infty}^{\infty}|\hat{g}(k-\omega)|^2\mathcal{E}(k)dk\ . \tag{6.29}$$

This equation shows that the common variance of the values of the CGT (when the frequency is frozen at ω) is a weighted average of the spectrum $\mathcal{E}(k)$ when k is near ω.[1] This suggests that the expected value of the square modulus of the Gabor transform could provide a reasonable estimate of the spectral density if one could find a way to compute this expectation from a sample realization of the signal. The previous discussion applies to any wide-sense stationary process of order 2. We now assume that the process is strictly stationary and ergodic. Using a simple ergodic argument as in the discussion of the Wiener's approach to spectral theory, we see that

$$\mathbb{E}\{|G_f(b,\omega)|^2\} = \lim_{B\to\infty}\frac{1}{B}\int_{-B/2}^{B/2}|G_f(b,\omega)|^2 db, \tag{6.30}$$

almost surely in the realization of the random function f. Consequently, in order to estimate the spectral density $\mathcal{E}_f(\omega)$ from a sample signal we choose a large B and we compute

$$V_B(\omega) = \frac{1}{B}\int_{-B/2}^{B/2}|G_f(b,\omega)|^2 db\ . \tag{6.31}$$

This estimator will be called the Gabor spectral function. It is biased since because of formula (6.29) we have

$$\begin{aligned}\mathbb{E}\{V_B(\omega)\} &= \frac{1}{B}\int_{-B/2}^{B/2}\mathbb{E}\{|G_f(b,\omega)|^2\}db \\ &= C_\omega(0) \\ &= \frac{1}{2\pi}\int_{-\infty}^{\infty}|\hat{g}(k-\omega)|^2\mathcal{E}(k)dk\ ,\end{aligned} \tag{6.32}$$

[1] Again, we assume that the Fourier transform of the window function g is localized near the origin of the frequency axis.

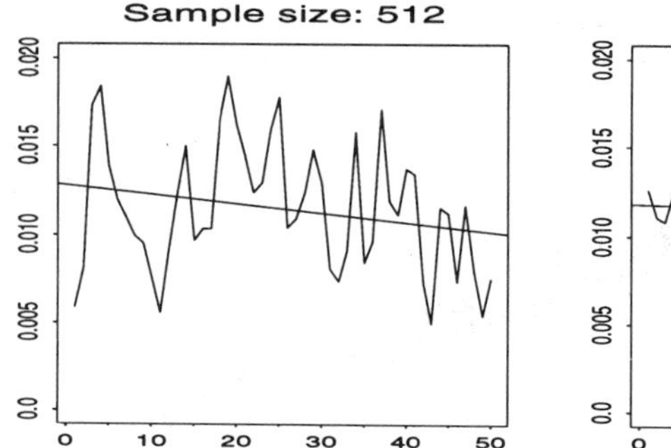

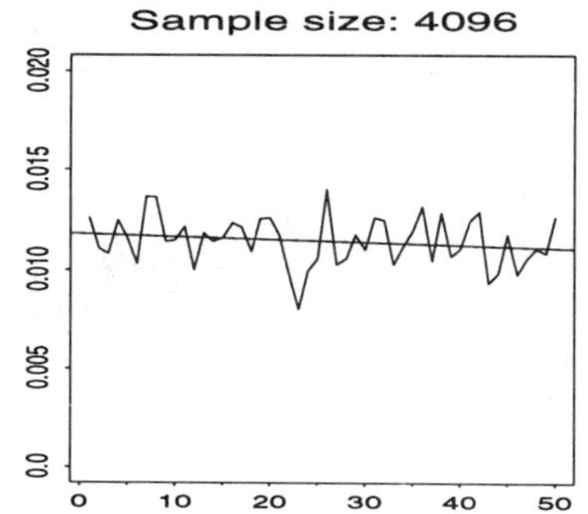

Figure 6.2. Plot of $V_B(\omega)$ defined in (6.31) as a function of ω, for Gaussian white noises with sizes 512 and 4096 (generated as in Example 6.3).

which is a weighted average of the values of $\mathcal{E}(k)$. Notice the similarity with the discussion of the statistical properties of the periodogram in Chapter 2. As in the case of the periodogram, such a smoothing reduces the variability of the estimate but it introduces a bias and a modicum of care is necessary to strike a reasonable balance between variance and bias of the estimate.

An example of power spectrum estimation in the case of white noise is given in Figure 6.2, where $V_B(\omega)$ is displayed as a function of ω for two very different values of the sample size. The plot in the case of the 512-sample example shows a more erratic behavior. We superimposed least square regression lines and the first one has a negative slope. This is another artifact of the sample size. The situation is much better in the case of a sample of size 4096. Indeed, not only are the fluctuations in the estimator much smaller, but the regression line is essentially horizontal (see following discussion for details). We shall use such estimations when discussing denoising algorithms.

Remark 6.6 Note, nevertheless, that the estimator $V_B(\omega)$ given in formula (6.31) turns out to be unbiased in some very specific situations. Indeed, let us assume that the spectral density $\mathcal{E}(k)$ is locally exponential in the sense that $\mathcal{E}(k) = Ae^{\lambda k}$ for some constants $A > 0$ and λ as long as $k \in \Omega = [\omega_1, \omega_2]$, and let us assume that $g(x)$ is band-limited with $Supp(\hat{g}) \subset [-k_0, k_0]$ with $k_0 \leq (\omega_2 - \omega_1)/2$. Then, if $\omega \in [k_0 - \omega_2, k_0 - \omega_1]$, we have

$$\mathbb{E}\{|G(b,\omega)|^2\} = e^{\lambda\omega} \int_{-k_0}^{k_0} |\hat{g}(k)|^2 e^{\lambda k} dk.$$

Therefore, the spectral estimator $V_B(\omega)$ can be unbiased in such a region. However, we have to stress that such a property only holds for a very limited

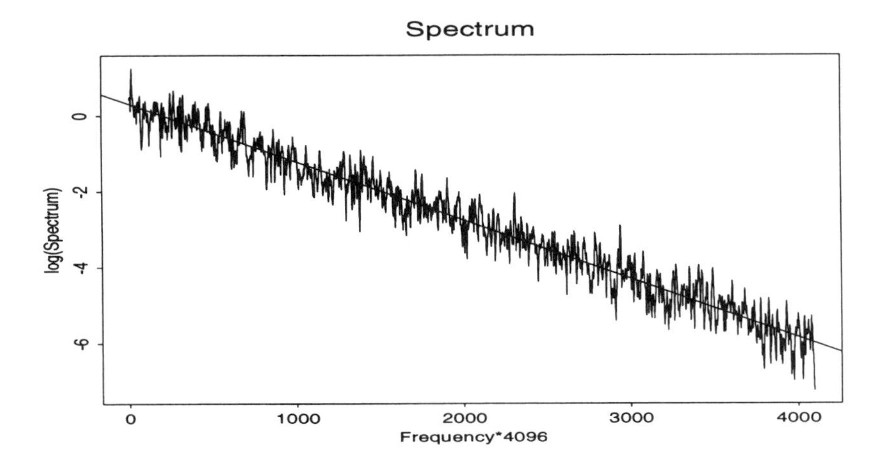

Figure 6.3. Periodogram of the simulated signal with an exponential spectrum.

class of spectral densities. Figure 6.3 gives an example of the estimation of the spectrum of such a process by the periodogram, while Figure 6.4 gives the plot of the "unbiased" Gabor spectrum estimator V_B.

Second-Order Moments of $V_B(\omega)$ for Gaussian Processes

As we noticed in our discussion of the standard spectral estimation reviewed in Chapter 2, the control of the bias is only half of the story: the control of the fluctuations (and in particular of the variance of the estimator) is the second major ingredient in the theory. In the present situation we are able to quantify these variations in a specific mode. Notice that

$$\mathbb{E}\{V_B(\omega)V_B(\omega')\} = \frac{1}{B^2}\int_{-B/2}^{B/2}\int_{-B/2}^{B/2} \mathbb{E}\{|G_f(b,\omega)|^2|G_f(b',\omega')|^2\}\, dbdb' \tag{6.33}$$

and using (6.26) one can see that the second moments of $V_B(\omega)$ can be computed if we can compute

$$\mathbb{E}\{W(dk)\overline{W(dk')}W(dk'')\overline{W(dk''')}\}. \tag{6.34}$$

This cannot be done in general, so from now on we assume that the process $f = \{f(x);\ x \in \mathbb{R}\}$ is Gaussian. In this case the $W(A)$'s form a (possibly complex) Gaussian system of mean zero random variables and we can use

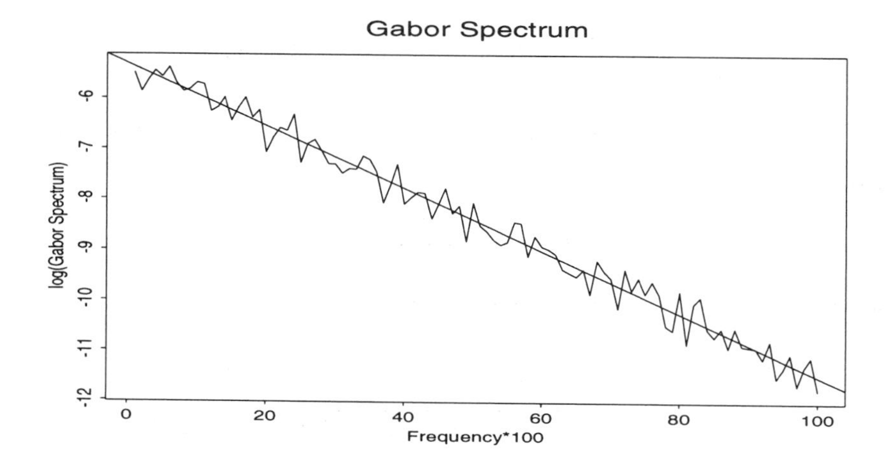

Figure 6.4. Gabor spectrum of the same signal as in Figure 6.3.

the fact that

$$\begin{array}{c}\mathbb{E}\{X_1X_2X_3X_4\} = \mathbb{E}\{X_1X_2\}\mathbb{E}\{X_3X_4\} + \mathbb{E}\{X_1X_3\}\mathbb{E}\{X_2X_4\} \\ +\mathbb{E}\{X_1X_4\}\mathbb{E}\{X_2X_3\}\end{array} \tag{6.35}$$

whenever X_1, X_2, X_3, and X_4 are mean zero jointly Gaussian complex random variables with an arbitrary covariance matrix. This formula is often referred to as Isserlis formula. Using this formula in (6.34) gives

$$\begin{aligned}&\mathbb{E}\{W(dk)W(dk')W(dk'')W(dk''')\} \\ &\quad = [\delta(k-k')\delta(k''-k''') + \delta(k-k'')\delta(k'-k''') \\ &\quad + \delta(k-k''')\delta(k'-k'')]dk\,dk'\,dk''\,dk'''\end{aligned} \tag{6.36}$$

and we can complete the computation of (6.33). In fact, a direct calculation shows that

$$\mathbb{E}\{V_B(\omega)\overline{V_B(\omega')}\} = C_\omega(0)C_{\omega'}(0) + \frac{1}{B^2}\int_{-B/2}^{B/2}\int_{-B/2}^{B/2} \left(|\mathbb{E}\{G_f(b,\omega)\overline{G_f(b',\omega')}\}|^2 + |\mathbb{E}\{G_f(b,\omega)G_f(b',\omega')\}|^2\right) dbdb' ,$$

and using the consequence of Isserlis formula discussed earlier, one sees that

$$\begin{aligned}&\frac{1}{B^2}\int_{-B/2}^{B/2}\int_{-B/2}^{B/2} |\mathbb{E}\{G_f(b,\omega)G_f(b',\omega')\}|^2\, dbdb' \\ &= \frac{1}{4\pi^2}\int K_B(\xi-\zeta)^2\mathcal{E}(\xi)\mathcal{E}(\zeta)\overline{\hat{g}(\xi-\omega)}\hat{g}(\zeta-\omega)\overline{\hat{g}(-\xi-\omega')}\hat{g}(-\zeta-\omega')d\xi d\zeta,\end{aligned}$$

$$\frac{1}{B^2}\int_{-B/2}^{B/2}\int_{-B/2}^{B/2}|\mathbb{E}\{G_f(b,\omega)\overline{G_f(b',\omega')}\}|^2\,dbdb'$$

$$=\frac{1}{4\pi^2}\int K_B(\xi-\zeta)^2\mathcal{E}(\xi)\mathcal{E}(\zeta)\overline{\hat{g}(\xi-\omega)}\hat{g}(\xi-\omega')\hat{g}(\zeta-\omega)\overline{\hat{g}(\zeta-\omega')}d\xi d\zeta,$$

where

$$K_B(\xi)=\frac{\sin(\xi B/2)}{\xi B/2} \tag{6.37}$$

is the cardinal function. In particular, we have

$$\mathrm{Cov}\{V_B(\omega)\overline{V_B(\omega')}\}=\frac{1}{4\pi^2}\int K_B(\xi-\zeta)^2\mathcal{E}(\xi)\mathcal{E}(\zeta)\overline{\hat{g}(\xi-\omega)}\hat{g}(\zeta-\omega')$$
$$\left(\hat{g}(\xi-\omega')\overline{\hat{g}(\zeta-\omega')}+\overline{\hat{g}(-\xi-\omega')}\hat{g}(-\zeta-\omega')\right)d\xi d\zeta.$$

We clearly see from such a result that the frequency localization of the estimator is fixed by the decay of the Fourier transform of the window, and not by that of the Fejér kernel as in the case of the periodogram. This is reminiscent of the situation encountered when we discussed the Welsh-Bartlett spectral estimators. The Gabor spectrum is very close to the Welsh-Bartlett estimator.

The Case of White Noise

We saw that, for each (fixed) frequency ω, the CGT of a stationary process is stationary, but, as the following example of a white noise shows, this CGT is correlated even when the original signal is uncorrelated. We will see in Subsection 6.2.2 that this property is shared by the continuous wavelet transform. This is an important feature of linear time-frequency representations.

The present discussion is only formal because of the fact that a continuous time white noise cannot have sample functions, but it is rigorous for finite sequences and consequently for samples of the theoretical white noise which we would like to consider. Anyway, at least at a formal level, we have

$$C_f(x)=\sigma^2\delta(x) \qquad \text{and} \qquad \mathcal{E}(k)\equiv\sigma^2\ ,$$

and since the power spectrum of white noise is constant, formula (6.27) implies that the power spectrum of $G_f(\cdot,\omega)$ is given by

$$\mathcal{E}_\omega(k)=\sigma^2|\hat{g}(k-\omega)|^2\ .$$

Consequently,

$$\mathbb{E}\{V_B(\omega)\}=\frac{1}{||g||^2}\mathbb{E}\{|G_f(b,\omega)|^2\}=\frac{\sigma^2}{2\pi||g||^2}\int|\hat{g}(k-\omega)|^2\,dk=\sigma^2\ ,$$

which shows that the spectral estimator $V_B(\omega)$ is unbiased in this case. In addition, the *cross covariance* $C_{\omega,\omega'}(b-b') = \mathbb{E}\{G_f(b,\omega)\overline{G_f(b',\omega')}\}$ is given by

$$C_{\omega,\omega'}(b-b') = \langle g_{(b',\omega')}, g_{(b,\omega)}\rangle = 2\pi\sigma^2 ||g||^2 \mathcal{K}_g(b,\omega;b',\omega') ,$$

which proves that the CGT of a white noise is a correlated process, and the correlations are governed by the reproducing kernel of the window $g(x)$ (see equation (3.6) for the expression of the reproducing kernel). Examples of power spectrum estimation with CGT are given in Figure 6.2 (white noise case) and Figure 6.4.

The discussion above contains the main reasons why statisticians have concentrated primarily on expansions in orthogonal bases and not on continuous transforms. Indeed, even though the latter are clearly important and well worth attention, their careful analysis is a difficult problem to tackle because of the correlations inherent in the nature of the continuous transforms.

6.2.2 Wavelet Analysis

The CWT shares some of the essential features of the CGT when it comes to the analysis of stationary processes. As in the Gabor case, we suppose $||\psi|| = 1$ to simplify the equations. Consider as before the example of a mean zero second-order stationary process $\{f(x); x \in \mathbb{R}\}$ with autocovariance function $C_f(x) = \mathbb{E}\{f(x)f(0)\}$ and common variance $\sigma^2 = C_f(0)$. Taking into account the spectral representation (2.16), it is easy to derive along the lines of the analysis of the CGT case, the following expressions for the spectral representation of $T_f(b,a)$:

$$\begin{aligned} T_f(b,a) &= \frac{1}{a}\int_{-\infty}^{\infty} \overline{\psi\left(\frac{x-b}{a}\right)} f(x)dx & (6.38)\\ &= \frac{1}{\sqrt{2\pi}}\int_{-\infty}^{\infty} e^{ikb}\overline{\hat{\psi}(ak)}\sqrt{\mathcal{E}(k)}\, W(dk) , & (6.39)\end{aligned}$$

and the autocovariance and common variance of $T_f(b,a)$:

$$C_{a,a'}(b-b') = \mathbb{E}\{T_f(b,a)\overline{T_f(b',a')}\} = \int_{-\infty}^{\infty} e^{ik(b-b')}\mathcal{E}(k)\overline{\hat{\psi}(ak)}\hat{\psi}(a'k)dk , \quad (6.40)$$

$$\mathbb{E}\{|T_f(b,a)|^2\} = \frac{1}{2\pi}\int_{-\infty}^{\infty} \mathcal{E}(k)|\hat{\psi}(ak)|^2 dk = \frac{1}{2\pi}\frac{1}{a}\int_{-\infty}^{\infty} \mathcal{E}\left(\frac{k}{a}\right)|\hat{\psi}(k)|^2 dk . \quad (6.41)$$

Again, the common variance is nothing but the wavelet spectrum introduced above in (6.21). Then, like the CGT, the fixed scale CWT of a stationary process is still stationary. In fact, formula (6.40) shows that the cross covariance of the CWT at two different scales depends only upon the difference $b-b'$. This property is usually called *cross stationarity*. A closer look at the derivation of (6.40) shows that this cross stationarity is in fact equivalent to the (weak) stationarity of the process f. But this transform is in general correlated, even when the process f is not. Take, for instance, the case of a white noise process with variance σ^2. In such a case, we have

$$\mathbb{E}\{T_f(b,a)\overline{T_f(b',a')}\} = \langle \psi_{(b',a')}, \psi_{(b,a)} \rangle = c_\psi \mathcal{K}_\psi(b,a;b',a') \qquad (6.42)$$

and as before, the correlations are governed by the reproducing kernel of the wavelet. But for fixed scale a (still in the case of white noise),

$$\sigma_a^2 = \mathbb{E}\{|T_f(b,a)|^2\} = \frac{\sigma^2}{a}\|\psi\|^2, \qquad (6.43)$$

which is not a constant any longer. This $1/a$ behavior is due to the fact that the frequency spread of wavelets is larger for high frequencies than for low frequencies, and to the normalization we have chosen for the wavelet transform.

As an example, we show in Figure 6.5 the continuous wavelet transform of a Gaussian white noise, generated in Example 6.4. It may be seen on the image of the modulus that the size of the correlations is a decreasing function of the scale (actually given by (6.41).)

Equations (6.40) and (6.41), together with standard ergodic arguments, suggest the use of

$$\tilde{V}_B(a) = aV_B(a) = \frac{a}{B}\int_{-B/2}^{+B/2} |T_f(b,a)|^2 db \qquad (6.44)$$

as an estimate for the power spectrum. This estimator $\tilde{V}_B(a)$ will be called the wavelet spectral function. We have that

$$\lim_{B\to\infty} \tilde{V}_B(a) = \frac{a}{2\pi}\int \mathcal{E}(k)|\hat{\psi}(ak)|^2 dk = \frac{1}{2\pi}\int_{-\infty}^{\infty} \mathcal{E}\left(\frac{k}{a}\right)|\hat{\psi}(k)|^2 dk \ .$$

Note that the right-hand side of this last equation is nothing but the wavelet spectrum introduced in (6.21), up to a constant factor (equal to one if one assumes $||\psi|| = 1$ as we did). We then have a "smoothed" but biased estimate. Here the smoothing reduces the variance of the estimate (as does standard tapering) but introduces bias. As usual, we face the classical problem of the bias-variance trade-off.

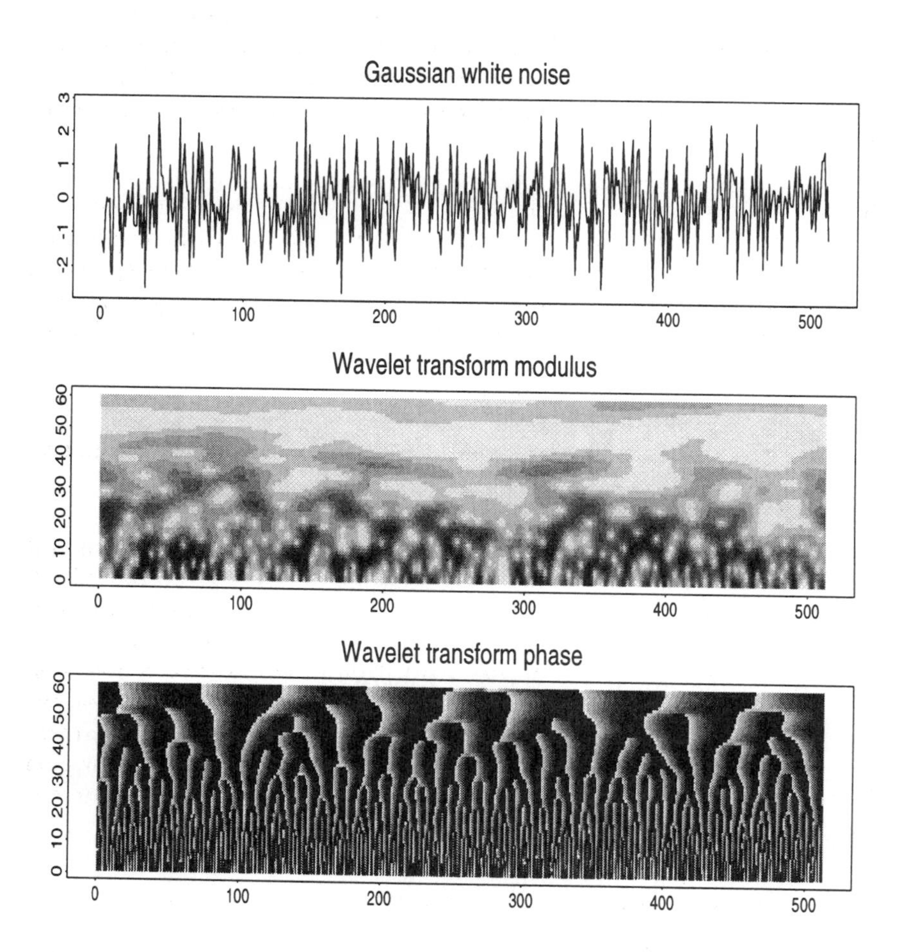

Figure 6.5. Modulus and phase of the CWT of a Gaussian white noise signal (with unit variance), generated in Example 6.4.

The plot of another white noise time series is given later, in the top left corner of Figure 6.8. On the same top row one can see the periodogram and the wavelet spectral estimate $V_B(a)$. The plot of the periodogram looks different from the plots given in Figure 2.3 of Chapter 2 because its values are plotted against the logarithm of the frequency ω. The wavelet spectral estimate is plotted against the logarithm of the scale, and since $V_B(a)$ is proportional to $1/a$ it is not a surprise to see a straight line of slope -1.

The commands needed to produce these three plots are given in Example 6.6.

Remark 6.7 Remark 6.6 has the following counterpart in the wavelet case. $V_B(a)$ (see Equation (6.44)) turns out to yield an unbiased estimator in specific situations. Indeed, assuming that $\mathcal{E}(k)$ is locally a power law spectrum in the sense that $\mathcal{E}(k) \sim k^\lambda$ for some constant λ for $k \in \Omega = [\omega_1, \omega_2]$ and that $\psi(x)$ is band-limited, then, for suitably chosen values of ω, we have

$$\mathbb{E}\{\tilde{V}_B(a)\} \sim a^\lambda ,$$

for a in a given interval $[a_1, a_2]$ whose bounds depend on ω_1 and ω_2 and on the support of $\hat{\psi}$.

Examples of wavelet spectrum estimation based on the wavelet spectral function $V(a)$ are given later in Figures 6.8 and 6.9. In the top row of 6.8, the analyzed signal was a Gaussian white noise, and we consider two different realizations of unequal lengths. The plot of the logarithm of $V(a)$ as a function of a shows the power law behavior, which is of course seen more easily in the case of the longer time series for which the fluctuations are smaller and for which the regression line gives a better fit. The comments made earlier in the discussion of the influence of the sample size on the properties of the CGT based spectral estimator apply to the present situation. A more quantitative analysis is given in Example 6.4.

Second-Order Moment of $\tilde{V}_B(a)$ for Gaussian Processes

As before, we compute the second-order statistics of our spectral estimator $\tilde{V}_B(a)$ when the process $f(x)$ is assumed to be Gaussian. A simple calculation, similar to the one performed in the Gabor case and therefore not reproduced here, shows that

$$\mathrm{Cov}\{\tilde{V}_B(a), \tilde{V}_B(a')\} = \frac{a^2}{4\pi^2} \int K_B(\xi - \zeta)^2 \mathcal{E}(\xi)\mathcal{E}(\zeta)\overline{\hat{\psi}(a\xi)}\hat{\psi}(a\zeta)$$
$$\left(\hat{\psi}(a'\xi)\overline{\hat{\psi}(a'\zeta)} + \hat{\psi}(-a'\xi)\overline{\hat{\psi}(-a'\zeta)}\right) d\xi d\zeta , \qquad (6.45)$$

and

$$\mathrm{Var}\{\tilde{V}_B(a)\} = \frac{a^2}{4\pi^2} \int K_B(\xi - \zeta)^2 \mathcal{E}(\xi)\mathcal{E}(\zeta)\overline{\hat{\psi}(a\xi)}\hat{\psi}(a\zeta)$$
$$\left(\hat{\psi}(a\xi)\overline{\hat{\psi}(a\zeta)} + \hat{\psi}(-a\xi)\overline{\hat{\psi}(-a\zeta)}\right) d\xi d\zeta \ . \tag{6.46}$$

Note again that the frequency resolution of such quantities are governed by the frequency localization of the wavelet instead of that of the cardinal function $K_B(\xi)$ defined in (6.37).

6.2.3 Self-Similarity of WAN Traffic

The development of models for local area network (LAN for short) traffic and their mathematical analysis has been a very important and very active field of research in the last decade. Some of the most spectacular progress in this area has been linked to the discovery of long-range dependence and self-similarity of the processes. Strong evidence of this self-similarity of the LAN traffic can be found in [204], for example. A stochastic process $f = \{f(x);\ x \in \mathbb{R}\}$ is said to be self-similar (with Hurst exponent H) if for every $\lambda > 0$ its distribution is the same as the distribution of the scaled process $f_\lambda = \{f_\lambda(x);\ x \in \mathbb{R}\}$ defined by

$$f_\lambda(x) = \lambda^{1-H} f(\lambda x).$$

This implies that, when f is stationary, its autocovariance function C_f must satisfy

$$C_f(x) = \lambda^{2(1-H)} C_f(\lambda x)$$

for all $x \in \mathbb{R}$ and all $\lambda > 0$. In turn, this scaling property of the autocovariance function implies that the spectral density $\mathcal{E}_f$ (whose existence we assume for the moment) is a power law in the sense that

$$\mathcal{E}_f(\xi) = c_f |\xi|^{1-2H} \tag{6.47}$$

for some constant $c_f > 0$ and all $\xi \in \mathbb{R}$. This is very disappointing because such a power function cannot be integrable both at ∞ and at 0 simultaneously. In other words, such a form (6.47) cannot give the spectral density of a bona-fide stationary process. Indeed, it is not possible to have a stationary process with a spectral density satisfying (6.47) and such that the sample realizations $x \hookrightarrow f(x)$ are functions. The only way to construct such a process would be to realize it in a space of Schwartz distributions, and this would take us too far from realistic applications. So we shall only require self-similarity near the origin of frequencies. In other words, we shall say that the process f is asymptotically self-similar if

$$\mathcal{E}_f(\xi) = c_f |\xi|^{1-2H} \qquad \text{as} \quad |\xi| \to 0. \tag{6.48}$$

Abelian-Tauberian arguments can be used to give an equivalent formulation in term of the decay as $|x| \to \infty$ of the autocovariance function $C_f(x)$. This is why the property (6.48) is sometimes called *long-range dependence with Hurst parameter H.*

Notice that this long-range dependence could be misinterpreted as a contradiction with the well-known independence assumption of most of the Poisson-based queuing models used in traffic networks. In fact, there is no dilemma because of the differences between the two phenomena which are modeled. The Poisson-based queuing models are used in telephony to explain the structure of (individual) call arrivals, whereas we are working here at the level of packet and byte traffic.

Note also that there is an equivalent form of the asymptotic self-similarity in terms of a scaling invariance in distribution very much in the spirit of the way strict self-similarity was defined earlier. The details are irrelevant to the present discussion, so we shall not elaborate on that equivalent form.

In [8], Abry and Veitch used orthonormal wavelet bases to check the asymptotic behavior (6.48) for measured Ethernet LAN traffic. Here we use the continuous wavelet transform and the wavelet spectral estimate $\tilde{V}_B$ introduced earlier to reproduce the results of the analysis of [127] on the scaling behavior of measured wide-area network (WAN for short) traffic traces. Over the past several years the World Wide Web (WWW for short) has become the main application for WAN traffic, and the analysis of the second data set is intended to illustrate the changes between the pre-WWW days and the post-WWW days, at least at the level of the scaling and self-similarity properties of the traffic data. Short forms (though long enough for our purpose) of the data sets are plotted in Figure 6.6. The two original data sets contain 5-hour-long packet level WAN traces from Bellcore measured in January 1990 and in December 1994, respectively. The 1990 trace data give the number of bytes each 1 msec. These measurements pre-date WWW, and `telnet` and `ftp` were the main applications at that time (also, the NSFnet backbone was T1-based, running at 1.5 Mbps). The 1994 WAN trace data give the number of bytes per 10 μsec. They contain approximatively 10% of Web traffic and represent measurements of Internet traffic as the Internet is transitioning to a 45 Mbps backbone. The results presented next are even more significant in more recent data where WWW is the main WAN application and makes up the major portion of the traffic (see [127] for details.)

For each of these time series we computed the periodogram, the Gabor spectrum $V_B(\omega)$, and the wavelet spectrum $V_B(a)$. The results are reproduced in Figure 6.7. The two leftmost columns contain the spectrograms and the Gabor spectra. They are difficult to read and interpret. On the other hand, the wavelet spectra are very clean (we used a (com-

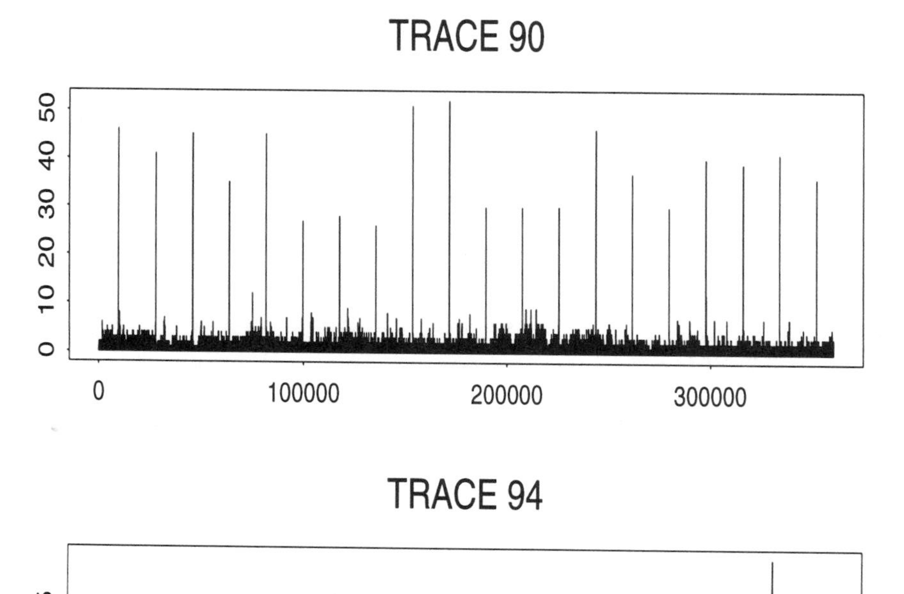

Figure 6.6. 1990 (top) and 1994 (bottom) network traces discussed in the text.

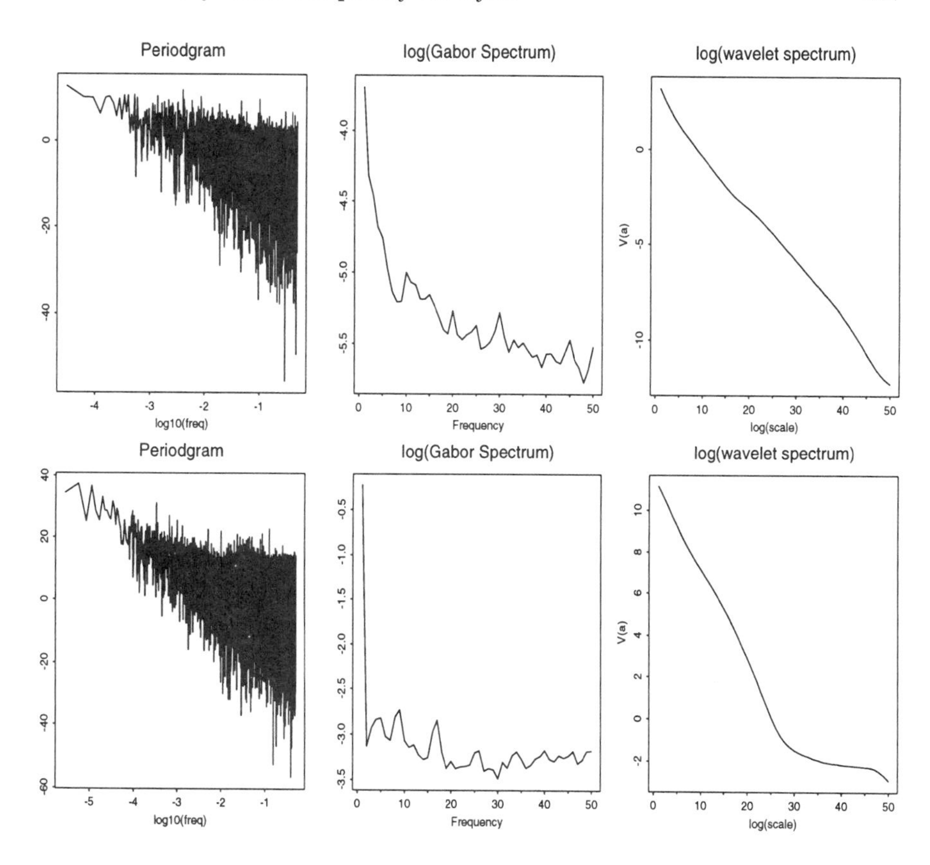

Figure 6.7. Spectral analysis of 1990 (top) and 1994 (bottom) network traces: periodogram(left), Gabor spectrum (middle), and wavelet spectrum (right.)

plexified) derivative of Gaussian). They show that the scaling law in the 1990 traces holds throughout the scale range. This clean linear spectrum makes it possible to estimate the Hurst parameter if one so desires. On the other hand, this spectrum shows a clear break in the case of the 1994 traces. The scaling exponent is different at large scales and at small scales (typically below a few hundreds of milliseconds). The breaking point of the *hockey-stick*-like spectrum is approximately at 500 msec. According to the discussion in [127] this value is robust and irrespective of whether we consider byte counts (as in the application discussed here) or packet counts. This break in the scaling properties of the traffic is a significant feature of the appearance of the Web traffic.

6.3 First Steps toward Non-stationarity

The versatility of wavelet analysis and more generally time-frequency analysis make it possible to consider the spectral analysis of different classes of nonstationary processes. The local nature of time-frequency analysis suggests that a first class of examples could be provided by processes with a spectrum slowly varying with time. A second class comprises processes with stationary increments. Indeed, because of the inherent *differentiation* nature of the wavelet transform, these processes can be analyzed with essentially the same ease as the stationary processes.

6.3.1 Locally Stationary Processes

After a long period of inactivity following the pioneering work of Priestley [248, 249], the problem of local spectrum of locally stationary processes has received a renewal of interest in recent years, mainly because of the growing success of time-frequency representations. We review the main elements of the theory originated by Priestley, and we add some of the more recent developments, at least those which fit nicely in the framework of time-frequency analysis.

Time-Dependent Spectral Representation

The first systematic account of a spectral theory for locally stationary processes is due to Priestley, who proposed considering random processes whose spectral density $\mathcal{E}(\omega) = \mathcal{E}(x,\omega)$ depends upon the time variable x. See, for example, [248, 249]. More precisely, Priestley proposed to study a class of stochastic processes $f = \{f(x);\, x \in \mathbb{R}\}$ which can be defined (at least in the distribution sense) by generalizing the spectral representation (2.16) formula by a formula of the type

$$f(x) = \frac{1}{\sqrt{2\pi}} \int_{-\infty}^{+\infty} e^{i\omega x} D_f(x,\omega) W(d\omega) \ , \qquad (6.49)$$

for some (possibly complex valued) function $D_f(x,\omega)$ of the time variable x and the frequency variable ω. Here, as in the case of the spectral representation of stationary processes, $W(d\omega)$ is a white noise random measure (see Chapter 2 for the notation). As usual, we assume that the mean of the process (which is now a possibly non-constant function of the time variable x) can be studied separately and we restrict ourselves to mean zero processes. The covariance of the process f is then given by the formula

$$\begin{aligned} C_f(x,y) &= \mathbb{E}\{f(x)\overline{f(y)}\} \\ &= \frac{1}{2\pi}\int\int e^{i\omega x}e^{-iy\omega'}D_f(x,\omega)\overline{D_f(y,\omega')}\mathbb{E}\{W(d\omega)\overline{W(d\omega')}\} \\ &= \frac{1}{2\pi}\int\int e^{i\omega x}e^{-iy\omega'}D_f(x,\omega)\overline{D_f(y,\omega')}\delta(\omega-\omega')d\omega d\omega' \\ &= \frac{1}{2\pi}\int_{-\infty}^{+\infty} e^{i\omega(x-y)}D_f(x,\omega)\overline{D_f(y,\omega)}\,d\omega. \end{aligned} \tag{6.50}$$

In particular,

$$\mathbb{E}\{|f(x)|^2\} = C_f(x,x) = \frac{1}{2\pi}\int_{-\infty}^{+\infty} |D_f(x,\omega)|^2 d\omega\ .$$

The positive quantity $|D(x,\omega)|^2$ is called the *evolutionary spectrum* by Priestley. It is still another candidate for a time-varying spectral density $\mathcal{E}(x,\omega)$. Unfortunately, the function $D_f(x,\omega)$ is not uniquely determined by the covariance kernel $C_f(x,y)$ (and hence the second-order statistics of the process). Note in particular that $|D_f(x,\omega)|^2$ differs from the Wigner-Ville spectrum $\mathcal{E}_0(x,\omega)$ we introduced in formula (6.23). For example, $|D_f(x,\omega)|^2$ is positive while $\mathcal{E}_0(x,\omega)$ is generally not. If we recall the definition (6.1) of the covariance operator $\mathcal{C}_f$ associated to the covariance function C_f,

$$[\mathcal{C}_f\varphi](x) = \int C_f(x,y)\varphi(y)dy,$$

and if we introduce the operator $\mathcal{D}_f$ associated to the function $D_f(x,\omega)$ by

$$[\mathcal{D}_f\varphi](x) = \frac{1}{\sqrt{2\pi}}\int D_f(x,\omega)e^{i\omega x}\hat{\varphi}(\omega)d\omega,$$

then formula (6.50) can be reinterpreted by saying that the operator $\mathcal{D}_f$ is a square root of the covariance operator $\mathcal{C}_f$ in the sense that $\mathcal{C}_f = \mathcal{D}_f\mathcal{D}_f^*$. Unfortunately, it is well known that a positive definite symmetric operator such as $\mathcal{C}_f$ may have infinitely many square roots in this sense. This lack of uniqueness led Priestley to many contortions in order to define an unambiguous choice of $D_f(x,\omega)$ as a function of the covariance structure of the process.

Under-spread Processes and Optimal Window Selection

A very different approach was developed by Kozek [196, 197]. The latter relies on a careful analysis of the *ambiguity function* of the process defined by

$$A(\tau,\xi) = \mathbb{E}\left\{ \int f\left(x+\frac{\tau}{2}\right) \overline{f\left(x-\frac{\tau}{2}\right)} e^{-i\xi x} dx \right\} . \tag{6.51}$$

This ambiguity function (sometimes called the *expected ambiguity function*) is nothing but the simplectic Fourier transform of the Wigner spectrum:

$$A(\tau,\xi) = \frac{1}{2\pi} \int \mathcal{E}_0(b,\omega) e^{i(\omega\tau - \xi b)} db d\omega . \tag{6.52}$$

This ambiguity function is essentially a measure of the spread of the process in the time-frequency plane (recall formula (1.50) for the ambiguity function in a deterministic context). Particularly interesting is the class of processes whose ambiguity function is compactly supported. If the support of the ambiguity function $A(\tau,\xi)$ is contained inside a rectangular domain, say, $[-\tau_0/2, \tau_0/2] \times [-\xi_0/2, \xi_0/2]$, such that $\xi_0\tau_0 \ll 2\pi$, the process is called an *under-spread process.*[2] Examples of under-spread processes are given in [196].

As we have seen, the Gabor spectrum may be considered as an appropriate tool for approximating the Wigner-Ville spectrum. Instead of being a shortcoming, the fact that it depends on the choice of a window $g(x)$ may be turned into an asset if the window is adapted to the process. We now describe two possible optimization procedures for choosing the best window g, for the case of under-spread processes. We do not elaborate on the mathematical details of the optimization. We only give formal calculations: the interested reader may check that all the integrals which are written here are well-defined because of the support properties of the ambiguity functions. We assume that the processes we are considering are under-spread, and we briefly review two possible ways of optimizing a window for local spectrum estimation.

Least Square Optimization Ideally, one would like to diagonalize the covariance operator of the process. This is not possible in the framework of Gabor analysis and the best one can expect is to make the covariance "as diagonal as possible" in the Gabor representation. We look for an "almost diagonalization" by stating an ansatz of the form

$$\mathcal{C}_f g_{(b,\omega)} = \lambda(b,\omega) g_{(b,\omega)} + r_{(b,\omega)} , \tag{6.53}$$

for some coefficients $\lambda(b,\omega)$ and for a remainder $r_{(b,\omega)}(x)$ which is to be chosen in such a way that $\langle g_{(b,\omega)}, r_{(b,\omega)} \rangle = 0$. Then, using the fact that

[2] Note that if this is indeed the case, the sampling theorem may be applied to the Wigner-Ville spectrum, yielding a decomposition in terms of time-frequency localized functions.

$\langle \mathcal{C}_f g_{(b,\omega)}, g_{(b,\omega)} \rangle = \mathcal{E}_{f,CGT}(b,\omega)$ (we recall that we have assumed $||g|| = 1$), we have

$$||r_{(b,\omega)}||^2 = ||\mathcal{C}_f g_{(b,\omega)}||^2 - \mathcal{E}_{f,CGT}(b,\omega)^2 . \tag{6.54}$$

In addition, we have that

$$\begin{aligned} \int ||\mathcal{C}_f g_{(b,\omega)}||^2 db d\omega &= \int C_f(y,x)\overline{C_f(y,x')} g_{(b,\omega)}(x)\overline{g_{(b,\omega)}(x')} dx\, dx'\, dy\, db\, d\omega \\ &= 2\pi\, ||\mathcal{C}_f||^2_{HS}\, ||g||^2 \\ &= 2\pi\, ||\mathcal{C}_f||^2_{HS} \end{aligned}$$

Therefore, the window selection criterion

$$\min_{g \in L^2(\mathbb{R}), ||g||=1} \int ||r_{(b,\omega)}||^2 db d\omega$$

is equivalent to

$$\max_{g \in L^2(\mathbb{R}), ||g||=1} \int \mathcal{E}_{f,CGT}(b,\omega)^2 db d\omega \tag{6.55}$$

and we shall call $g_{\text{opt},1}$ the (or at least one) argument of this optimization procedure. Using the relationship between the Gabor spectrum and the Wigner spectrum given in (6.24) and the connection to the ambiguity functions given in formula (6.52), the problem (6.55) is easily seen to be equivalent to

$$g_{\text{opt},1} = \arg \max_{g \in L^2(\mathbb{R}), ||g||=1} \langle |A_0|^2, |A_g|^2 \rangle , \tag{6.56}$$

i.e., the least square optimal window is that whose squared ambiguity function matches best that of the process.

Minimum Bias Optimization As an alternative, Kozek [196] proposes to search for the window which minimizes the bias:

$$\min_{g \in L^2(\mathbb{R}), ||g||=1} \sup_{(b,\omega)\in\mathbb{R}} |\mathcal{E}_0(b,\omega) - \mathcal{E}_{f,CGT}(b,\omega)| . \tag{6.57}$$

This expression may be bounded from above in several ways. For example,

$$\begin{aligned} |\mathcal{E}_0(b,\omega) - \mathcal{E}_{f,CGT}(b,\omega)| &\le \frac{1}{2\pi} \int |A_0(\xi,\tau)(1 - A_g(\xi,\tau)|\, d\xi d\tau \\ &\le \frac{1}{2\pi} ||A_0||\, ||\mathbf{1}_S - A_g|| , \end{aligned}$$

where $\mathbf{1}_S(\xi, \tau)$ is the indicator function of the support of the ambiguity function $A_0(\xi, \tau)$. This suggests the following criterion for window selection:

$$g_{opt,2} = \arg \min_{g \in L^2(\mathbb{R}), ||g||=1} ||\mathbf{1}_S - A_g|| \; . \tag{6.58}$$

Again, this is a criterion based on support properties of the ambiguity function. We refer to [196] for further details on the analysis of its properties.

Families of Windows In practice, it is impossible to search for the optimal window within an infinite family, and one has to restrict the search to easily generated windows. A simple way of doing so amounts to starting from a unique window $g(x)$ and considering other windows generated from $g(x)$ using simple transformations. The simplest such transformations are presumably dilations. Looking for the optimal scale for the window is a data-driven way of deciding whether to use broad-band or wide-band Gabor analysis (see the discussion in Chapter 3). Among other possible transformations, one may think of all natural transformations of the time-frequency plane, including, for example, time-frequency rotations. These natural transformations actually form a group, called the *metaplectic group.*

Locally Stationary Processes

The window selection procedures we just reviewed are of a global nature in the sense that the same window is used to estimate a local spectrum $\mathcal{E}_0(b, \omega)$ (or $\mathcal{E}_{f,CGT}(b, \omega)$) for all values of the time variable b. However, this is not necessarily the most convenient solution, for the rate of change of the local spectrum $\mathcal{E}_0(b, \omega)$ may vary as a function of b. Such a problem was studied in [213] by Mallat, Papanicolaou, and Zhang. The heuristic is the following. As we have seen, the diagonalization of the covariance operator (or more precisely of $\mathcal{C}_f\mathcal{C}_f{}^*$) is generally not an easy task. However, suppose that the function $C_0(u, v)$ is slowly varying in u when v is fixed, and that it is rapidly decaying in v for u fixed. More precisely, suppose that for any x_0, the Wigner spectrum $\mathcal{E}_0(x, \omega)$ varies very little within an interval $[x_0 - \delta/2, x_0 + \delta/2]$. Such a parameter $\delta > 0$ is called the *stationarity length.* In general, its value depends upon x_0.

Let us now consider a C^∞-window function g, compactly supported in the interval $[-\delta/2, \delta/2]$. Then in the same spirit as before, an easy formal computation shows that

$$\mathcal{C}_f g_{(x_0,\omega_0)} \approx \mathcal{E}_0(x_0, \omega_0) g_{(x_0,\omega_0)} \; .$$

Therefore, a suitably chosen Gabor function appears as an *almost eigenfunction* of the covariance operator, which suggests that general Gabor functions may achieve an *almost diagonalization* of the covariance operator and therefore "almost solve" the spectral representation problem for that class of stochastic processes.

However, the stationarity length δ is generally a function of the time variable x_0, which suggests that we should consider more general Gabor functions, for example, functions of the form

$$\frac{1}{\beta(b)} e^{i\omega(x-b)} g\left(\frac{x-b}{\beta(b)}\right),$$

or trigonometric functions involving sine and cosine functions in which the scale is matched to the stationarity length of the process.

Since the authors of [213] were interested in the estimation of the covariance operator and the spectral representation of locally stationary processes, they suggested using the families of local cosine bases of $L^2(\mathbb{R})$ derived by Coifman and Meyer, i.e., families of functions of the form

$$u_{k\nu}(x) = \sqrt{\frac{2}{\ell_k}}\, w_k(x) \cos\left(\frac{\pi(\nu+1/2)}{\ell_k}(x-a_k)\right), \quad k \in \mathbb{Z}, \nu = 0, 1, \ldots.$$

(See the discussion in Section 5.5.2 of Chapter 5 and Theorem 5.8 for more details on the construction of such bases).

In this context a *locally stationary process* is defined to be a process such that there exists a local trigonometric basis such that

1. The matrix elements $\langle C_f u_{k\nu}, u_{k'\nu'}\rangle$ have fast decay with respect to the two indices.
2. The lengths ℓ_k entering the definition of the local trigonometric basis are slowly varying, i.e., such that

   $$\max(\ell_k, \ell_{k'}) \leq C|k-k'|^\alpha \min(\ell_k, \ell_{k'}),$$

   for some constants $C > 0$ and $\alpha < 1$.

It may be shown that non-stationary processes generated through a time-dependent spectral representation of the form (6.49) are locally stationary as long as the symbol $D_f(b,\omega)$ satisfies some smoothness estimates of the type

$$\partial_b^k \partial_\omega^l D_f(b,\omega) \leq C_{kl}\ell(b)^{l-k}$$

for some function $\ell(b)$ satisfying $\inf \ell(b) > 0$ and $|\ell(b) - \ell(b')| \leq C|b-b'|^\mu$ and where $C > 0$ and $\mu \leq 1/2$ are uniform constants.

The estimation procedure proposed in [213] is very much in the spirit of (6.55). It is in fact a local version of the optimization problem (6.55), solved using the "best basis strategy" described in the previous chapter (see Section 5.5.3). Starting from a family (a library) of local trigonometric bases associated with a binary tree (see the discussion in Section 5.5.2), one looks for the one which maximizes an ℓ^2 norm of the matrix elements of the covariance operator $\mathcal{C}_f$ in the local trigonometric basis:

$$\max \sum_{k,\nu} |\langle \mathcal{C}_f u_{k\nu}, u_{k\nu} \rangle|^2. \tag{6.59}$$

As in the best basis search problem which we discussed in Chapter 5, the maximization may be achieved through a dynamic programming algorithm: first estimate the matrix elements of the covariance with respect to all the bases in the tree, then run a "bottom to top" search of the optimal basis. Examples of estimation of local spectrum are given in [213], but the reader used to the standard spectral estimation of stationary processes has to be warned: more than one sample realization of the whole process is needed for the estimation procedures to be used.

6.3.2 Processes with Stationary Increments and Fractional Brownian Motion

An important class of nonstationary processes is provided by the processes with stationary increments. In a certain sense, these processes should be understood as antiderivatives (and more generally *fractional antiderivatives*) of stationary processes. Many of these processes are of great importance because of their scaling properties. Unfortunately, because of these scaling properties, the spectral densities of their derivatives are not integrable and these derivatives have to be interpreted in a *generalized* sense. In other words, the derivatives in question do not make sense in the sense of the classical theory of functions. They have to be understood in the sense of Schwartz distributions and the analysis cannot reduce (modulo integration or derivation) to the analysis of a bona fide stationary process. We first present some theoretical results on fractional Brownian motion (definitely the best known of these processes) and we give a detailed analysis of two examples. The first one is academic. It deals with numerical simulations of a sample path of a standard Brownian motion and of a sample path of a fractional Brownian motion with given Hurst exponent. The second example concerns data from the time evolution of a pixel value in a 128 × 128 Amber AE 4128 InSB focal plane array. These data have been presented in [162, 163] by Hewer and Kuo as a possible candidate for tests of the fractional Brownian motion model, and we reproduce some of the results of [162, 163] while adding the results of our analysis using the machinery of the wavelet spectral estimator developed in this chapter.

Some Theoretical Results

The epitome of processes with stationary increments is the so-called *fractional Brownian motion* (fBm.) A fractional Brownian motion is a mean zero (non-stationary) real Gaussian process $\{f(x);\ x \in \mathbb{R}\}$ with stationary increments, and an autocovariance function of the form

$$\mathbb{E}\{f(x)f(y)\} = \frac{\sigma^2}{2}\left\{|x|^{2h} + |y|^{2h} - |x-y|^{2h}\right\}\ , \tag{6.60}$$

where $\sigma > 0$ and $0 \leq h \leq 1$ is a characteristic exponent of the process. This exponent is often called the Hurst exponent of the process. It is important to control the value of this exponent because it determines the fractal (Haussdorff) dimension of the sample paths of the process. Setting $h = 1/2$ in the definition formula (6.60) gives

$$\mathbb{E}\{f(x)f(y)\} = \sigma^2 \min\{x, y\}, \tag{6.61}$$

which shows that the classical Brownian motion is an fBm with Hurst exponent $h = 1/2$.

In the same way the "process of white noise" is shown to be the derivative of the process of Brownian motion, it is possible to show that the fBm process f has a derivative (in the generalized sense of Schwartz distributions), say, $f' = \{f'(x);\ x \in \mathbb{R}\}$ which can be considered as a stationary mean zero Gaussian process with autocovariance function

$$C_{f'}(x-y) = \sigma^2 h(2h-1)|x-y|^{2h-2}, \qquad x, y \in \mathbb{R},$$

which should be interpreted as

$$C_{f'}(x-y) = \sigma^2 \delta(x-y), \qquad x, y \in \mathbb{R},$$

in the case $h = 2$ of the standard Brownian motion. Equivalently, the distribution of the fractional white noise f' can be characterized by the spectral density:

$$\mathcal{E}_{f'}(\omega) = |\omega|^{1-2h}. \tag{6.62}$$

This form of the spectral density is responsible for the terminology $1/f$ process (or $1/f$ spectrum), the use of which is widespread in the electrical engineering literature. Formula (6.62) can be made rigorous with the appropriate mathematical apparatus. It is of crucial importance for most of the statistical procedures used to estimate the Hurst exponent h are based, in one form or another, on this formula. We shall use it hereafter in its integrated form.

Under some mild assumptions on the analyzing wavelets, it is readily shown that the CWT of a fBm with exponent h satisfies

$$\mathbb{E}\{T_f(b,a)\overline{T_f(b',a')}\} = -\frac{\sigma^2}{2}a'^{2h}\int |\tau|^{2h}\overline{\psi(u)}\psi\left(\frac{a}{a'}u+\tau+\frac{b-b'}{a'}\right)dud\tau\ . \tag{6.63}$$

Again it is interesting to make a few remarks.

1. If we fix $a'=a$, then $\mathbb{E}\{T_f(b,a)\overline{T_f(b',a)}\}$ is a function of $b'-b$ only. More precisely, we have

$$\mathbb{E}\{T_f(b,a)\overline{T_f(b',a)}\} = -\frac{\sigma^2}{2}a^{2h}\int |\tau|^{2h}\overline{\psi(u)}\psi\left(u+\tau+\frac{b-b'}{a}\right)dud\tau\ .$$

   Hence, the fixed scale CWT of a fractional Brownian motion is stationary. In fact, the whole autocovariance is a fixed function of $(b'-b)/a$. This is a reflection of the *self-similarity* of the increments of the fBm.

2. If we now fix $b'=b$, we then get

$$\mathbb{E}\{T_f(b,a)\overline{T_f(b,a')}\} = -\frac{\sigma^2}{2}a'^{2h}\int |\tau|^{2h}\overline{\psi(u)}\psi\left(\frac{a}{a'}u+\tau\right)dud\tau\ .$$

   With a different normalization, i.e., setting $\tilde{T}_f(b,a)=a^hT_f(b,a)$, we see that the autocovariance function is a function of a'/a only. Then, $T_f(b,a)$ appears as a *multiplicatively stationary* function of the scale, or equivalently stationary as a function of $\log a$.

3. If we now consider the case $(b',a')=(b,a)$, we obtain

$$\mathbb{E}\{|T_f(b,a)|^2\} = -\frac{\sigma^2}{2}a^{2h}\int |\tau|^{2h}\overline{\psi(u)}\psi(u+\tau)\,dud\tau\ .$$

   We get a power law behavior as a function of a. Notice also that the variance is independent of b, as in the case of stationary processes. Although there is no *stricto sensu* spectrum associated with fBm (remember that this process is not stationary), one sometimes speaks of a *pseudo-spectrum* of the form $\sigma^2\xi^{-2h-1}$.

Remark 6.8 In fact, putting together the preceding results, it is possible to prove that if f is a fBm of Hurst exponent h, and if we set

$$\tilde{T}_f(b,a) = a^{-h}T_f(b,a),$$

where the wavelet transform is computed with respect to a suitably chosen wavelet, then $\tilde{T}_f(b,a)$ is stationary with respect to the action of the affine group G_{aff} (see the remark on group covariance in Chapter 4), in the sense that we have

$$\mathbb{E}\{\tilde{T}_f(b,a)\overline{\tilde{T}_f(b',a)}\} = \Gamma\left((b',a')^{-1}\cdot(b,a)\right)$$

for some positive definite function $\Gamma(b,a)$ on the affine group.

Application to Spectral Estimation

One of the most important problems for processes with "power-law-type" spectral density is the estimation of the Hurst exponent from one (or several) realization(s) of the process. Namely, given a (pseudo) spectrum of the form

$$\mathcal{E}_f(\omega) = C|\omega|^{-2h-1}\ , \tag{6.64}$$

one wants to estimate the exponent $\alpha = -2h-1$. Classical methods involve standard and tapered periodograms (see the discussion in Chapter 2) and related tools. However, it follows from the preceding discussion that wavelet transforms provide very convenient and efficient alternatives. Indeed, with a spectrum of the type (6.64), one has

$$\mathbb{E}\{|T_f(b,a)|^2\} = C'a_{-\alpha-1}$$

with

$$C' = \frac{C}{2\pi}\int|\hat{\psi}(\omega)|^2|\omega|^\alpha d\omega\ .$$

Therefore, as soon as C' is finite, wavelet coefficients may be used to construct unbiased estimators for α. For example, the functions $V_B(a)$ defined in (6.44) exhibit the same behavior:

$$\mathbb{E}\{V_B(a)\} = \frac{C'}{||\psi||^2}a^{-\alpha-1}\ , \tag{6.65}$$

and a linear fit on

$$\log V_B(a) = \log\left(\frac{C'}{||\psi||^2}\right) - (\alpha+1)\log a \tag{6.66}$$

yields estimates for α.

Remark 6.9 In fact, such estimators may be shown to be unbiased and of minimum variance [6].

Remark 6.10 Once the constant α (and therefore the Hurst exponent h) has been estimated, the other constant of the problem, namely, C, may be estimated as well. Indeed, $\log C'$ is the intercept of the regression line in (6.66), and C is obtained from C' via a multiplication by a constant, depending only on α and the analyzing wavelet $\psi(x)$.

Simulation Tests

The middle (resp. bottom) row of Figure 6.8 gives the plots of a sample path with 8192 (resp. 4096) samples of a Brownian motion process (resp. a fBm process with Hurst exponent $h = 0.2$) on the left, the raw periodogram plotted against the logarithm of the frequency in the middle, and the plot of the wavelet spectral estimator $V_B(a)$ still plotted against the logarithm of the frequency on the right. The S-commands needed to produce the plots of the top row are given in Example 6.6, while the commands for the middle and bottom rows are given in Example 6.7.

The Hurst exponent (which is 1/2 for Brownian motion and which was 0.2 in the case of the fBm simulated for this purpose) is given by the slopes of linear fits. For example, the spectral representation of fBm implies that

$$2h + 1 = -\text{slope}$$

for the slope of the linear fit of the raw periodogram as plotted in these logarithmic axes. Similarly, using the wavelet spectral function, the Hurst exponent can be estimated by the value of the slope in the rightmost plot. Notice that the wavelet spectral estimator is much smoother than the periodogram and that the linear fit is easier and more reliable (even though some boundary effects may occur as in the case of fBm in the bottom row). This is especially true in the light of the large fluctuations in the raw periodogram at high frequencies. In fact, the high-frequency part of the periodograms could not be used for the linear fit. Including these frequencies does cause wrong estimates of h. We used only the first forth of the frequencies for the linear fit. The estimate $\hat{h}$ of the Hurst exponent which we found was $\hat{h} = 0.482$ in the case of the Brownian motion simulation (this is not too far from the target value of 1/2) and $\hat{h} = 0.196$ in the case of fBm (which is reasonably close to 0.2). We used the DOG (derivative of a Gaussian) wavelet to estimate the wavelet spectral functions and, ignoring scales too close to the boundary, we obtained $\hat{h} = 0.495$ in the case of Brownian motion and $\hat{h} = 2.02$ in the case of fBm. These results seem to indicate that the wavelet spectral estimate is a better predictor of the Hurst exponent than the periodogram, but most importantly, it is obviously more robust in that it does not require an ad hoc selection of an interval of low frequencies.

Remark 6.11 For some applications, the behavior of $\hat{\psi}(\omega)$ at low frequencies (i.e., as $\omega \to 0$) is not a crucial issue (essentially when one is interested in intermediate frequencies in signals). This is why the Morlet's wavelet is often used in practice, even though it is not strictly speaking a wavelet. This is *not* the case here, which explains our choice of a different wavelet (here the LOG wavelet).

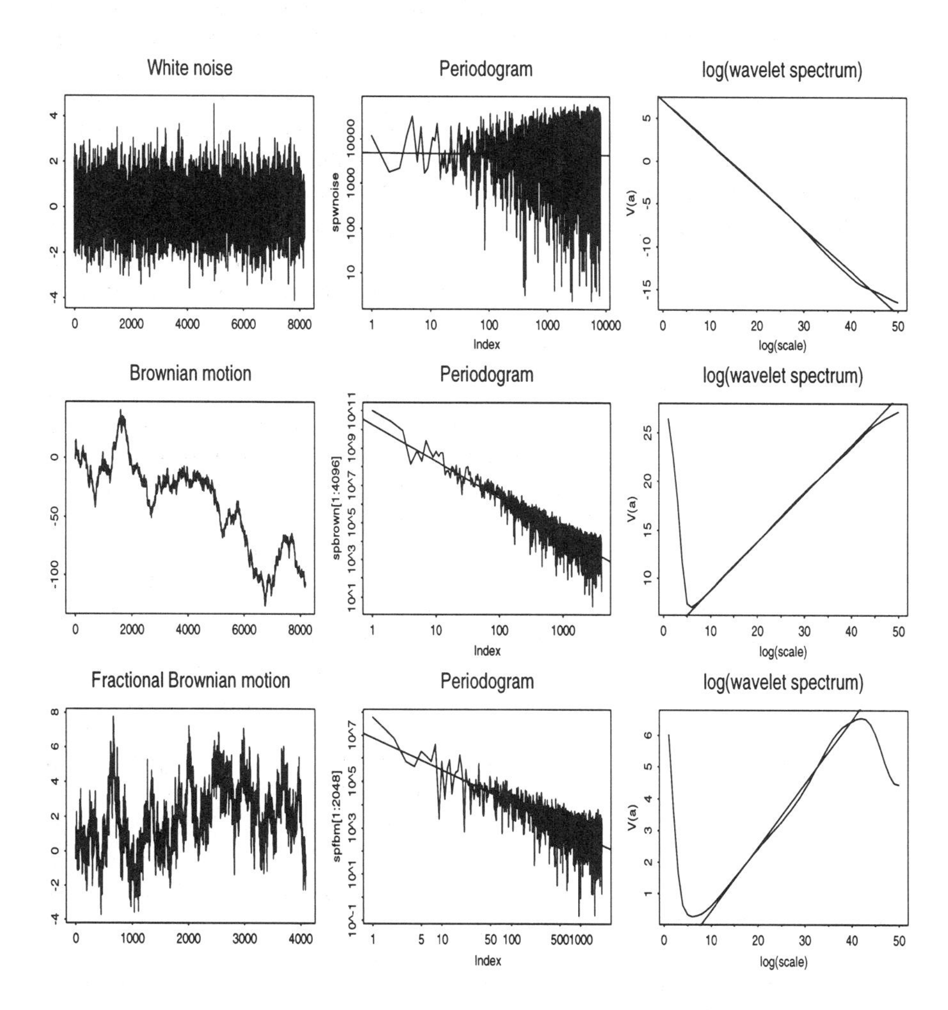

Figure 6.8. Spectral analysis for white noise (top), Brownian motion (middle), and fractional Brownian motion (bottom): time series (left), periodogram (middle), and wavelet spectrum (right.)

More Examples

Both signals analyzed in this subsection were provided by Dr. Gary Hewer from the NAWCWPNS China Lake (California) who analyzed these data in [162] and [163]. The first one is shown in Figure 6.9. It represents 7000 time-sample outputs from an individual pixel on a 128 × 128 Amber AE 4128 InSB focal plane array at a 54-Hz sampling rate.

It was suggested in [162, 163] that a fBm embedded in white noise could be a good model for this data. Indeed, the time series plotted in the leftmost column of Figure 6.9 show some of the features one typically finds in the simulations of fBm. We compute the raw periodogram (which we plot against the logarithm of the frequency) and the wavelet spectral estimator (again plotted against the logarithm of the scale) to try to validate such a model and possibly estimate the Hurst exponent. These plots are given in the middle and rightmost column of Figure 6.9 (see Example 6.9 for the S commands used to generate the plots).

The plots of the periodogram are consistent with the hypothesis that the power spectrum may be modeled as the sum of two power law spectra, one for the fBm and the other for the independent white noise. However, for each of the three time series, estimating the slope of the low-frequency component (as well as that of the high-frequency component) turns out to be fairly tricky because of the irregularity of the spectrum. Moreover, the location of the breakpoint between the two components is extremely difficult to choose, and unfortunately this choice has a great influence on the estimations.

The wavelet spectral functions are much smoother. The interpretation of the spectrum as a sum of two components is not obvious from the wavelet spectrum, since different regions of the spectrum may be observed. For each pixel (i.e., for each time series), we divided the scale axis into three relatively homogeneous intervals and we computed in each of them an estimation for the Hurst exponent. We chose the regions $j \in 5:25$, $j \in 25:35$, and $j \in 39:45$, where the spectrum is relatively well approximated by a power law behavior. Recall that with our convention, $a = 2 \times 2^{j/n_v}$, where n_v is the number of intermediate scales per octave. See Example 6.9 for the S commands used to generate the plots.

Pixels 7 and 9 exhibit a behavior close to white noise $\hat{h} \approx -0.5$ at high frequencies, and a similar behavior in the range 35 : 45. Pixel 8 has a different behavior, as noted by Hewer and Kuo. In all cases, the behavior at low frequency differs from the one obtained using the periodogram, and published in [162, 163]. We could not find a breakpoint which would give us the same estimators $\hat{h}$ as in [162, 163]. However, the estimation from the wavelet spectrum in the scale interval 35 : 50 yields an estimate $\hat{h}$ close to the one obtained using the periodogram. But the power law behavior in

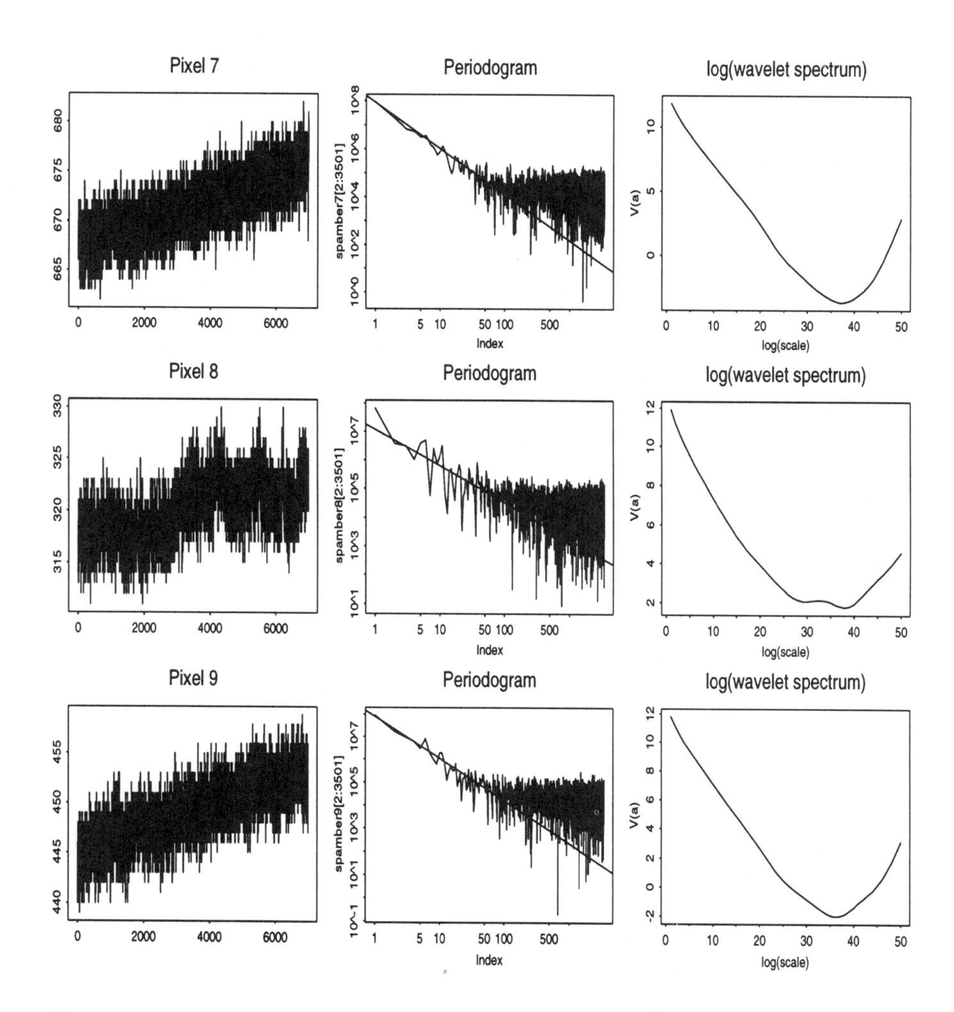

Figure 6.9. Spectral analysis for time series taken from three different pixels of amber camera: time series (left), periodogram (middle), and wavelet spectrum (right.)

that range is somewhat questionable, and at these frequencies, boundary effects are likely to be important.

Summarizing all that, the analysis clearly indicates the presence of at least two components in the power spectrum. The samples are presumably too short to allow the claim that the low-frequency component has a power law spectrum, or even that it is a single component. However, as a time series of length 7000, it may be modeled like that. We summarize the numerical results in the following tables.

Estimates Based on the Raw Periodogram

Frequency range	2:100
pixel 7	-0.46
pixel 8	-0.14
pixel 9	-0.42

Estimates Based on the Wavelet Spectral Function

Scale range	5:25	25:35	39:45	35:50
pixel 7	-0.47	-0.32	0.37	0.42
pixel 8	-0.34	-0.05	0.23	0.19
pixel 9	-0.45	-0.24	0.29	0.33

6.4 Examples and S-Commands

The examples of this chapter were produced with the following S-commands. First we show an example of Gabor analysis of a white noise signal together with the behavior of the associated Gabor spectral function $V_B(\omega)$. The example makes use of the function `tfmean`, which computes the function $V_B(\omega)$.

Example 6.2 White Noise Gabor Transform
Compute the Gabor transform of a white noise sequence, and display modulus and phase. We use $n_f = 50$, $\delta_f = 0.2$, *and scale=25.*

```
> gnoise <- rnorm(512)
> par(mfrow=c(1,3))
> tsplot(gnoise)
> title(``White Gaussian noise'')
> cgtgnoise <- cgt(gnoise,50,.2)
> image(Arg(cgtgnoise))
> title(``Phase of Gabor transform'')
```

Example 6.3 White Noise Gabor Spectral Function
Estimate $\mathbb{E}[|G_f(b,\omega)|^2]$.

```
> par(mfrow=c(1,2))
> gspec <- tfmean(Mod(cgtgnoise)*Mod(cgtgnoise), plot=F)
> tsplot(gspec,ylim=c(0,.02))
> lsline <- lsfit(1:50,gspec)
> abline(lsline$coef)
> title(''Sample size:  512'')
```

Same thing for a longer time series.

```
> gnoise <- rnorm(4096)
> cgtgnoise <- cgt(gnoise,50,.02,25,plot=F)
> gspec <- tfmean(Mod(cgtgnoise)*Mod(cgtgnoise), plot=F)
> tsplot(gspec,ylim=c(0,.02))
> lsline <- lsfit(1:50,gspec)
> abline(lsline$coef)
> title(''Sample size:  4096'')
```

The results are displayed in Figure 6.2. As usual, the larger the sample, the better the estimate.

We now consider the wavelet transform of a white noise signal.

Example 6.4 White Noise Wavelet Transform
Compute the wavelet transform of a white noise signal.

```
> gnoise <- rnorm(512)
> par(mfrow=c(1,3))
> tsplot(gnoise)
> title(''White Gaussian noise'')
> cwtgnoise <- cwt(gnoise,5,12)
> image(Arg(cwtgnoise))
> title(''Phase of wavelet transform'')
```

The results are displayed in Figure 6.5.

Example 6.5 Exponential-type Spectrum: The following commands are used to generate a stationary signal with an exponential spectrum, to estimate this spectrum with the periodogram, and to fit a straight line to the logarithm of this spectrum. The results have been plotted in Figure 6.3 in the text. This example makes use of the utilities `SampleGen`, which generates a stationary time series with prescribed spectrum, and `SpecGen`, which does spectral estimation.

Generate process and estimate spectrum using the periodogram.

```
> lambda <- -1
> expspec <- exp(lambda*2*pi*(0:4095)/4096)
> expran <- SampleGen(expspec)
> estspec <- SpecGen(expran)
> tsplot(log(estspec))
> lsline <- lsfit(0:4096,log(estspec))
> abline(lsline$coef)
> lsline$coef
```

The following commands are used to produce the result of the unbiased estimation procedure based on the computation of $V_B(\omega)$. The result was given in Figure 6.4 in the text.

Estimate the spectrum using CGT.

```
> tfrep <- Mod(cgt(expran[1:4096],100,0.01,100,plot=F))
> gspec <- tfmean(tfrep*tfrep,plot=F)
> tsplot(log(gspec))
> lsgline <- lsfit(1:100,log(gspec))
> abline(lsgline$coef)
```

Example 6.6 White Noise Hurst Exponent: The plots on the top row of Figure 6.8 were produced by the following S-commands. These make use of the two functions `Hurst.est`(estimation of Hurst exponent from CWT) and `wspec.pl` (display wavelet spectrum).

Compare the periodogram and the wavelet spectral estimate.

```
> wnoise <- rnorm(8192)
> tsplot(wnoise)
> spwnoise <- fft(wnoise)
> spwnoise <- Mod(spwnoise)
> spwnoise <- spwnoise*spwnoise
> plot(spwnoise[1:4096],log="xy",type="l")
> lswnoise <- lsfit(log10(1:4096),log10(spwnoise[1:4096]))
> abline(lswnoise$coef)
> cwtwnoise <- DOG(wnoise,10,5,1,plot=F)
> mcwtwnoise <- Mod(cwtwnoise)
> mcwtwnoise <- mcwtwnoise*mcwtwnoise
> wspwnoise <- tfmean(mcwtwnoise,plot=F)
> wspec.pl(wspwnoise,5)
> hurst.est(wspwnoise,1:50,5)
```

The estimation of the Hurst parameter from the periodogram gave the value $\hat{h} = -0.494$ (which is a fairly good approximation of the true value $-1/2$), while the estimation from the wavelet spectral function gave $\hat{h} = -0.498$, which is closer to the true value. Note that the fit was not perfect in the large scale region.

Example 6.7 fBm Hurst Exponent
Periodogram estimate for Brownian motion.

```
> tsplot(brown)
> spbrown <- fft(brown)
> spbrown <- Mod(spbrown)
> spbrown <- spbrown*spbrown
> plot(spbrown[1:4096],log="xy",type="l")
> lsbrown <- lsfit(log10(1:2000),log10(spbrown[1:2000]))
> lsbrown$coef
> abline(lsbrown)
```

The high frequency part of the spectrum could not be used for the regression, because it leads to wrong estimate of h. The result was $\hat{h} = 0.482$.
Wavelet spectral estimate for Brownian motion.

```
> cwtbrown <- DOG(brown,10,5,1,plot=F)
> mcwtbrown <- Mod(cwtbrown)
> mcwtbrown <- mcwtbrown*mcwtbrown
> wspbrown <- tfmean(mcwtbrown,plot=F)
> wspec.pl(wspbrown,5)
> hurst.est(wspbrown,10:40,5)
```

The fine scales could not be used for the fit, because of the importance of boundary effects (strong singularity at boundaries). The result was $\hat{h} = 0.495$.
Periodogram estimate for fBm.

```
> tsplot(fbm)
> spfbm <- fft(fbm)
> spfbm <- Mod(spfbm)
> spfbm <- spfbm*spfbm
> plot(spfbm[1:2048],log="xy",type="l")
> lsfbm <- lsfit(log10(1:500),log10(spfbm[1:500]))
> lsfbm$coef
> abline(lsfbm)
```

The high frequency part of the spectrum could not be used for the regression, because it leads to wrong estimate of h. We obtained $\hat{h} = 0.196$.
Wavelet spectral estimate for fBm.

```
> cwtfbm <- DOG(fbm,10,5,1,plot=F)
> mcwtfbm <- Mod(cwtfbm)
> mcwtfbm <- mcwtfbm*mcwtfbm
> wspfbm <- tfmean(mcwtfbm,plot=F)
> wspec.pl(wspfbm,5)
```

```
> hurst.est(wspfbm,10:40,5)
```

Again, the fine scales could not be used for the fit, because of the importance of boundary effects (strong singularity at boundaries). Our result was $\hat{h} = 0.202$.

Example 6.8 WAN Traces Self-Similarity

Plots of the trace time series.

```
> par(mfrow=c(2,1))
> tsplot(TRACE90)
> title("TRACE 90")
> tsplot(TRACE94)
> title("TRACE 94")
```

The results are shown in Figure 6.6.

Wavelet spectrum of the 1990 trace.

```
> wtSTRACE90 <- DOG(TRACE90[1:(2^15)],10,5,1,plot=F)
> mcwtSTRACE90 <- Mod(cwtSTRACE90)
> mcwtSTRACE90 <- mcwtSTRACE90 * mcwtSTRACE90
> wspSTRACE90 <- tfmean(mcwtSTRACE90, plot=F)
> wspec.pl(wspSTRACE90,5)
```

Wavelet spectrum of the 1994 trace.

```
> cwtSTRACE94 <- DOG(TRACE94[1:(2^15)],10,5,1,plot=F)
> mcwtSTRACE94 <- Mod(cwtSTRACE94)
> mcwtSTRACE94 <- mcwtSTRACE94 * mcwtSTRACE94
> wspSTRACE94 <- tfmean(mcwtSTRACE94, plot=F)
> wspec.pl(wspSTRACE94,5)
```

Gabor spectrum of the 1990 trace.

```
> cgtSTRACE90 <- cgt(TRACE90[1:(2^15)],50,0.02,25,plot=F)
> gspec90 <- tfmean(Mod(cgtSTRACE90)*Mod(cgtSTRACE90),
+ plot=F)
> tsplot(log(gspec90))
```

Gabor spectrum of the 1994 trace.

```
> cgtSTRACE94 <- cgt(TRACE94[1:(2^15)],50,0.02,25,plot=F)
> gspec94 <- tfmean(Mod(cgtSTRACE94)*Mod(cgtSTRACE94),
```

```
+ plot=F)
> tsplot(log(gspec94))
```

Spectrogram of the 1990 trace.

```
> s90 <- spec.pgram(TRACE90[1:(2^15)])
> plot(log10(s90$freq),s90$spec,type="l",xlab="log10(freq)",
+ ylab="")
```

Spectrogram of the 1994 trace.

```
> s94 <- spec.pgram(TRACE94[1:(2^15)])
> plot(log10(s94$freq),s94$spec,type="l",xlab="log10(freq)",
+ ylab="")
```

Figure 6.7 contains the results of the preceding computations. It was created with the following commands.

Putting everything together.

```
> par(mfrow=c(2,3))
> plot(log10(s90$freq),s90$spec,type="l",xlab="log10(freq)",
+ ylab="")
> title("Periodogram")
> tsplot(log(gspec90),xlab="Frequency")
> title("log(Gabor Spectrum)")
> wspec.pl(wspSTRACE90,5)
> plot(log10(s94$freq),s94$spec,type="l",xlab="log10(freq)",
+ ylab="")
> title("Periodogram")
> tsplot(log(gspec94),xlab="Frequency")
> title("log(Gabor Spectrum)")
> wspec.pl(wspSTRACE94,5)
```

Example 6.9 Amber Camera Data

Read signals (here only pixel 7) from disk and plot.

```
> par(mfrow=c(3,3))
> amber7 <- scan("./signals/amber16/pixel_8.7")
> tsplot(amber7)
> title("Pixel 7")
```

Same commands for pixels 8 and 9.

Periodogram-based spectral estimation.

```
> spamber7 <- Mod(fft(amber7))
> spamber7 <- spamber7*spamber7
```

```
> plot(spamber7[2:3501],log="xy",type="l")
> title("Periodogram")
> lsamber7 <- lsfit(log10(2:100),log10(spamber7[2:100]))
> abline(lsamber7$coef)
```

Wavelet-based spectral estimation.

```
> cwtamber7 <- DOG(amber7,10,5,1,plot=F)
> mcwtamber7 <- Mod(cwtamber7[501:6500,])
> sp7 <- tfmean(mcwtamber7*mcwtamber7,plot=F)
> wspec.pl(sp7,5)
```

To get the values of exponents in several parts of the frequency domain:

```
> hurst.est(sp7,5:25,5,plot=F)
> hurst.est(sp7,25:35,5,plot=F)
> hurst.est(sp7,39:45,5,plot=F)
```

Pixels 8 and 9 are processed the same way. See the plots in Figure 6.9.

6.5 Notes and Complements

The wavelet transform of second-order stationary processes has been extensively studied and further details on the abstract considerations presented in this section can be found in the article [173] by C. Houdré or in O. Zeitouni's paper [253] for example. Another instance of the use of the continuous wavelet transform for spectral estimation can be found in [174] by Frehe, Mayer, and Hudgins where an object very similar to the estimator $\tilde{V}_B$ of the *scale spectrum* of the wavelet transform was considered, whereas spectral estimation using the theory of wavelet orthonormal bases as presented in Chapter 5 is discussed by Gao in [140].

The self-similarity of network traffic (and the pervasive presence of heavy tails) has been documented and analyzed in detail. We refer the interested reader to the review article [296] by Willinger, Paxson, and Taqqu and to the references therein. Wavelet analysis is particularly efficient when it comes to analyzing the scaling properties of data, and this is why we concentrated on the self-similarity of network traffic. Evidence of self-similarity for local-area networks (LANs) was given in [295], and the case of the Internet was discussed in [296]. The application to the self-similarity of wide-area network (WAN) traffic which we presented in the text is borrowed from the work [127] of Feldman, Gilbert, Willinger, and Kurtz. Inspired by the work [8] of Abry and Veitch, they used the energy distribution across the scales of the wavelet decomposition (in a wavelet orthonormal basis) of the data to illustrate the scaling properties of the traffic. This energy distribution is nothing but the analog in the case of a wavelet basis of the spectral density $\tilde{V}_B(a)$.

Silverman, and Mallat, Papanicolaou and Zhang, and Kozek independently proposed different ways of studying locally stationary process. See [196, 213, 264]. Silverman's analysis was restricted to the case of processes whose Wigner spectrum is a separable function:

$$\mathcal{E}_0(b,\omega) = m(b)\mathcal{E}(\omega) \ ,$$

i.e., essentially products of stationary processes with a deterministic function. In such a case, the analysis is greatly simplified because standard results from the spectral theory of stationary processes may be used. There are very few textbooks dealing with the spectral theory of non-stationary processes from a probabilistic and/or statistical point of view. Priestley's book [251] is the only exception we know of. This book of Priestley comes after a series of works by the author (see, for example, [248, 249]) which remained mostly ignored by the community at large. The article of Mallat, Papanicolaou, and Zhang [213] initiated a renewal of interest in the work of Priestley. The small number of publications on the spectral theory of nonstationary processes does not mean that specific classes of models have not been investigated. Indeed, processes with stationary increments, and especially self-similar Gaussian processes (the most famous case being of course fractional Brownian motion fBm), have attracted the attention of many investigators. P. Flandrin was one of the first time-frequency investigators to revisit spectral estimation of random processes in general and of fBm in particular. See, for example, the articles [130, 131, 132] or [5, 6] and his book [Flandrin93]. Wornell considered the properties of the wavelet transform of fBm. See, for example, [297, 298]. So did Masry [221], Ramanathan and Zeitouni[253], and many others. Wornell also considered the statistical problem of the estimation of $1/f$ processes from noisy measurements [299]. So did Hwang [176].

Fractional Brownian motion was proved to play a very important role in communication networks modeling by Beran *et al.* in [40]. But fBm is far from being the only example of interesting stationary increments processes which have been studied by wavelet tools. Indeed, Benassi, Jaffard, and Roux have considered more general classes of Gaussian processes, defined by a covariance operator which is the Green's function of an elliptic operator (possibly of low regularity). Such processes are called *elliptic Gaussian processes* and have interesting local self-similarity properties. Benassi, Jaffard, and Roux have shown in particular how to estimate the principal symbol of the covariance from a single realization of the process. We shall not go further in that subject, which is far beyond the scope of this text. See [34, 35], for example.

Chapter 7

Analysis of Frequency Modulated Signals

Throughout the first chapters of this volume, we have focused on simple examples of CGT and CWT in order to emphasize certain characteristic behaviors. In particular, we concentrated on the qualitative properties of the continuous wavelet and Gabor transforms of functions of the form

$$f(x) = \sum_{\ell=1}^{L} A_\ell(x) \cos \phi_\ell(x) \ . \tag{7.1}$$

Our main remark was that if the Fourier transform of an analyzing wavelet $\psi(x)$ is sharply concentrated near a fixed value $\xi = \omega_0$ of the frequency, the continuous wavelet transform has a tendency to "concentrate" near a series of curves, called the *ridges* of the transform. The same holds in the Gabor case.

We shall describe in this chapter how to turn this remark into a series of innocent-looking though very powerful numerical methods for characterizing important features of the original signal from the contribution of these ridges. Our strategy will essentially be as follows:

- Associate with the signal a time-frequency representation, for example, a continuous wavelet or Gabor transform.
- Use the time-frequency representation to estimate ridges. In other words, associate a series of *curves* with a function of two variables.
- For each "component" estimate the signal which produced this ridge component and reconstruct the signal accordingly.

In this chapter, we shall focus on the first two parts of this program. The problem of "reconstruction from a ridge" will be addressed in the next chapter.

7.1 Generalities on Asymptotic Signals

Before entering the subject, we need to recall a few concepts and definitions from standard signal analysis. We briefly describe here the basic definitions and properties of the so-called *asymptotic signals*. They are the signals for which the methods we shall describe in this chapter are adequate.

7.1.1 The Canonical Representation of a Real Signal

For the sake of completeness we review some of the material already introduced in Subsection 1.1.2 of the first chapter.

An arbitrary real signal f can always be represented in terms of instantaneous modulus and argument, in the form

$$f(x) = A(x)\cos\phi(x)\ . \tag{7.2}$$

However, such a representation is far from unique since infinitely many pairs (A, ϕ) may be associated with a given real signal f. In order to illustrate this last point, let us consider a simple example. Given a signal as in Equation (7.2), suppose that $|A|$ is a bounded function and let us set $A = \sup|A(x)|$ for its sup-norm. Let us now consider the function $\tilde{f}(x) = f(x)/A$. Clearly, $|\tilde{f}(x)| \leq 1$ for all x, and one may write $\tilde{f}(x) = \cos\tilde{\phi}(x)$, for some function $\tilde{\phi}$. Then we get another representation of the same kind as (7.2), namely, $f(x) = A\cos\tilde{\phi}(x)$. This is the simplest evidence of the non-uniqueness of the representation (7.2).

Several authors (see, e.g., [242, 243, 288]) have noticed that among these pairs, it is convenient to specify a particular one, called the *canonical pair*, defined as follows. By definition, the *analytic signal* Z_f associated with f is obtained by a linear filtering of f canceling its negative frequencies. It may be expressed by making use of the Hilbert transform (see Section 1.1.2):

$$Z_f(x) = [I + iH]f(x)\ .$$

The function Z_f has an analytic continuation to the upper half (complex) plane $\Pi_+ = \{z = x + iy;\ y \geq 0\}$ (hence the term "analytic signal"). Then Z_f is completely characterized by the pair (A_f, ϕ_f) via the formula

$$Z_f(x) = A_f(x)\exp[i\phi_f(x)]\ , \tag{7.3}$$

if one assumes that A_f is non-negative and that ϕ_f takes its values in the interval $[0, 2\pi)$. The pair (A_f, ϕ_f) is called the canonical pair associated to f. Obviously,

$$\Re(Z_f(x)) = f(x) \tag{7.4}$$

may be written as

$$f(x) = A_f(x)\cos(\phi_f(x)) , \tag{7.5}$$

which defines the *canonical representation* of f. The canonical representation allows the introduction of the instantaneous frequency $\omega_f(x)$ of $f(x)$, defined by

$$\omega_f(x) = \frac{1}{2\pi}\frac{d\phi_f(x)}{dx} . \tag{7.6}$$

Oddly enough, a pair of functions (A, ϕ) is not necessarily a canonical pair; in other words, a signal of the form $A(x)\exp[i\phi(x)]$ is not necessarily an analytic signal. We refer to [243] for several counterexamples and an elaboration of this point.

Remark 7.1 Note that although the definition of the instantaneous frequency always makes sense mathematically, its physical significance can be doubtful in some particular situations. This is especially true when $f(x)$ is not oscillating enough, i.e., when $\phi_f(x)$ varies slowly compared to $A_f(x)$, or when the frequency $\omega_f(x)$ itself has fast variations.

7.1.2 Asymptotic Signals and the Exponential Model

Let $f \in L^2(\mathbb{R})$ be a real signal of finite energy of the form (7.2) with $A(x) \geq 0$ and $\phi(x) \in [0, 2\pi)$ for all $x \in \mathbb{R}$. It is tempting to state that its analytic signal equals $A(x)\exp[i\phi(x)]$. Unfortunately, this is wrong in general (consider, for example, the case $f(x) = \sin(1/x)$, whose Hilbert transform is $\cos(1/x) - 1$ and not $\cos(1/x)$ as one might have expected). However, this result is approximately true for a certain class of signals, called asymptotic signals. More precisely, we have the following:

Lemma 7.1 *Let λ be a (large) positive number, and let $f \in L^2(\mathbb{R})$ be of the form $f(x) = A(x)\cos[\lambda\phi(x)]$, where A and ϕ are twice and four times continuously differentiable functions, respectively. Then*

$$Z_f(x) = A(x)e^{i\lambda\phi(x)} + O\left(\lambda^{-3/2}\right) \tag{7.7}$$

as $\lambda \to \infty$.

The proof of the lemma is obtained by using twice the stationary phase method (see [91, 106, 172]) to evaluate the two Fourier transforms involved in the Hilbert transform.

Of course, such a result is not directly applicable since in practice one does not have a parameter λ to play with. However, it may be used at least in an approximate way. We shall say that a signal f as in (7.2) is *asymptotic* if it is oscillatory enough so that we can approximate its associated analytic signal in the form

$$Z_f(x) \approx A(x)e^{i\phi(x)} . \tag{7.8}$$

Even though we do not attempt to make more precise the sense in which this approximation has to be understood, we can state that it essentially means that the oscillations coming from the phase term $\phi(x)$ are much faster than the variations coming from the amplitude term $A(x)$.

The asymptotic signals turn out to have a simple behavior when represented in a time-frequency domain. All the results we present are based on a specific estimate which expresses the localization properties of the transform. We first give the estimate in the case of the wavelet transform (a similar result holds in the Gabor case), before explaining the practical consequences.

Proposition 7.1 *Let λ be a (large) positive number, let*

$$f(x) = A(x)\cos\left[\lambda\phi(x)\right] \tag{7.9}$$

be a signal with $A \in C^2$ and $\phi \in C^4$, and let

$$\psi(x) = A_\psi(x)e^{i\lambda\phi_\psi(x)} \in H^2(\mathbb{R}) \tag{7.10}$$

be a progressive wavelet. We assume that for each fixed $(b,a) \in \mathbb{R} \times \mathbb{R}_+^*$ (or a sub-domain of interest of the time-frequency plane,) the equation

$$\phi'(x_0) = \frac{1}{a}\phi_\psi'\left(\frac{x_0 - b}{a}\right) \tag{7.11}$$

has a unique solution $x_0 = x_0(b,a)$, and that this solution satisfies

$$\phi''(x_0) \neq \frac{1}{a^2}\phi_\psi''\left(\frac{x_0 - b}{a}\right) .$$

Then as $\lambda \to \infty$, the CWT of f may be approximated as

$$T_f(b,a) = \frac{e^{i\frac{\pi}{4}\mathrm{sgn}(\Phi''_{(b,a)})}}{\sqrt{2\pi\lambda a^2\left|\Phi''_{(b,a)}\right|}}\overline{\psi\left(\frac{x_0 - b}{a}\right)}Z_f(x_0) + O\left(\lambda^{-3/2}\right) , \tag{7.12}$$

where we used the notation

$$\Phi_{(b,a)}(x) = \phi(x) - \phi_\psi\left(\frac{x - b}{a}\right) . \tag{7.13}$$

Again, the proof follows from a standard stationary phase argument [102]. The stationary points of the integral defining the wavelet transform

$$T_f(b,a) = \frac{1}{2a}\int A(x)A_\psi\left(\frac{x-b}{a}\right)e^{i\lambda\Phi_{(b,a)}(x)}dx \tag{7.14}$$

(where the function $\Phi_{(b,a)}$ is given in (7.13)) satisfy (7.11) and are assumed to be unique and of first order. Then the application of the stationary phase method directly yields (7.12).

In practice, the best we can hope for is to have a parameter λ large enough that the approximation provided by the stationary phase method is reliable. Let us now assume that this is true. The stationary points $x_0 = x_0(b,a)$ are the points where the frequency of the scaled wavelet coincides with the local frequency of the signal. This is how the fact that they provide a significant contribution to the wavelet coefficient may be understood intuitively. It follows from (7.12) that the following set of points plays a particular role, in the sense that the energy of the wavelet transform tends to "concentrate" around it:

$$R = \{(b,a) \mid x_0(b,a) = b\} \ . \tag{7.15}$$

Putting this definition into (7.11) yields

$$a = a_r(b) = \frac{\phi'_\psi(0)}{\phi'(b)} \ . \tag{7.16}$$

This curve is called a *ridge of the wavelet transform* and plays a major role in the analysis of the behavior of the whole wavelet transform. Examples illustrating this fact and the localization properties of asymptotic signals in the time-frequency domain may be found in Figures 4.5 and 4.6 in Chapter 4.

Remark 7.2 These approximations have to be compared with those given in Lemma 4.1 in Chapter 4. The conclusion is the same (i.e., localization near ridges), but the approximation is different. Here, the quantity $\phi'_\psi(0)$ replaces the "central frequency" ω_0 used in Chapter 4 as reference frequency for the wavelets. In addition, in Proposition 7.1, the spectral information is carried by the stationary points $x_0 = x_0(b,a)$.

Similar approximations may be derived in the case of the Gabor transform, leading to conclusions similar to those of Chapter 3. Let us suppose that we are given a window which is a low-pass filter, i.e., such that its Fourier transform has fast decay away from the origin of the frequency axis. Then the CGT of an asymptotic signal of the form (7.2) "concentrates" near a ridge of the form

$$\omega = \omega_r(b) = \phi'(b) ,$$

which provides information about the instantaneous frequency of the signal. For an illustration, see Figure 3.5 in Chapter 3.

Remark 7.3 *The case of noisy signals:* The most important point about frequency modulated signals is that their wavelet transform is "concentrated" in a well-defined region in the time-frequency plane. Let us assume now that the considered frequency modulated signal is embedded in noise. As we saw in the previous chapters, the wavelet transform of the noise is spread in the whole time-frequency space. Then near the ridge(s) of the wavelet transform, the contribution of the signal may be expected to be much larger than that of the noise. We shall see applications of this fact later. For an illustration, see in Figure 7.1, later, the case of the superposition of a chirp signal, a sine wave, and an additional white Gaussian noise. The deterministic part of the signal does not appear in the plot at the top of the figure. Nevertheless, two well-defined ridges appear in the modulus of the CWT, and the CWT of the noise is spread all over the time-frequency domain which is displayed.

7.2 Ridge and Local Extrema

Now that we know how to analyze the behavior of the wavelet transform of asymptotic signals, let us turn to the following general problem. Given a signal f which we want to model as in (7.1,) how can we determine numerically the local amplitudes and frequencies?

Depending on the situation (signals with a unique component or several components, pure or noisy signals ...), there are several ways to answer such a question. We shall describe here a certain number of possible answers, keeping in mind that we want algorithms that are robust to noise.

Throughout this section, we shall restrict ourselves to the case of the CWT. But it should be clear that the algorithms can easily be adapted with minor modifications to the case of the CGT. From now on, we shall assume that ψ is a progressive wavelet with good frequency localization (the Morlet wavelet is a typical example).

7.2.1 "Differential" Methods

Let us start with a set of methods that may be termed *differential methods*, since they are based on a local analysis of the extrema of the modulus of the wavelet transform. In the simple case where the signal possesses a unique ridge, it is in general sufficient to look for the maxima in the scale variable of the wavelet transform, for fixed values of the position variable b. This yields an estimate for the ridge, which is precise enough for most of the applications.

This fact may be turned into an explicit algorithm in several ways. The simplest one is a search for the global maxima of the wavelet transform modulus. For each value b of the time variable we look for the (hopefully unique) value of the scale, say a_b, such that

$$|T_f(b, a_b)| = \max_a |T_f(b, a)| \ .$$

The ridge is then given by the graph of the function $b \hookrightarrow a_b$ (see Example 7.1 for the S commands used to search for global maxima).

But if more than one ridge is present, such a procedure can only yield one of them in the best case. In general it will give small segments of ridges, those segments where the modulus of the transform is the highest. An alternative is to search for the local maxima (still with respect to the scale variable).

However, such a method turns out to be unstable in the presence of noise; the noise creates additional local maxima, and to determine the ridge, one has to discriminate between the local maxima which come from a "true" ridge and those that are noise artifacts.

As an example, we show in Figure 7.1 the wavelet transform modulus and the corresponding local maxima for the sum of a chirp, a sine wave and an additive Gaussian white noise (see Example 7.2 for the S-commands used to generate the figure). Clearly, the local maxima reproduce the ridges of the two deterministic components, but also produce a lot of spurious local extrema. In general, an additional chaining stage is needed in order to reorganize the maxima into ridges (i.e., one-dimensional objects) and to get rid of spurious local extrema produced by the method. But as can be seen in Figure 7.1, even the local maxima produced by the noise look like one-dimensional objects, which makes the chaining difficult.

We now discuss an approach which generally gives good results when the noise component is relatively small in the regions occupied by the ridges. We replace the search for local extrema by a numerical resolution of the equation

$$\partial_a |T_f(b, a)| = 0 \ , \tag{7.17}$$

by an iterative scheme, say, Newton's method. Iterative schemes need an initial guess to start the process and in our case, for each b, we choose to take $\varphi(b-1)$ as an initial guess for $\varphi(b)$. In other words, we try to draw

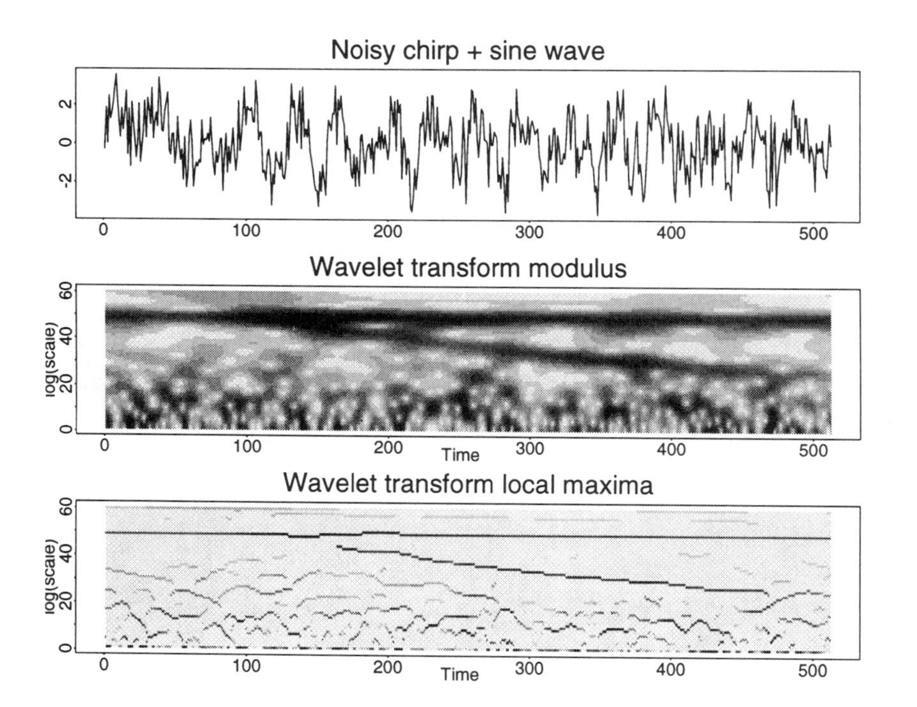

Figure 7.1. Modulus and local maxima of the CWT of a chirp, generated in Example 7.2.

the graph of a given ridge from the left to the right, trying to determine $\varphi(b)$ from $\varphi(b-1)$.

In any case, it follows from the estimate (7.12) that the location of the modulus maximum of the wavelet transform does not exactly coincide with the ridge, because of the possibly non-constant denominator. If more precision is needed, let us quote the so-called *Marseille method*, based on the use of the phase of the CWT. More precisely the numerical resolution of equation (7.17) is replaced with that of

$$\partial_b \arg T_f(b,a) = \frac{\omega_0}{a} , \tag{7.18}$$

i.e., the search for the points (b,a) where the frequency of the wavelet transform coincides with that of the scaled wavelet. Again, such a search may be done numerically, using Newton's method for example. An interesting aspect of the two methods just mentioned is that they require no discretization of the CWT, since Equation (7.18) may be solved without any prior computation of the CWT. This makes it possible to evaluate the CWT at only a few (representative) points. We refer to [102, 272] for a more complete description of this method.

Let us stress that such "local" methods lead in general to ill-posed problems, and thus to algorithms which can be extremely unstable in the presence of significant noise.

In addition, let us also notice that in both cases, not all the available information has been explicitly used in the estimation of the ridge. Namely, we have not taken into account the fact that, at least for the class of signals we are interested in, the ridge is a smooth function of the time variable b. We shall describe now alternative methods which incorporate this type of information explicitly. These are based on post-processing of wavelet transform and are closely related to image processing methods [142, 191].

7.2.2 "Integral" Methods: Spaces of Ridges

To start with, let us revisit the approach described earlier. We think of a ridge as a set of points in the time-frequency plane (more precisely at this stage we think of a ridge as the graph of a function in this plane,) and we look for those points $(b, \varphi(b))$ using a standard optimization technique. Let us stress that when doing so, we never use explicitly the fact that such a set of points is supposed to yield a smooth curve.

Moreover, in many instances, the signal from which one wants to extract amplitude and frequency components is corrupted by noise. In such cases, the differential methods discussed earlier may fail to provide reasonable and stable answers, and it is necessary to turn to more robust methods. In particular, it becomes necessary to use more of the *a priori* information which is available about the ridges (i.e., their smoothness,) and to include them explicitly in the estimation algorithm.

Let $\Omega \subset \mathbb{R}^2$ be the domain in the time-frequency plane under consideration. Points in Ω are labeled by $(b,\omega) \in \mathbb{R}^2$, and we denote generically by $M(b,\omega)$ the time-frequency representation under consideration. We will sometimes call $M(b,\omega)$ a time-frequency energy density, or an energy for short. In the present discussion, we focus on two specific cases:

1. The CGT modulus: in this case ω is the frequency variable, and
$$M(b,\omega) = |G_f(b,\omega)|^2 \ .$$
2. The CWT modulus: in this case it is natural to introduce the variable $\omega = \log(a)$ (we recall that in discrete wavelet transforms, the scale variable is sampled in the form $a_j = a_1^j$, which amounts to a regular sampling of the variable ω), and
$$M(b,\omega) = |T_f(b,e^{\omega})|^2 \ .$$

More generally, $M(b,\omega)$ may also be any other energetic time-frequency or time-scale representation.

Remark 7.4 *The Case of Noisy Signals:* In real-life applications, signals are not in general "impure" and most often embedded in noise. And although the methods just described are in general quite robust, the presence of a strong noise may degrade the quality of results. However, when some information on the noise is available, it may be taken into account in order to enhance the performances of the algorithm.

Consider the example of the signal-plus-noise model introduced in Chapter 6. The signal $f(x) = f_0(x) + \epsilon(x)$ is the sum of a (true) deterministic signal f_0 embedded into a random, mean zero stationary process $\epsilon = \{\epsilon(x)\}$ with autocovariance function
$$C_\epsilon(\tau) = \mathbb{E}\{\epsilon(x)\epsilon(x+\tau)\}$$
and common variance $C_\epsilon(0) = \sigma_\epsilon^2$. Let us denote by
$$\mathcal{E}(\xi) = \int C_\epsilon(\tau) e^{-i\xi\tau} d\tau$$
the power spectrum of the noise. Then because wavelets are of zero integral we clearly have
$$\mathbb{E}\{|T_f(b,a)|^2\} = |T_{f_0}(b,a)|^2 + \mathbb{E}\{|T_\epsilon(b,a)|^2\}$$

(this is true when the mean is a constant, but let us stress that for processes whose mean varies slowly, at scales larger than a, $\mathbb{E}\{T_\epsilon(b,a)\}$ is the wavelet transform of the mean, which is numerically negligible). Now, as we have seen before, we have that

$$\mathbb{E}\{|T_\epsilon(b,a)|^2\} = \int |\hat{\psi}(a\xi)|^2 \mathcal{E}(\xi) d\xi \ .$$

Assuming that the autocovariance function of the noise is known (in some cases, it may be estimated directly from the data, as it is the case in underwater acoustics where part of the signal is set aside to this effect), this suggests modifying the time-frequency energy by taking into account the noise term, replacing $|T_f(b,a)|^2$ by $|T_f(b,a)|^2 - \mathbb{E}\{|T_n(b,a)|^2\}$.

We may then choose

$$M(b,\omega) = |T_f(b,e^\omega)|^2 - \mathbb{E}\{|T_\epsilon(b,e^\omega)|^2\}$$

instead of $M(b,\omega) = |T_f(b,e^\omega)|^2$.

Similar arguments may be developed in the context of the CGT. There is nevertheless an important difference: the Gabor functions are not of zero integral, and the argument which we developed in the case of the CWT will hold only approximately, and for large values of the frequency.

Remark 7.5 This a priori information about the noise may also be taken into account directly when evaluating the time-frequency representation. Supposing for example that the noise is stationary, with spectral density $\mathcal{E}(\omega)$; one may introduce a *pre-whitened wavelet transform* as follows:

$$W_f(b,a) = \langle \mathcal{C}^{-1/2} f, \psi_{(b,a)} \rangle \tag{7.19}$$

$$= \frac{1}{2\pi} \int e^{ikb} \frac{\hat{f}(k)\overline{\hat{\psi}(ak)}}{\sqrt{\mathcal{E}(k)}} dk \ . \tag{7.20}$$

As usual, $\mathcal{C}$ and $\mathcal{E}(k)$ stand for the covariance operator and the spectral density of the noise, respectively. Recall the discussion on pre-whitening in Chapter 2. Such a pre-whitened wavelet transform was used in [178] for the problem of gravitational waves detection (see the discussion in Section 4.5.3).

7.2.3 Ridges as Graphs of Functions

Let us first consider the simple case where the ridge (assumed to be unique for the sake of simplicity) is modeled as the graph of a function

$$b \hookrightarrow \varphi(b) \ .$$

A classical approach is to use a penalization method, i.e., to introduce a natural (as natural as possible at least) penalty function on the set of possible ridges, and to define the ridge as the one that minimizes the penalty. In order to realize such a program, the penalty function has to take into account two basic facts:

1. According to Proposition 7.1, the time-frequency representation has a tendency to "localize" the energy near the ridges.
2. In the models of interest to us and in particular in the case of frequency modulated signals, the ridges are smooth and slowly varying functions.

Let us examine both facts separately and show how they can be used to define an appropriate penalty function.

1. It is a traditional technique in time-frequency signal analysis to use the energy integrated over the ridge. Then, given a ridge candidate φ, the quantity $\int M(b,\varphi(b))db$ attains its maximal value when $b \hookrightarrow \varphi(b)$ is the "true" ridge. It is then "natural" to use
$$\mathcal{E}_1(\varphi) = -\int M(b,\varphi(b))db \tag{7.21}$$
as a first term in the penalty function. In fact, it is easy to see (at least formally) that minimization of $\mathcal{E}_1(\varphi)$ with respect to φ is equivalent to maximizing $|T(b,a)|$ with respect to a for all b independently.

2. To take care of the smoothness requirement for the ridge, it is a standard approach to introduce terms of the form
$$\mathcal{E}_2(\varphi) = \lambda \int |\varphi'(b)|^2 db + \mu \int |\varphi''(b)|^2 db\ , \tag{7.22}$$
and minimizing such terms will penalize non-smooth ridges.

All together, we end up with the following penalty function on the set of ridge candidates:
$$\mathcal{E}(\varphi) = -\int M(b,\varphi(b))db + \lambda \int |\varphi'(b)|^2 db + \mu \int |\varphi''(b)|^2 db\ . \tag{7.23}$$

The ridge detection problem is then reformulated as a minimization problem
$$\hat{\varphi} = \arg\min\ \ \mathcal{E}(\varphi)\ , \tag{7.24}$$
which is to be attacked numerically.

There are several ways to attack the optimization problem. The obvious one is to write and solve the Euler-Lagrange equations associated with the problem (7.24). In the present situation these equations reduce to a single equation which can be written formally as

$$\mu\varphi^{(4)}(b) - \lambda\varphi''(b) = \frac{1}{2}\partial_a M(b, \varphi(b)) \ . \tag{7.25}$$

However, the numerical algorithms solving this equation are known to be unstable in general, and in any case, likely to lead to local extrema of the penalty function. Another reason why we shall refrain from using such algorithms is our interest in noisy signals. We describe alternative approaches in Section 7.3.

7.2.4 Ridges as "Snakes"

The previous approach has several drawbacks. One of the shortcomings is that, when ridges are considered as functions, the domain of definition of these functions becomes a problem and the boundaries of the ridge require a special treatment. Fixing a common domain for all the ridges limits the class of signals, while allowing the domain of definition to vary with the ridge causes problems with the detection algorithms. Notice also that viewing ridges as functions of the time variable b breaks the symmetry between the time and the frequency variables b and ω. However, this problem will not be a hindrance for most of the types of signals we consider. But the most important drawback is the fact that signals with multiple ridges, which are typical of speech signals, for example, cannot be handled by this approach.

Except perhaps for the problem of the multiple ridges, one can remedy these shortcomings by modeling the ridge as a parametric curve:

$$s \in [0,1] \hookrightarrow ((\rho(s), \varphi(s)) \in \Omega \ ,$$

which we shall call *snake* and to penalize it in order to force it to trace the ridge. Using the ideas introduced before to justify the penalty functional (7.23,) we set

$$\begin{aligned} \mathcal{E}(\rho,\varphi) &= -\int M(\rho(s),\varphi(s))ds + \lambda_a \int |\varphi'(s)|^2 ds + \mu_a \int |\varphi''(s)|^2 ds \\ &\quad +\lambda_b \int |\rho'(s)|^2 ds + \mu_b \int |\rho''(s)|^2 ds \ , \end{aligned} \tag{7.26}$$

and the snake estimate of the ridge is obtained by minimizing this penalty. Again, one possible option is to write the Euler-Lagrange equations associated with the minimization problem, and solve them by conventional methods. The remarks made earlier to explain why this approach is not always appropriate also apply to the snake approach and we shall describe other algorithmic procedures below.

Let us also stress that in the snake approach, since the boundaries are free, they will adjust dynamically to the true boundaries of the ridge.

7.2.5 Bayesian Interpretation

Before going deeply into the details of the minimization algorithms which we recommend for the frequency modulated signals, we would like to shed some light on the formulation of the optimization problems introduced earlier.

It is possible to give a Bayesian interpretation for the optimization problem from which we derive the estimate of the ridge. We give it here for the sake of illustration. Our argument is perfectly rigorous in the case of (finite) discrete signals. It requires sophisticated measure theoretic arguments in the continuous case. We refrain from providing the details even though we opted to present our discussion with the convenient notations of integration and differentiation for continuous signals.

We focus on the case of ridges modeled as graphs of a function (the discussion is strictly identical in the case of snakes). Let us denote by Φ the space of all the possible ridge functions $b \hookrightarrow \varphi(b)$. Let us set

$$\mu_{\text{prior}}(d\varphi) = \frac{1}{Z_1} \; e^{-\frac{1}{T}\int\left[\mu_a|\varphi'(b)|^2+\lambda_a|\varphi''(b)|^2\right]db} \;\; "d\varphi" \, , \tag{7.27}$$

where the normalization constant Z_1 is chosen so that μ_{prior} is a probability measure on Φ. In that formula, T is a constant *temperature parameter*, introduced for convenience. The infinitesimal $"d\varphi"$ has an obvious (rigorous) meaning in the case of a discrete variable b. Its meaning when b varies continuously is not as clear. In fact, the whole formula defining the measure μ_{prior} has to be reinterpreted as the definition of a Gaussian probability measure on an infinite dimensional space (without using the density given by formula (7.27) above). In any case, the probability measure $\mu_{\text{prior}}(d\varphi)$ should be viewed as a model for the prior knowledge (or lack of knowledge) about the kind of ridge we are looking for.

Given an element $\varphi \in \Phi$, let us consider the probability $\mu_\varphi(df)$ defined by

$$\mu_\varphi(df) = \frac{1}{Z_2} \; e^{\frac{1}{T}\int M(b,\varphi(b))db} \;\; "df" \tag{7.28}$$

on the space $\mathcal{S}_2$ of signals of finite energy. As before, Z_2 is a normalization constant whose role is to guarantee that the total mass of the measure is 1. Intuitively speaking, $\mu_\varphi(df)$ gives the probability, knowing that the ridge is given by the function φ, that the true signal is in the infinitesimal "df" in the space $\mathcal{S}_2$. Notice that the probability $\mu_\varphi(df)$ is large when the modulus of the wavelet transform of f is large along the function φ. This is consistent with the intuitive interpretation of this probability.

Given an observation of a signal f of finite energy, the posterior probability of the ridge is given by the classical Bayes formula:

$$\mu_{\text{posterior}}(d\varphi) = \frac{1}{Z_f} e^{-\frac{1}{T}[\int (\mu_a|\varphi'(b)|^2+\lambda_a|\varphi''(b)|^2)db - \int M(b,\varphi(b))db]} \text{ "}d\varphi\text{"} , \tag{7.29}$$

for some normalizing constant Z_f depending on the observation f. The maximum a posteriori likelihood principle suggests to choose the value of φ which maximizes the *density* of this distribution. In other words, given the observation f, the estimate $\hat{\varphi}$ of the ridge should be the solution of the minimization problem

$$\varphi_0 = \arg\min \left(\int (\mu_a|\varphi'(b)|^2 + \lambda_a|\varphi''(b)|^2)\, db - \int M(b,\varphi(b))db \right) , \tag{7.30}$$

which is indeed our choice for the estimation procedure.

7.2.6 Variations on the Same Theme

Many other choices of penalty functions are possible. For the record we mention those which we constructed along the same lines as in the previous discussion and which we implemented in the Swave package.

In order to justify the first one of them, we come back to the situation of Section 7.2.3 and we make the following remark. The penalty function described in Equation (7.23) is based on the balance between two terms: an "internal" ridge energy term which controls the ridge's rigidity, and a second "external" term which "pushes" the ridge toward regions where $M(b,\omega)$ is large. However, far away from such regions, the external term is small, so that essentially only the smoothness penalty is present. As a result, the estimation algorithm may fail to produce a reasonable estimate if initialized too far away from the "energetic regions" of $M(b,\omega)$. Indeed it is likely to produce a function $b \hookrightarrow \varphi(b)$ which is very smooth and whose graph is contained in a region where the modulus of the transform is small! This suggests locally renormalizing the smoothness penalty, and introducing the new penalty function

$$\mathcal{E}(\varphi) = \int M(b,\varphi(b)) \left[-1 + \lambda|\varphi'(b)|^2 + \mu|\varphi''(b)|^2\right] db \, . \tag{7.31}$$

A similar argument leads to the following modification of the snake penalty function (7.26.):

$$\begin{aligned} \mathcal{E}(\rho,\varphi) &= \int M(\rho(s),\varphi(s))\Big[-1+\lambda_a|\varphi'(s)|^2+\mu_a|\varphi''(s)|^2 \\ &\quad +\lambda_b|\rho'(s)|^2+\mu_b|\rho''(s)|^2\Big]\,ds\ . \end{aligned} \tag{7.32}$$

Ridge estimation procedures based on the minimization of the functionals (7.31) and (7.32) are implemented in the package Swave (see the later examples).

7.3 Algorithms for Ridge Estimation

Let us now turn to the discussion of alternative methods for minimizing the penalty functions introduced in equations (7.23) and (7.26.)

7.3.1 The "Corona" Method

As an alternative to the direct approach discussed earlier we propose to use combinatorial optimization techniques which carry the guarantee of eventually reaching the global minimum. Simulated annealing algorithms are presumably the best known of these algorithms. We refer the interested reader to [142, 143, 201] for detailed discussions of the various properties of these algorithms. Simulated annealing algorithms are known for being difficult to fine tune in order to make sure that a global minimum of the cost function (or at least a reasonable approximation of this global minimum) is attained in a reasonable number of iterations. Nevertheless, the simulated annealing procedure turned out to be quite efficient for the ridge detection problem. In this context it was first proposed in [67, 68].

Let us describe the algorithm in the practical case of a discrete time-frequency representation $M(k,j)$, in the particular case where the ridge is modeled as the graph of a function. Then the ridge is a discrete sequence $\varphi(0),\ldots,\varphi(B-1))$ with values into a finite set denoted by $\{0,1,\ldots,A-1\}$ for simplicity. The algorithm is based on repeated Monte Carlo simulations of the measure in (7.29) with decreasing temperature T, and goes as follows.

1. Set an initial temperature T_0.
   Choose an initial guess

   $$\varphi_0 = \{\varphi_0(0),\varphi_0(1),\ldots,\varphi_0(B-1)\}\ ,$$

   Compute the corresponding value of the penalty function $\mathcal{E}(\varphi_0)$.

2. At each time step t of the algorithm:
   Select randomly an integer $b = 0, \ldots, B-1$.
   Select randomly a potential move $\epsilon = \pm 1$.
   The new ridge candidate φ_t^c is given by

$$\varphi_t^c = \{\varphi_{t-1}(0), \ldots, \varphi_{t-1}(b) + \epsilon, \ldots, \varphi_{t-1}(B-1)\} .$$

   Compute the corresponding penalty $\mathcal{E}(\varphi_0)$.

   - If $\mathcal{E}(\varphi_t^c) < \mathcal{E}(\varphi_{t-1})$ (good move), set $\varphi_t = \varphi_t^c$.
   - If not (adverse move), then set randomly $\varphi_t = \varphi_t^c$ with probability

$$p_t = \exp[(\mathcal{E}(\varphi_{t-1}^c) - \mathcal{E}(\varphi_t))/T_t\} , \qquad (7.33)$$

   and $\varphi_t = \varphi_{t-1}$ with probability $1 - p_t$.

   Update the temperature T_t.
3. Iterate until the ridge does not move for a certain number of time steps.

The last point to be specified in the algorithm is the temperature schedule, i.e., the rate at which the parameter T_t goes to 0 when t increases. The goal is to lower the temperature T_t to freeze the system, but we do not want the temperature to go to zero too fast because we do not want the system to get frozen in a local minimum of the penalty function. It has been proven rigorously that the choice

$$T_t = \frac{T_0}{\log_2 t} \qquad (7.34)$$

guarantees the convergence to a global minimum, provided the initial temperature is large enough. Other choices are possible, but we shall not use them. See [201] for a discussion of the properties of the various temperature schedules.

An example of application of this corona method is given in Figure 7.2. See the details later.

Obviously, the snake penalty function can also be minimized by a similar combinatorial optimization procedure. A discrete snake is a series of nodes:

$$(\rho(n), \varphi(n)) \in \{0, 1, \ldots, B-1\} \times \{0, 1, \ldots, A-1\}, \; n = 0, \ldots, N-1,$$

and a discrete version of the penalty function is easily derived. We refer the reader to [67, 68] for details on this implementation.

7.3.2 The "ICM" Method

A classical alternative to simulated annealing algorithms is provided by the so-called ICM (or Iterated Conditional Modes) algorithms [45], which may be seen as a way of approaching steepest descent methods. In this approach again, one starts with an initial guess $b \hookrightarrow \varphi_0(b)$ for the ridge function, and one searches iteratively the minimizer of the penalty function. The iterative scheme goes as follows:

1. Start with an initial guess

$$\varphi_0 = \{\varphi_0(0), \varphi_0(1), \ldots, \varphi_0(B-1)\}$$

   and compute the corresponding value of the penalty function $\mathcal{E}[\varphi_0]$. Choose a maximal number of iterations $t_{\max}$.

2. For $t = 0$ to $t = t_{\max}$ do
   - For $b = 0$ to $b = B - 1$, do
     - Set $\tilde{\varphi}(k) = \varphi_t(k)$ for $k = 0, \ldots, b-1$ and $\tilde{\varphi}(k) = \varphi_{t-1}(k)$ for $k = b+1, \ldots, B-1$.
     - $\varphi_t(b) = \arg\min_{\tilde{\varphi}(b)} \mathcal{E}(\tilde{\varphi})$
   - If $\varphi_t = \varphi_{t-1}$, return.

Such a method produces good results as long as the SNR is large enough (i.e., the noise is relatively weak).

7.3.3 Choice of Parameters

The choice of the parameters in the penalty functions, as well as in the temperature schedule, is traditionally a touchy business. Various things, for example, the required precision and the noise level, can be very influential in the success or failure of the procedure. However, there are some general rules of thumb that should be followed.

Let us consider first the parameters introduced in the smoothness part of the penalty function. Clearly, they have to be matched to the function $M(b, \omega)$: they should be proportional to some characteristic value of $M(b, \omega)$, for example the average.

Let us now consider the parameters specific to the simulated annealing algorithm. The most important one is the initial temperature T_0 introduced in Equation (7.34). Again, it follows from formula (7.33) that T_0 must be matched to $M(b, \omega)$ in the same way as the smoothness parameters. Theoretical results (see, for example, [201]) show that it is sufficient to choose T_0 at least as large as the highest hill to climb in the energy landscape (hence $\max(M) - \min(M)$ is a rough upper bound for this initial

T_0). We shall not elaborate here on more specific details of the parameter choice problem. This would require sophisticated arguments from the fine analysis of the minimization problem and the algorithms used.

7.3.4 Examples

We illustrate the use of the two annealing-based ridge detection procedures on real life signals.

The Corona Method

We first consider the example of a sonar signal emitted by certain species of bats. This signal is remarkably "clean." It presents a main frequency modulated component, whose instantaneous frequency is close to an hyperbolic function of time. We display in Figure 7.2 the bat sonar signal (top) and its wavelet transform modulus (middle.) We choose Morlet's wavelet for the analysis and the transform is computed on 4 octaves, with 15 intermediate scales per octave. The modulus shows a sharp concentration around a ridge, which is displayed at the bottom of the figure. The ridge is estimated using the simulated annealing method (i.e., the corona method.) See Example 7.3 for the list of S commands needed to produce these results with the Swave package. Since this is an easy case, the ICM method produces the same result (not presented here).

The same experiment was run with the same signal after adding a white Gaussian noise, with input signal to noise ratio equal to −5 dB. As can be seen in Figure 7.3, the presence of a coherent structure in the wavelet transform modulus is hardly distinguishable. Nevertheless, the annealing algorithm was able to get a relatively precise estimate for the ridge, except at its boundaries where the signal's energy becomes too small. See example 7.4 for the list of S commands used to generate that example.

The Snake Method

We now revisit the acoustic backscattering signals we already discussed in Chapter 3. We consider the CWT of such signals as an illustration for the snake method. The CWT was computed on 6 octaves, with 8 intermediate scales per octave, and the snake algorithm was used twice, to estimate the frequencies of the two chirps. The S commands used are given in Example 7.6, and the results are displayed in Figure 7.4. Clearly, the snake algorithm had no difficulty to give a precise estimate of the ridges since the signal was relatively "clean." Notice also that the boundaries of the structures have been properly identified.

As another example, we show in Figure 7.5 the results obtained with the snakes algorithm in the case of the noisy bat signal discussed earlier. Again

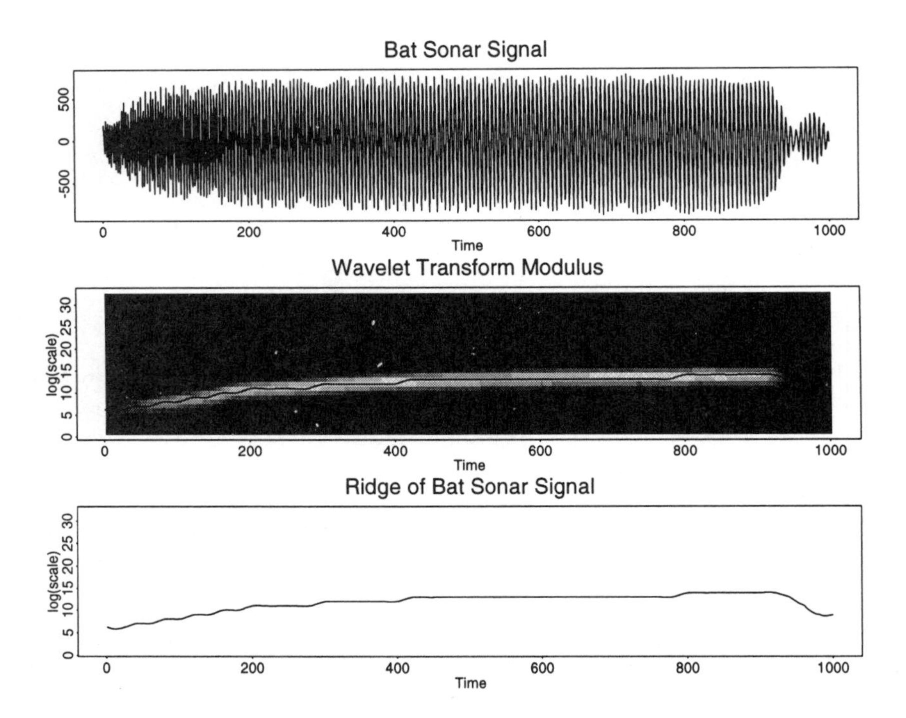

Figure 7.2. Bat sonar signal, its CWT modulus and the corresponding ridge estimate given by the corona algorithm.

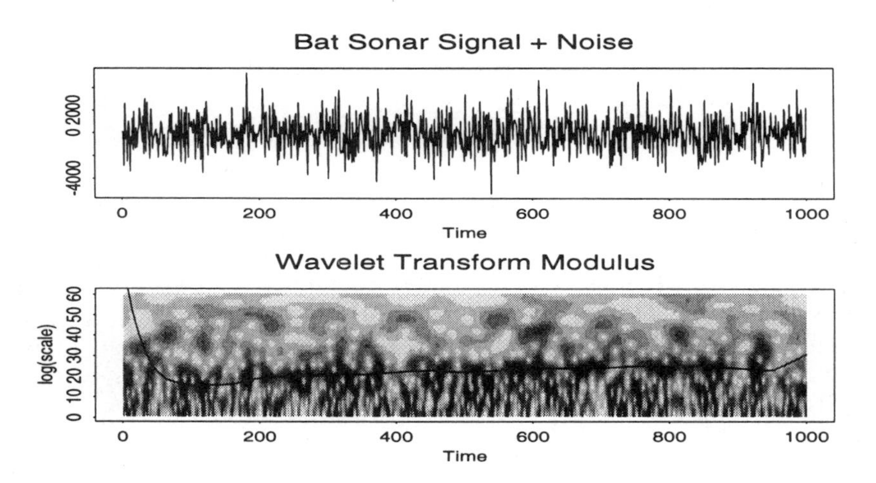

Figure 7.3. Noisy bat sonar signal (SNR=−5 dB), its CWT modulus, and the corresponding corona ridge estimate (see Example 7.4).

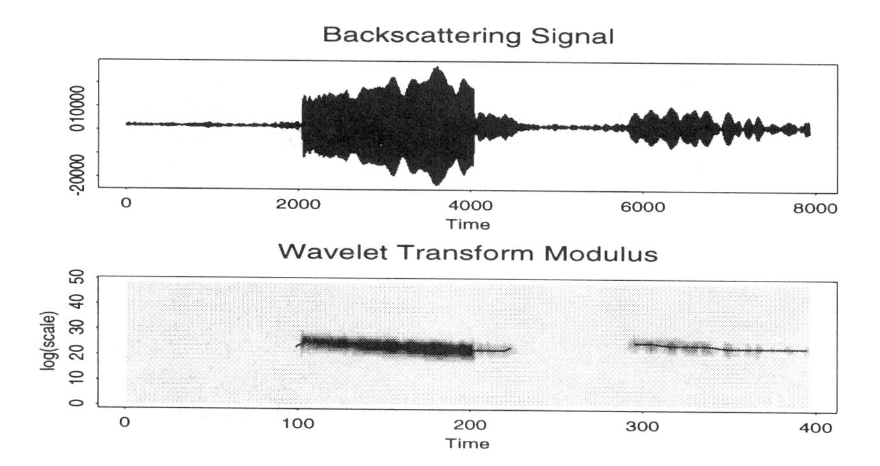

Figure 7.4. Acoustic backscattering signal, and corresponding snake estimates of the ridges (see Example 7.6).

the ridge is correctly estimated, except for its boundaries. The details of the computation are given in Example 7.5. Notice that the values of the smoothness penalties and of the initial temperature constant have to be taken larger than they were in the non-noisy case. This was needed because the SNR is quite low in that particular example.

7.4 The Multi-ridge Problem and the "Crazy Climbers" Algorithm

In many instances, the signal to be analyzed is characterized not only by one but by several frequency and amplitude modulated components. This is the case for signals well approximated by the expression given the model (7.1) when $L > 1$. The importance of this model comes from the fact that it is regarded as a good model for voiced speech [222]. The methods described earlier are clearly not adapted to such problems, and it is necessary to turn to better suited procedures capable of handling multiple ridges. We shall describe here one of these, namely the *crazy climber* algorithm. Like the simulated annealing-based ridge estimation methods described above, it relies on Monte Carlo Markov Chain (MCMC for short) simulation ideas. But its philosophy is very different: in particular, it does not attempt to minimize a penalty function of any kind. The stochastic system is introduced to generate (weighted) occupation densities which (hopefully) draw the ridges.

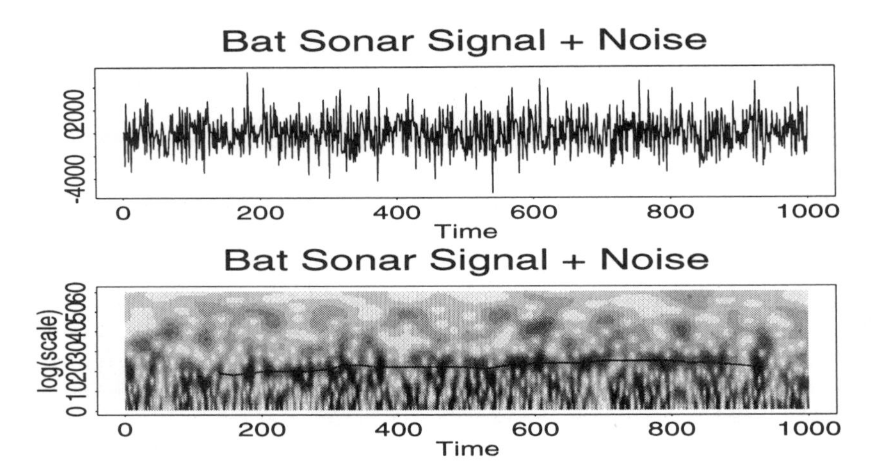

Figure 7.5. Noisy bat sonar signal (SNR=−5 dB) its CWT modulus and the corresponding ridge, estimated with the snake method (see Example 7.5).

Before we engage in the presentation of this algorithm, we spend some time putting the previously described methods into perspective. Except for the *Marseille method* described in Section 7.2.1, which may be implemented with a random access to the time-frequency plane, they all deal with *post-processing* of a time-frequency distribution. More precisely, the time-frequency representation is used as some sort of energy landscape with an attractive "potential," and the curves are determined by some sort of least action principle. The crazy climber algorithm follows the same philosophy. The main idea is to use the time-frequency representation to generate a random walk on the time-frequency plane, in such a way that the random walker is attracted by the ridges of the hills (hence the name "climber"). In addition, the random walk is done at a given "temperature" which as before, changes with time. The lower the temperature, the stronger the attraction of the ridge. The temperature is gradually decreased in the process, as in simulated annealing methods, but contrary to the simulated annealing procedure, the motion of the walker is unrestricted in one direction and the walker is never stuck on a ridge: the walker will always escape a ridge if the latter does not disconnect the part of the time-frequency plane on which the walk takes place. But because of the temperature schedule, each climber is expected to spend most of his time walking along one ridge or another. Therefore, if instead of following the walk of one climber we keep track of a large number of climbers and if we assign to each point of the time-frequency domain the number of times it has been "visited" by a climber, one obtains an *occupation measure*, which is expected to be sharply localized near the ridges. Before we turn to a more precise description of the algorithm, we want to emphasize one more important difference between the crazy climber algorithm and the simulated annealing procedures used so far. It is the complexity of the space in which the random walk takes place. The crazy climbers are evolving in a sub-domain of the two-dimensional time-frequency plane and especially after discretization, this space is very simple (and even small): we need a large number of climbers because we want to plot occupation densities, but each individual walk is very simple. On the other hand, the random walks which we introduced in the MCMC simulations leading to the corona and the snake methods were taking place in a very large set, the space of all the possible graphs of functions or the space of all the possible snakes. One uses only one sample realization of the walk, but this walk takes place in a very complex space.

We shall use the following notation: Let $f \in L^2(\mathbb{R})$, and $M(k,j), j = 0,\ldots,A-1, k = 0,\ldots,B-1$, be a discrete time-frequency representation of f. The typical examples are given by the discretized square modulus of the CGT or the CWT of f. We shall term "vertical" the A direction, and "horizontal" the B direction. The algorithm involves two stages, a "stochastic relaxation" stage and a "chaining" stage which we describe independently.

7.4.1 Crazy Climbers

We describe first the stochastic part of the algorithm, namely the generation of the moves of the climbers. We shall come to the construction of the occupation measures later on.

Since the climbers are going to move independently of each other and since several climbers can occupy the same site at the same time, it is enough to describe the statistical motion of one single climber. The first stage is an iterative method; as before, we denote by t the time variable of the iteration.

1. At $t = 0$, choose randomly (with uniform probability distribution) the position $X(0)$ of the climber on the grid,

   $$\Gamma = \{(k,j);\ k = 0,\ldots,B-1,\ j = 0,\ldots,A-1\},$$

   and initialize the temperature of the climber to your favorite temperature T_0.

2. Given the position $X(t) = (k,j)$ at time t, the position $X(t+1) = (k',j')$ at time $t+1$ is determined according to the following rule.

   (a) • If $k \geq 1$ and $k \leq B-2$, then $k' = k+1$ with probability 1/2 and $k-1$ with probability 1/2.

- If $k = 0$ then $k' = 1$.
- If $k = B - 1$ then $k' = B - 2$.

(b) Once the move in the k (i.e., time) direction has been made, the possible vertical move is considered. The climber tries to move up ($j' = j + 1$) or down ($j' = j - 1$). But unlike the horizontal move, the vertical move is not necessarily made: the probability of the vertical move depends on the temperature T_t.

- If $M(k', j') > M(k', j)$, then the vertical move is done: $X(t + 1) = (k', j')$.
- If $M(k', j') < M(k', j)$, then $X(t + 1) = (k', j')$ with probability

$$p_t = \exp\left[\frac{1}{T_t}\left(M(k', j') - M(k', j)\right)\right] ,$$

and $X(t + 1) = (k', j)$ with probability $1 - p_t$.

Update the temperature.

3. The climber "dies" when its temperature is below a certain threshold.

Again it is necessary to set a temperature schedule in such a way that when the temperature goes to zero, the trajectories concentrate on the ridges.

Let us turn to the generation of the occupation measures. These are (discrete) measures on the domain $\{0, \ldots, B - 1\} \times \{0, \ldots, A - 1\}$ constructed as follows: Each time a site is visited by a climber, the corresponding value of the measure is incremented appropriately. More precisely, at each time t we consider the two occupation measures:

$$\left\{ \begin{array}{lcl} \mu^{(0)}(t) & = & \frac{1}{N}\sum_{\alpha=1}^{N} \delta_{X_\alpha(t)} , \\ \mu(t) & = & \frac{1}{N}\sum_{\alpha=1}^{N} M(X_\alpha(t))\, \delta_{X_\alpha(t)} . \end{array} \right. \tag{7.35}$$

The first one is obtained by putting at the location of each climber a mass equal to $1/N$, and the second one by replacing the uniform mass by the value of the time-frequency representation at the location of the climber. These occupation measures are random and we would like to work with their expected (average) values. A typical ergodic argument can be used to justify the fact that one way to get these expected occupation measures is enough to compute the time averages of a single realization. In other words, the quantities of interest to use will be the integrated occupation measures, defined by

$$\left\{ \begin{array}{lcl} \mu_T^{(0)} & = & \frac{1}{T}\sum_{t=1}^{T} \mu_t^{(0)} , \\ \mu_T & = & \frac{1}{T}\sum_{t=1}^{T} \mu_t . \end{array} \right. \tag{7.36}$$

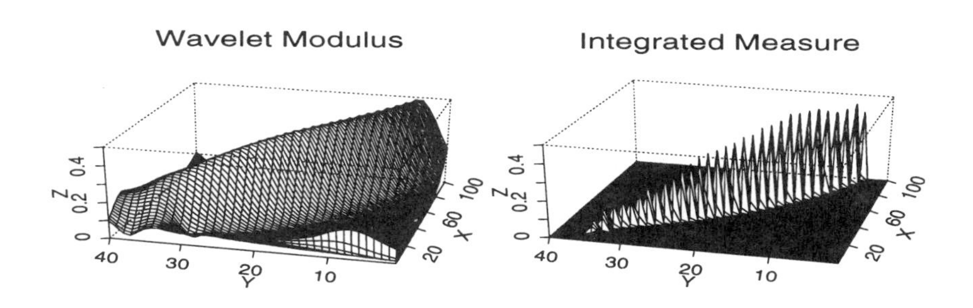

Figure 7.6. CWT modulus and integrated occupation measure (output of the crazy climbers algorithm) for an academic chirp signal.

As the time t increases, the temperature decreases and the climbers spend more time on or near the ridges, so that the occupation measures tend to be sharply localized near ridges.

As an example, we display in Figure 7.6 perspective views of the CWT modulus and the integrated occupation measure for the chirp signal described in Example 4.5. As can be seen on the two plots, the smooth modulus has been turned into a "singular" measure, concentrated on a curve.

7.4.2 Chaining

The output of the crazy climbers stage is a discrete measure on the time-frequency plane. The measure is expected to concentrate around one-dimensional structures (ridges). Let $\mu(k,j)$ denote the mass affected by either one of the two measures in (7.36) to the point (k,j) of the time-frequency plane. The second stage transforms the measure $\mu(k,j)$ into ridges. This may be achieved through a classical chaining algorithm, based on the following two steps.

- Thresholding of the density function, output of the crazy climbers stage: The values of the density function below a certain fixed threshold τ are forced to zero. Let us denote by $\rho(k,j)$ the thresholded measure:
$$\rho(k,j) = \begin{cases} \mu(k,j) & \text{if } \mu(k,j) \geq \tau \\ 0 & \text{otherwise .} \end{cases}$$
- "Propagation" in the horizontal (left and right) direction. We choose a parameter b_{step}. This parameter b_{step} essentially gives the mini-

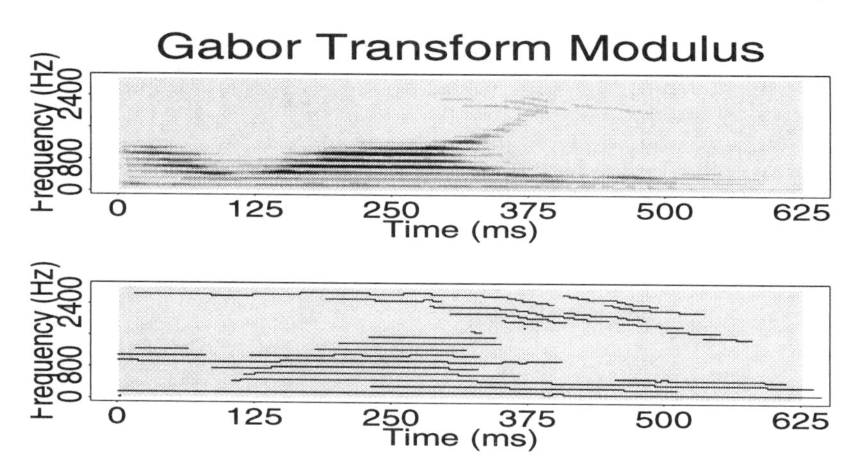

Figure 7.7. Modulus of the CGT (top) and ridge estimates (bottom) of the speech signal /How are you ?/ as produced by the crazy climber algorithm. The Swave commands needed to produce these results are given in Example 7.7.

mal length of a gap between two ridges. Let us describe the "right-propagation" (the "left-propagation" is identical). Given a (discrete) ridge $R = \{(k, \omega_k);\ k = k_m, \dots, k_M\}$, check whether there is a value $k_R \in \{k_M + 1, \dots, k_M + b_{\text{step}}\}$ such that $\rho(k, \omega_{k_M} + \delta) \neq 0$, where $\delta = \pm 1$. If yes, extend the ridge to $(k_R, \omega_{k_M} + \delta)$. If not, the ridge stops at (k_M, ω_{k_M}).

All together, the result is a series of ridges $R_1, \dots, R_n$.

7.4.3 Examples

Crazy Climbers for the Gabor Transform of a Speech Signal

As illustration, we display in Figure 7.7 an example of a Gabor transform and the associated series of ridges given by the crazy climbers algorithm. The (narrow band) Gabor transform was computed for 70 values of the frequency variable, between 0 and 2400 Hz, and processed by the crazy climbers method. As can be seen in Figure 7.7, the algorithm produced a series of 26 ridges. Remarkably enough, the ridge structure obtained by this method is completely different from the results obtained in [99, 208, 209]. This effect is due to the fundamentally different description of the time-frequency plane given in these two works (essentially, wavelets and narrow-band Gabor). We shall revisit this example a last time when we illustrate the reconstruction algorithms in the next chapter.

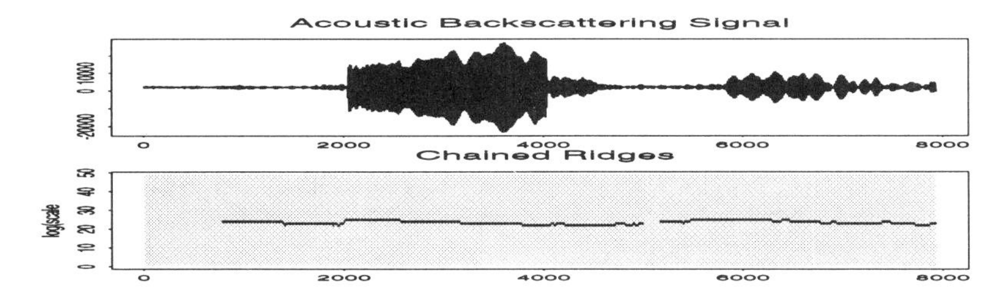

Figure 7.8. Acoustic backscattering signal, and output of the crazy climber method.

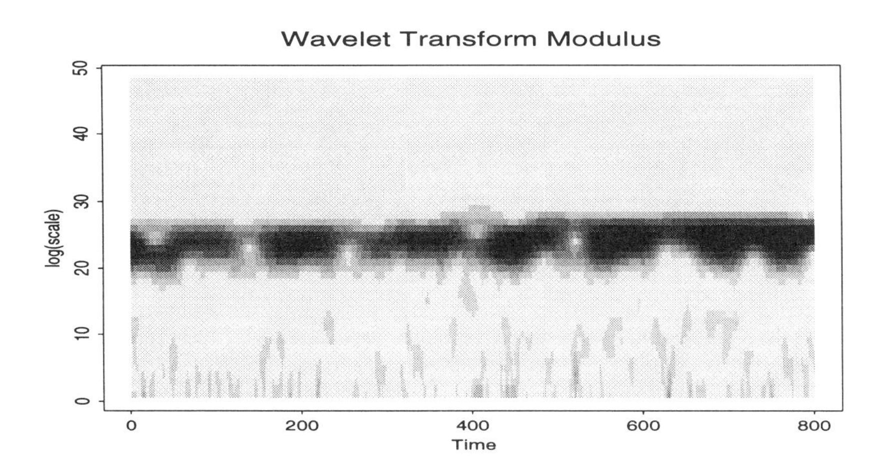

Figure 7.9. CWT of the space between the two chirps in the acoustic backscattering example.

Climbers and Wavelets: Acoustic Backscattering Revisited

Let us now revisit for the last time the acoustic backscattering signal we have analyzed throughout this volume. Running the crazy climbers algorithm on the same data as before, we find (see Figure 7.8) only one long ridge, instead of two as expected from the previous analysis. This motivates a closer look at the space in between the two chirps. This is done in Figure 7.9, in which one can clearly see that there is still a frequency modulated component in the region previously thought to be ridge free. Such a component is of much smaller amplitude than the two chirps, which explains why it was not detected in the previous analyses.

7.5 Reassignment Methods

Let us conclude this chapter with a family of time-frequency methods aimed at "improving the readability" of time-frequency representations, and hopefully facilitating the salient features extraction problem. For this reason, reassignment methods could be conceived as a pre-processing operation preceding the implementation of the procedures described earlier in this chapter, and that is why we consider them here. But let us also stress that they may as well be used for different purposes, and this should be clear from the discussion of a speech example.

7.5.1 Overview

In the cases of the CGT and the CWT, the introduction of a window function allows a more precise time localization of the signals. But the unavoidable price to pay is a weakening of the frequency localization. For instance, the CGT of a sine wave (see Figure 3.4) is indeed localized around a horizontal (i.e., fixed frequency) line, but has a certain frequency width which depends on the window's bandwidth. In some sense, one may say that a part of the energy is not located where it should be.

The goal of the reassignment algorithms is to move the coefficients of a time-frequency representation in the time-frequency plane, so as to improve the readability of a transform, changing the energy distribution so doing. The general scheme can be described as follows. Let $\mu(b,\omega)$ be a time-frequency representation, viewed as a *measure* on the time-frequency plane $\mathbb{R}^2$. Notice that we use the letter μ instead of the letter M used thus far. This is to emphasize the fact that we do not have the modulus (or the square of the modulus) of a transform in mind. So μ can very well be complex valued. The whole procedure depends upon the choice of a map $\mathcal{R}:\mathbb{R}^2 \to \mathbb{R}^2$ which we will call a *reassignment map*. Then the reassigned representation takes the form of the measure defined by

$$\mu_r(\Omega) = \mu(\{(b,\omega) \in \mathbb{R}^2;\, \mathcal{R}(b,\omega) \in \Omega\}), \qquad \Omega \subset \mathbb{R}^2 .$$

In general, the reassignment map $\mathcal{R}$ is chosen so as to "reassign" the energy (and more precisely the values of the representation) where its "true location" is *expected* to be. Since this "true location" is of course signal dependent, we are naturally led to nonlinear transformations.

For the sake of illustration we outline the reassignment strategy used in [23]. The input time-frequency representation, say $\mu_0(b,\omega)$, is first used to compute at each point (b,ω) in the time-frequency plane a local time average $\langle b\rangle(b,\omega)$ and a local frequency average $\langle\omega\rangle(b,\omega)$. This fixes the reassignment map $\mathcal{R}(b,\omega) = (\langle b\rangle(b,\omega), \langle\omega\rangle(b,\omega))$ and, according to the strategy described earlier, the coefficient $\mu_0(b,\omega)$ of the representation is "moved" to the point $(\langle b\rangle(b,\omega), \langle\omega\rangle(b,\omega))$. The results of the implementation of several variations on this reassignment procedures are given in [23] for the analysis of various bilinear (and more general nonlinear) time-frequency representations.

The Crazy Climber Algorithm as Reassignment It is interesting to point out the similarity existing between the general reassignment philosophy and the crazy climber algorithm. The latter does not deal with the actual values of the time-frequency representation, and it will not be possible to interpret the output in terms of a representation. Nevertheless, if one thinks of the modulus (or more precisely of the square of the modulus) of the transform as a distribution (in the probabilistic sense) of energy over the two-dimensional plane, the choice of the random mechanism governing the statistics of the motions of the climbers is to reach an equilibrium distribution given by the (square) modulus as rescaled vertically (i.e., for b fixed) according to the value of the temperature. The effect of the lowering of the temperature is to squeeze the equilibrium distribution toward the high values, in other words, the ridges. So in some sense, the crazy climbers try to rearrange themselves in the time-frequency plane to give a statistical distribution which is rearranged by their statistical motion and which is squeezed into "delta" like distribution on the ridges.

7.5.2 The Synchrosqueezed Wavelet Transform

Reassignment methods have been described in great generality in [23], but we shall now concentrate on the simple version which Daubechies and Maes described in [99, 208] in the case of the CWT, and which we implemented in the Swave package.

Synchrosqueezing This version of the reassignment strategy is very much in the spirit of the differential algorithm described in Section 7.2.1, i.e., the algorithm provided by the numerical resolution of Equation (7.18.) The main remark is that when we compute the phase derivative

$$\omega(b,a) = \partial_b \arg T_f(b,a)$$

of the CWT at some point (b,a), we actually compute a new frequency, and thus a corresponding new scale:

$$\tilde{a} = \frac{\omega_0}{\omega(b,a)} ,$$

which is naturally associated to the coefficient $T_f(b,a)$. This yields a natural candidate for the reassignment map:

$$\mathcal{R}(b,\omega) = \left(b, \frac{\omega_0}{\partial_b \arg T_f(b,a)}\right) . \tag{7.37}$$

As noticed earlier in the discussion of the differential approach to the ridges, this computation is done for each fixed value of the time variable b.

More precisely, the synchrosqueezing may be described as follows: Let $f \in L^2(\mathbb{R}))$ and let $T_f(b,a)$ be its CWT. Given a value $\tilde{a}$ for the scale variable, introduce the following subset of the scale semi-axis:

$$\Omega_{\tilde{a}}(b) = \left\{a \in \mathbb{R}^+ ;\ \partial_b \arg T_f(b,a) = \frac{\omega_0}{\tilde{a}}\right\} . \tag{7.38}$$

Then the reassigned CWT, or *synchrosqueezed wavelet transform*, is defined by

$$T_f^r(b,\tilde{a}) = \frac{1}{c_\psi} \int_{u \in \Omega_{\tilde{a}}(b)} T_f(b,u) \frac{du}{u} . \tag{7.39}$$

We used the integral notation, which is more suggestive when it comes to computations, even though this integral does not always make perfectly good sense. A rigorous definition can be obtained in term of change of measure as in our original definition.

Remark 7.6 As explained in the beginning of the chapter, the technical difficulty is the computation of the phase derivative. The latter may be implemented through finite differences, but this creates a difficult phase unwrapping problem. It is often much more convenient to remark that the phase derivative of $T_f(b,a)$ is nothing but the derivative of the imaginary part of $\log T_f(b,a)$. This can only be done when $|T_f(b,a)| \neq 0$, of course, but the phase derivative would make no sense anyway! Then a simple alternative to finite differences amounts to computing two wavelet transforms, one with ψ and the other with ψ', and using them to compute $\partial_b \log T_f(b,a)$.

Remark 7.7 It is interesting to compare the synchrosqueezing to the differential algorithm discussed in Section 7.2.1. The latter is based on the numerical resolution of Equation (7.18) and it simply amounts to looking for the fixed point set of the reassignment transform $\Omega_{\tilde{a}} \to \tilde{a}$. Moreover, even though the synchrosqueezed transform is not the wavelet transform of a signal any more, it is interesting to notice that, by the very nature of the construction of the reassignment, the simple reconstruction formula given in Chapter 4 is still valid after the synchrosqueezing.

Remark 7.8 Clearly, the same method may be applied to the Gabor transform. Given the CGT $G_f(b,\omega)$ of $f \in L^2(\mathbb{R})$, it suffices to set

$$G_f^r(b,\tilde{\omega}) = \int_{u\in\Omega_{\tilde{\omega}}(b)} G_f(b,u)du \ , \tag{7.40}$$

where the sets $\Omega_{\tilde{\omega}}(b)$ are now defined as

$$\Omega_{\tilde{\omega}}(b) = \{\omega \in \mathbb{R}^+ ;\ \partial_b \arg G_f(b,\omega) = \tilde{\omega}\} \ . \tag{7.41}$$

In this case, the reassignment map is even simpler since it is given by

$$(b,\omega) \to \mathcal{R}(b,\omega) = (b, \partial_b \arg G_f(b,\omega) \ .$$

Chaining It is important to notice that synchrosqueezing does not do feature extraction by itself. Like the crazy climbers algorithm, it yields a measure in the time-frequency domain, and such a measure has to be post-processed by a chaining algorithm as before. The chaining method is still adapted to some situations, but we refer to [209] for the description of an alternative better suited to speech analysis by synchrosqueezing.

An example To illustrate the synchrosqueezing algorithm, we display in Figure 7.10 the CWT and the synchrosqueezed CWT of the speech signal we already studied with CGT and the crazy climbers method. Both transforms were computed on 6 octaves (i.e., 6 powers of 2), with 8 intermediate scales between 2 consecutive powers of 2. Obviously, the description of the signal given by the wavelets is completely different from that given by the narrow band Gabor transform. As we have seen in Figure 3.11, the signal is an harmonic signal (at least locally) in the sense that it contains a fundamental (slowly varying) frequency and a few harmonics. Figure 7.10 may be interpreted as follows. At the scale corresponding to the fundamental frequency, the wavelet's bandwidth is such that there is only one harmonic component (the fundamental) in the analyzed frequency range. In contrast, at the scale corresponding to the first harmonic, the wavelet's bandwidth is twice the previous one, and the wavelet can hardly discriminate between harmonics 1 and 2. This produces interferences which are clearly seen on the images. The effect of squeezing is simply to improve the concentration, but it is not enough to produce ridges or one-dimensional structures. There, a robust additional chaining step is needed.

Remark 7.9 Synchrosqueezing a narrow-band Gabor transform would have produced a result similar to that produced by the crazy climbers algorithm in Figure 3.11.

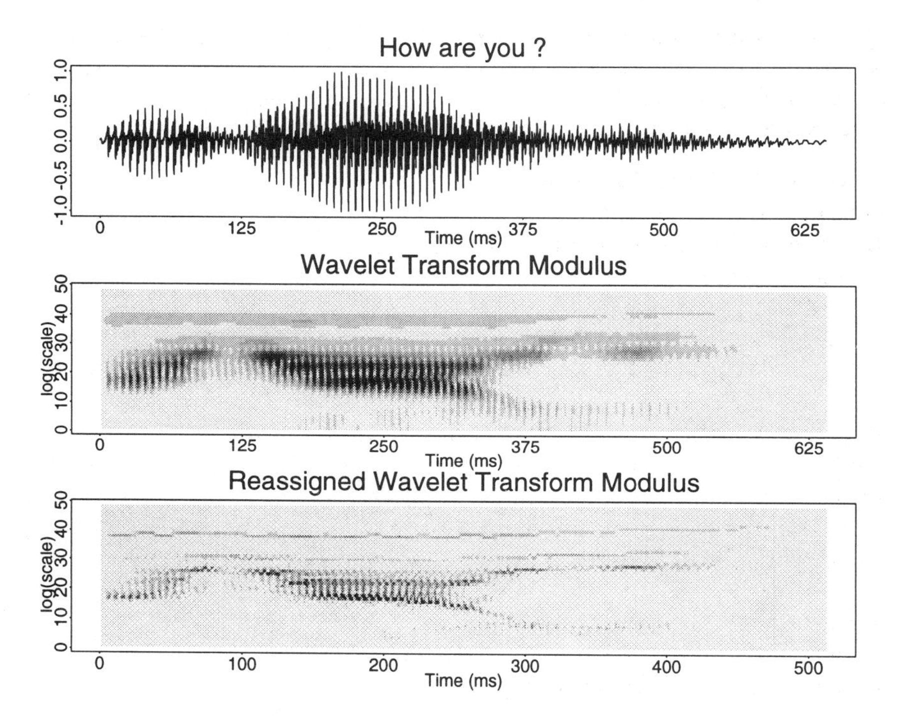

Figure 7.10. Modulus of the CWT and of the reassigned CWT of the speech signal /How are you ?/. The top plot is the original signal; the middle plot shows the modulus of the CWT, and the bottom plot, the modulus of the reassigned CWT as generated in Example 7.9.

7.6 Examples and S-Commands

We give new examples illustrating the algorithms just described, and we describe the Swave command needed to obtain the results and to produce the plots of this chapter.

7.6.1 Transform Modulus Maxima

Two S-plus functions permit to determine the global and local maxima of a time-frequency representation (maxima with respect to the frequency or scale variable, for fixed values of time variable): `tfgmax` (global maxima) and `tflmax` (local maxima).

Example 7.1 Chirp (continued)
Generate the chirp and compute the continuous wavelet transform (see Example 4.5) and compute the global maxima:

```
> image(Mod(cwtchirp))
> title(‘‘Wavelet transform modulus’’)
> chirpgmax <- tfgmax(Mod(cwtchirp))
> title(‘‘Global maxima’’)
```

Example 7.2 Chirp + Sine Wave + Gaussian White Noise
Generate the chirp and the sine wave, compute their continuous wavelet transforms (see Examples 4.5 and 4.3: the wavelet transform being a linear transform, the transform of the sum equals the sum of transforms) and compute the local maxima:

```
> par(mfrow=c(3,1))
> gnoise <- rnorm(512)
> cwtgnoise <- cwt(gnoise,5,12,plot=F)
> ncs <- chirp + sinwave + gnoise
> tsplot(ncs)
> title(‘‘Noisy chirp + sine wave’’)
> cwtncs <- cwtsinwave + cwtchirp + cwtgnoise
> image(Mod(cwtncs))
> title(‘‘Wavelet transform modulus’’)
> cslmax <- tflmax(Mod(cwtncs))
> title(‘‘Wavelet transform local maxima’’)
```

7.6.2 Ridge Detection Methods

Before describing the S functions of the Swave package, let us start with some general remarks on practical implementations of the ridge detection methods.

Any implementation of the methods described earlier has to deal with discrete signals, discrete (wavelet and Gabor) transforms, and thus discrete ridges. For the sake of the present discussion we shall restrict ourselves to the case of the wavelet transform and we shall assume that the modulus of the transform is given in a discretized form $M(b,a)$, $b = 0,\dots,B-1$, $A = 0,\dots,A-1$ (typically, B equals a few hundreds and $A \approx 50$). Let us consider the implementation of the "corona" method first. Since a ridge is assumed to be a slowly varying function, it does not necessarily make sense to search in the whole (finite but large) space of ridges whose cardinality is A^B. Most of the time, it is convenient to deal with lower dimensional subspaces, in other words, to "sample" the ridge with a sampling period smaller than that of the signal. In the implementation of the Swave package, such a "subsampling" is accompanied with a smoothing of the transform modulus (with respect to the time variable b).

The situation with the "snake" approach is similar, except that the dimensionality of the space in which the snake is searched can even be much larger: for example, the space of discrete snakes with n nodes has cardinality $(AB)^n$. Again, it is often convenient to (smooth and) subsample the transform in the b direction.

Swave has several S functions to solve the problem of ridge detection, under the hypothesis of existence of a single ridge. These functions implement the algorithms that have been developed within this chapter, namely the corona method (function `corona`) and its variant described in Section 7.2.6, Equation (7.31) (function `coronoid`), the ICM method (function `icm`), the snake method (function `snake`), and the corresponding variant (see equation (7.32) in Section 7.2.6) (function `snakoid`).

All these function require a certain number of arguments, including a time-frequency representation `tfrans` and an initial guess for the ridge. The initial guess argument, say `guess`, is a 1D array when the ridge is modeled as the graph of a function (`corona`, `coronoid`, and `icm`), and a pair of 1D arrays for the snake case (`snake`, `snakoid`). Also to be specified are the smoothness penalty constants μ and λ in Equations (7.23), (7.31), and (7.23) (these come by pairs in the snake case), the subsampling rate `subrate` alluded to earlier, and the initial temperature for the simulated annealing method (`temprate`). The functions return a ridge in the form of a 1D array (a pair of arrays in the snake case) and the values of the penalty as a function of time (the time of annealing), subsampled by a factor `costsub`.

Let us consider the example displayed in Figures 7.2 and 7.3, to illustrate the use of the functions `corona` and `coronoid`. We now give the S commands used to generate these figures.

Example 7.3 Bat Sonar Signal
Compute the CWT of the signal:

```
> cwtbat <- cwt(bat,4,15)
```

Estimate the ridge:

```
> guess <- numeric(1000)
> guess[] <- 30
> rbat <- corona(cwtbat[201:1200,],guess,temprate=3,subrate=20,
+         mu=10,lambda=10)
```

The same problem may be attacked using the function `coronoid`. *With that function, the initial guess may be taken far away from the true ridge:*

```
 >guess[] <- 50
> tmp <- coronoid(cwtbat[201:1200,],guess,temprate=3,subrate=20,
+         mu=10,lambda=1)
```

Example 7.4 Noisy Bat Sonar Signal
Compute the CWT of the signal:

```
> tsplot(nbat5[201:1200])
> cwtnbat <- cwt(nbat5,4,8)
```

Estimation of the ridge (the initial temperature parameter and the smoothness penalty have to be raised in order to accommodate the presence of noise):

```
> guess <- numeric(1000)
> guess[] <- 30
> rnbat5 <- corona(cwtnbat[201:1200,],guess,subrate=50,
+         temprate=5,mu=100)
```

Example 7.5 Noisy Bat Sonar: Snake Estimation
Enter initial guess and estimate the ridge:

```
> guessA <- numeric(30)
> guessA[] <- 30
> guessB <- 50 + 900*(1:30)/30
> snakenbat <- snake(Mod(cwtnbat[201:1200,]),guessA,guessB,
+         temprate=10,muA=2000,muB=200)
```

Example 7.6 Backscattering Signal with Snakes
Preparation: compute the CWT of the signal, and prepare two initial guesses for the two snakes to be estimated:

```
> back180 <- scan(''./signals/backscatter.1.180'')
> tsplot(back180)
> cwtback <- cwt(back180,6,8)
> guessA <- numeric(40)
> guessA[] <- 25
> guessB <- 1900 + 2600*(1:40)/40
> guessC <- 5900 + 2000*(1:40)/40
```

Estimation of snakes:

```
> backsn1 <- snake(cwtback,guessA,guessB,plot=F)
> backsn2 <- snake(cwtback,guessA,guessC,plot=F)
> lines(backsn1$B,backsn1$A)
> lines(backsn2$B,backsn2$A)
```

7.6.3 The Multiridge Problem

Swave contains a Splus function which implements the crazy climbers method (function `crc`) and the corresponding chaining method (function `cfamily`). The function `crc` needs three arguments: `tfrep`, the output of `cwt`, `cgt`, or related functions; `nbclimb`, the number of climbers; and `rate`, the initial temperature for the search. It returns a 2D array containing the chosen occupation measure (by default the weighted one).

The function `cfamily` chains the output of `crc` into a family of ridges. It outputs a sophisticated data structure containing information about the chained ridges. We refer to Part II for a more precise description of this data structure. Let us mention the utility `crfview`, which makes it possible to display the chained ridges either as an image or as a series of curves.

As an example, let us consider the CGT of the speech signal *How are you?* displayed in Figure 3.11 (top image.) The crazy climbers algorithm produces a series of 26 ridges, as follows.

Example 7.7 : CGT Analysis of *How are you?* (continued)
As before, the signal's name is HOWAREYOU, and its CGT is named cgtHOWAREYOU:

```
> par(mfrow=c(2,1))
> image(Mod(cgtHOWAREYOU))
```

Run the crazy climbers and chain:

```
> clHOWAREYOU <- crc(Mod(cgtHOWAREYOU),nbclimb=1000)
> cfHOWAREYOU <- cfamily(clHOWAREYOU)
> crfview(cfHOWAREYOU,twod=F)
```

Example 7.8 Acoustic Backscattering Signal (continued)
Reconsider the signal and its wavelet transform studied in Example 7.6, and estimate ridges using crazy climbers method, with very small threshold (this is possible since the noise is completely insignificant here):

```
> crcback <- crc(Mod(cwtback), nbclimb = 2000, bstep = 100)
> fcrcback <- cfamily(crcback,ptile=0.0005,bstep=10)
> npl(2)
> tsplot(back180)
> crfview(fcrcback)
```

Study the `cwt` *between the two chirps: there is indeed a frequency modulated component whose amplitude is very small, but still bigger than the noisy background:*

```
> image(Mod(cwtback[4700:5500,]))
```

7.6.4 Reassignment

The Splus function `squiz` implements the reassignment algorithm described in Section 7.5 (i.e., the "synchrosqueezing" time-frequency method). See Remark 7.6 for details on the implementation. The syntax is exactly the same as that of `cwt`. The results are displayed in Figure 7.10 and commented in Section 7.5.

Example 7.9 : *How are you?* revisited
As before, the signal's name is HOWAREYOU:

```
> par(mfrow=c(3,1))
> tsplot(HOWAREYOU)
> cwtHOWAREYOU <- cwt(HOWAREYOU,6,8)
> sqHOWAREYOU <- squiz(HOWAREYOU,6,8)
```

7.7 Notes and Complements

Most if not all the signal analysis algorithms are based on the detection and the identification of the salient features of a signal in a transform domain. The detection of the ridges in the energy landscape given by the (square) modulus (or the phase) of the continuous wavelet and Gabor transforms is an example of these general problems. See, for example, the papers by Delprat *et al.* [102] or by Tchamitchian and Torrésani, [272] or the more recent article [156] by P. Guillemain and R. Kronland-Martinet for an account of the differential methods alluded to in Section 7.2.1. The algorithms presented in this chapter have been introduced by the authors in [67] and [69] (see also [68] for another application). The penalty functions which we use are mainly based on the maximization of the time-frequency representation along ridges. This is a natural and already old idea, which has found several applications (see for example Flandrin's discussions in [129]). Similar ideas have more recently been used by Innocent and Torrésani in the context of gravitational waves detection.

Various forms of reassignment have been proposed since the original work of Kodera [194]. This idea was rediscovered and reintroduced more recently in works by Flandrin and Auger [23] in a general context, and Daubechies and Maes [99, 208, 209] in a speech processing context.

Chapter 8

Statistical Reconstructions

As part of our "signal analysis program," we devoted the previous chapters to methods able to identify and extract salient features of signals from their time/frequency transforms. We now turn our attention to another aspect of this program: the reconstruction of signals from the features previously identified. Since the part of the signal which is not reconstructed is often called noise, in essence what we are doing is denoising signals. We review some of the approaches which have been recently advocated. In particular, we review thresholding methods and the cases of reconstruction from the local extrema of the dyadic wavelet transform and from ridges of continuous wavelet and Gabor transforms, and we try to give a unified form of these algorithms. In order to put these methods in perspective, we start with generalities on non-parametric regression and spline smoothing methods.

8.1 Nonparametric Regression

The goal of nonparametric regression is to estimate a function f from a finite set of (possibly noisy) observations $\{y_j;\ j = 1, \ldots, n\}$. The typical situation faced by the statistician concerns observations of direct evaluations of the function. In this case the model is

$$y_j = f(x_j) + \epsilon_j, \qquad j = 1, \cdots, n. \tag{8.1}$$

In the best-case scenario, the specific sample values $x_1, \ldots, x_n$ of the time variable are equally spaced (the signal analyst would say *regularly sampled*) and the ϵ_j's are independent identically distributed (i.i.d. for short) mean zero random variables of order 2 with common variance σ^2. More generally, the observations can be of the values of linear functionals $L_1, \dots, L_n$ when computed on the unknown function f. This generalization will be of crucial importance in the examples treated later. The model (8.1) becomes

$$y_j = L_j f + \epsilon_j, \qquad j = 1, \cdots, n. \tag{8.2}$$

Obviously, the model (8.1) can be viewed as a particular case of the model (8.2) by merely choosing the linear functions L_j as the evaluations $L_j f = f(x_j)$. In the applications discussed later, the linear functionals L_j will be given by the values of the wavelet transform of f at points (b_j, a_j) of the time/scale plane, i.e., $L_j f = T_f(b_j, a_j)$, or by the value of the Gabor transform of f at points (b_j, ω_j) of the time/frequency plane, i.e., $L_j f = G_f(b_j, \omega_j)$. Notice that these linear functionals are complex valued. We shall eventually restrict the analysis to *real* observations, and for this reason, we shall have to double the number of linear functionals and choose for L_j the real part and the imaginary part of the transform in question separately.

There is an immense literature on the many nonparametric techniques of function estimation. See, for example, [57, 157, 159, 219, 233, 276] and especially [Hardle95] for an implementation in S. Most of these works concern the theory and the implementation of tree regressions, the nearest neighbor method, and the kernel regression. Some of them discuss expansion methods. The last is particularly well suited when it is possible to assume that the unknown function belongs to a Hilbert space for which a complete orthonormal basis is known, say $\{e_n;\, n \geq 0\}$. Repeating part of the introductory discussion of linear approximations given in Subsection 6.1.3, we use the fact that the exact value

$$f(x) = \sum_{n=0}^{\infty} \langle f, e_n \rangle \, e_n(x)$$

is first approximated by the value of a finite sum,

$$f^{(N)}(x) = \sum_{n=0}^{N} \langle f, e_n \rangle \, e_n(x),$$

and then, the coefficients $\langle f, e_n \rangle$ used in this finite sum are estimated from the data and the final function estimate is chosen in the form

$$\hat{f}(x) = \sum_{n=0}^{N} \hat{a}_n \, e_n(x).$$

This time-honored approach has been used originally with trigonometric, Hermite, Legendre, ... bases. Its performance depends upon the properties of the specific function f and how well (in other words, how fast) it can be approximated by finite linear combinations of the basis functions e_n and how accurately the coefficients

$$\langle f, e_n \rangle = \int f(x)\overline{e_n(x)}\, dx$$

can be estimated. It is obvious from the preceding discussion that this nonparametric estimation method is not restricted to orthonormal bases. Indeed, it can be used with biorthogonal systems, frames, or even Schauder bases (and especially unconditional bases) in more general Banach spaces. We shall not consider this level of generality here.

8.2 Regression by Thresholding

Expansion estimates can be implemented with wavelet bases. But as discussed in Subsection 6.1.3 of Chapter 6, a nonlinear approach is usually preferred. Instead of keeping the N first coefficients of the expansion (after all, there is not a single reason why these particular coefficients should be the most important), it is more natural to compute (estimate) all the coefficients of the expansion and to keep the coefficients (estimates) with the largest absolute values (setting all the other ones to zero). This estimation method is called estimation by coefficient thresholding. It has been implemented with wavelet orthonormal bases in several forms, both in the context of the approximation theory of stationary random functions discussed in Chapter 6 and in the present context of nonparametric regression. One form could be to keep only the N coefficients with the largest absolute values or to keep all the coefficients (or estimates) which are significantly different from 0. See, for example, [Bruce94] and the references therein. We discuss this last method in more detail in the next section.

In the present chapter we are interested in specific classes of signals (such as transients, fractals, chirps, and more general frequency modulated signals) embedded in noise with very low SNRs. In these situations, the performance of the orthogonal wavelet estimation procedures described earlier and in the next section do not compare well with the results of other denoising methods we present in this chapter. The latter are derived from the smoothing splines methodology (as presented in [122] and especially [289]) after pre-processing of a time-frequency/time-scale transform and a special sampling of the high energy regions of the time-frequency/time-scale plane.

8.2.1 A First Example

We first consider a simple (academic) example which should help illustrate some of the ideas of coefficient thresholding regression. We choose for

$\{e_n\}_n$ the trigonometric basis, and as a consequence, the thresholding is done in the (Fourier) transform domain. We revisit the sine-wave signal f_0 introduced in Example 3.3. We assume that the original signal is perturbed by a Gaussian white noise with variance $\sigma^2 = 1$. The corresponding signal,

$$f(x) = f_0(x) + \epsilon_x, \qquad x = 0, 1, \ldots, 511,$$

is shown in the top part of Figure 8.1 while the modulus of its FFT is shown in the middle part of the same figure. In signal processing, smoothing usually is achieved by low-pass filtering. This is done by setting to zero the high frequency coefficients and by coming back to the time domain by inverse FFT. The bottom part of Figure 8.1 shows the result of this procedure when the Fourier coefficients of the modulus of the Fourier transform outside the ranges 1:25 and 488:512 were set to 0. The reconstruction obtained by inverse Fourier transform after thresholding is superimposed on the top of the original signal. This reconstruction is reasonably good because the original signal is very sharply concentrated in the Fourier domain and the noise is not strong enough to mask the high energy concentrations.

Unfortunately, these two important features are not present in the applications we describe in this chapter, and this justifies the use of different methods.

The results of the reconstruction vary significantly with the choice of the set of coefficients forced to zero. This sensitivity to the choice of a parameter in the smoothing algorithm is not specific to the preceding example. It is fair to say that it is the main problem of the nonparametric regression procedures. Their results depend strongly upon the choice of a parameter which is usually called *the smoothing parameter* of the model.

In some cases, there are methods for estimating the optimal value of a smoothing parameter for a given problem. However, the optimal value often depends on a priori assumptions on the signal, which are generally hard to verify.

8.2.2 Thresholding Wavelet Coefficients

We consider now the thresholding of coefficients in the time domain based on the expansion of the input signal in a wavelet basis. To conform with the notation system used for wavelet bases, we use the notation $\{\psi_{jk}\}_{jk}$ instead of $\{e_n\}_n$. The method is based on the standard coefficient thresholding, i.e., setting to 0 the coefficients smaller than a pre-assigned level. It is especially well adapted to the wavelet bases because, in most decompositions on wavelet bases, the bulk of the coefficients are small (and hence can be set to 0 without affecting much the reconstructed signal). In other

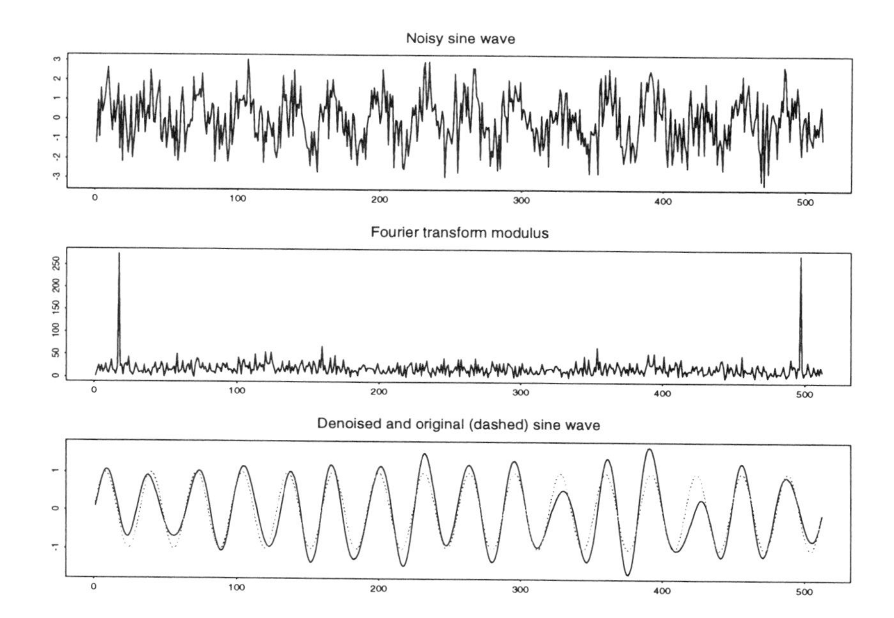

Figure 8.1. Denoising by smoothing. (Top) Plot of a noisy sine wave. (Middle) Plot of the modulus of the Fourier transform of the noisy sine wave after the coefficients corresponding to the high frequencies have been set to 0. (Bottom) Reconstruction (solid line) by inverse Fourier transform (from the thresholded set of coefficients) and original signal (dotted line).

words, the estimated signal is given by

$$\hat{f}(x) = \sum_{j,k\in\mathbb{Z}} \tau(\hat{a}_{jk})\psi_{jk}(x), \qquad x \in \mathbb{R},$$

where the $\hat{a}_{jk}$ are the estimates of the wavelet coefficients of the true signal f_0 in the wavelet orthonormal basis we work with, say, $\{\psi_{jk};\ j \in \mathbb{Z}, k \in \mathbb{Z}\}$ (see Chapter 5 for the definitions and notations). We use the wavelet coefficients of the observation f as (unbiased) estimates; in other words, we choose $\hat{a}_{jk} = t_k^j = \langle f, \psi_{jk}\rangle$. The *thresholding function* $\tau(a)$ is given by

$$\tau(a) = \begin{cases} a & \text{if} \quad |a| > \epsilon, \\ 0 & \text{otherwise.} \end{cases} \tag{8.3}$$

In other words, in the decomposition of the (unknown) signal in the wavelet orthonormal basis $\{\psi_{jk};\ j \in \mathbb{Z}, k \in \mathbb{Z}\}$, we use only the coefficients whose estimates are greater than ϵ in absolute value. Here ϵ is a positive number to be determined. It is the smoothing parameter of the reconstruction algorithm.

Wavelet Shrinkage

The thresholding method just presented was greatly enhanced by Donoho and Johnstone in a series of papers, among them [112, 113]. See also [109, 110, 111, 115]. The thrust of their work is the implementation of an old idea of Stein to *shrink* the coefficients toward 0 in order to get better results. Typically, they interpret the thresholding function τ defined in (8.3) as a *hard shrinkage* function and they replace it by a *soft shrinkage* thresholding function τ_{ss} defined by

$$\tau_{ss}(a) = \begin{cases} \operatorname{sign}(a)(|a| - \epsilon) & \text{if} \quad |a| > \epsilon, \\ 0 & \text{otherwise.} \end{cases} \tag{8.4}$$

The mathematical analysis of the corresponding estimators is very interesting, and Donoho and Johnstone have proven among other things that they are optimal min-max estimators.

8.3 The Smoothing Spline Approach

The material presented in this section owes much to Chapter I of Grace Wahba's monograph [289]. For the sake of simplicity we shall refrain from considering the Bayesian interpretation of the solution. This is unfortunate, for such a discussion would be enlightening. The reader is referred to [289] for a detailed discussion of this interpretation.

The purpose of this section is to derive the formulas for the construction of an approximation of the original signal from the observations of the values of the transform at sample points of the time/scale plane or the time/frequency plane. For the sake of the present discussion we use the notation $T_f(b,a)$ for the transform of the signal f. This notation is used throughout this book for the wavelet transform, but all the considerations of this section apply as well to the Gabor transform $G_f(b,\omega)$ of the signal f.

We assume that we have observations $f(x)$ of an unknown signal f_0 in the presence of an additive noise ϵ with mean zero. In other words, we work with the model

$$f(x) = f_0(x) + \epsilon(x)$$

and we assume that the noise is given by a mean zero stationary process with covariance

$$\mathbb{E}\{\epsilon(x)\epsilon(y)\} = C(x-y) ,$$

with the same notations as before. The case $\mathcal{C} = I$ (i.e., $C(x-y) = \delta(x-y)$) corresponds to the case of an additive white noise. We transform the observations and choose a discrete set $\{(b_1,a_1),\ldots,(b_n,a_n\})$ of sample points in the time/frequency-time/scale plane where the highest energetic concentration is found. These points are chosen with the reconstruction of the unknown signal from the values $T_f(b_j,a_j)$ of the transform at these points in mind. See the next section for details on the choices of these sample points in the case of the various transforms which we consider in this work. We assume that the observations follow the usual linear model

$$z_j = T_f(b_j,a_j) + \epsilon'_j,$$

where the computational noise terms ϵ'_j are assumed to be identically distributed and uncorrelated between themselves and with the observation noise terms ϵ. Hence, the final model is of the form

$$z_j = L_j f_0 + \epsilon_j, \qquad j = 1,\ldots,n , \tag{8.5}$$

where L_j is the linear form representing the value of the transform at the point (b_j,a_j) and where

$$\epsilon_j = T_\epsilon(b_j,a_j) + \epsilon'_j . \tag{8.6}$$

The assumption that the two sources of noise are uncorrelated implies that the covariance matrix[1] Σ of the ϵ_j is the sum of the covariance of the

[1] To avoid confusion, we use the symbol Σ to denote covariance in the wavelet domain, and reserve the symbol C for covariance in the time domain.

$T_\epsilon(b_j, a_j)$ and the covariance of the ϵ'_j. The latter being of the form $\sigma'^2 I$, we have

$$\Sigma = \sigma'^2 I + \Sigma^{(1)} , \tag{8.7}$$

where the entries of the matrix $\Sigma^{(1)}$ are given by the formula

$$\Sigma^{(1)}_{j,k} = \langle \psi_j, C\psi_k \rangle = \int \psi_j(x) C(x-y) \overline{\psi(y)} \, dxdy . \tag{8.8}$$

Here we have introduced the notation

$$\psi_j(x) = \frac{1}{a_j} \psi\left(\frac{x - b_j}{a_j}\right) \tag{8.9}$$

for the copies of the wavelet ψ shifted by b_j and scaled by a_j. An expression similar to (8.8) holds in the case of the Gabor transform.

The reconstruction algorithm is formulated as the solution of the minimization problem

$$\hat{f} = \arg \min_f \left[\frac{1}{n} \left| \Sigma^{-1/2} [Z - T_f(\cdot, \cdot)] \right|^2 + \mu \langle Qf, f \rangle \right] , \tag{8.10}$$

where $\langle Qf, f \rangle$ is a quadratic form in the candidate function f, Z denotes the n-vector of observations z_j, $T_f(\cdot, \cdot)$ denotes the n-vector of values of the transform of the function f at the points (b_j, a_j), and $| \cdot |$ denotes the Euclidian norm in $\mathbb{C}^n$. The constant $\mu > 0$ is introduced to balance the effects of the two components of the penalty (see [289] for a discussion of the choice of such a parameter).

The first term is a measure of the fit of the vector Z of observations to the transform of the candidate f sampled at the points (b_j, a_j). Notice that the measure of the fit which we use is a weighted sum of squares where the weights are given by the inverse of the covariance matrix. Statisticians know this measure as the *Malahanobis distance.* The second term is to be interpreted as a smoothness penalty. Its purpose it to prevent the candidate f from oscillating too wildly. These irregularities of f are typical consequences of the contribution of the first term of the cost function to minimize, since the latter tries to make sure that the values of the transform $T_f(b, a)$ at the sample points (b_j, a_j) remain close to the observations z_j. The choice of the matrix Q (and the corresponding quadratic form) is application dependent. See the next section for examples of this choice.

The following result is easily derived (see, for example, Theorem 1.3.1 of [289] for details):

Theorem 8.1 *The solution $\hat{f}$ of the optimization problem* (8.10) *is given by*

$$\hat{f}(x) = \sum_{j=1}^{n} \lambda_j \tilde{\psi}_j(x), \tag{8.11}$$

where the dual wavelets/atoms are defined by

$$\tilde{\psi}_j = Q^{-1}\psi_j \ , \tag{8.12}$$

the wavelets ψ_j being those defined in Equations (8.9)*. The coefficients λ_j of the linear combination* (8.11) *are the components of a one-dimensional vector Λ, given by the formula*

$$\Lambda = A^{-1}\Sigma^{-1/2}Z \ , \tag{8.13}$$

where the matrix A is defined as $A = n\mu I + \tilde{\Sigma}$ and the matrix $\tilde{\Sigma}$ is defined by $\tilde{\Sigma}_{j,k} = \langle \tilde{\psi}_j, \psi_k \rangle$.

Remark 8.1 Notice that we did not use the full generality of the smoothing spline problem as defined in [289]. Indeed, we could have chosen a quadratic penalty of the form $\|Q^{1/2}P_1 f\|^2$ where P_1 is the projection onto the orthogonal complement of a subspace of finite dimension. In this generality it is possible to avoid penalizing special subspaces of functions (for example, the space of polynomial functions of degree smaller than a fixed number). Since the form of the solution is much more involved and since we shall not consider applications at this level of generality, we decided to use the smoothing spline approach in our simpler context. The reader interested in this extra feature of the smoothing splines technique can consult [289].

Remark 8.2 The approach presented here was alluded to as a possible extension to the reconstruction algorithm derived and used in [64] and [67]. The latter corresponds to the case where the knowledge of the wavelet transform of the unknown signal is assumed to be perfect, in other words, to the case where both C and σ'^2 are assumed to be zero. It is easy to see that, under these extra assumptions, the reconstruction procedure given by the preceding minimization problem reduces to the minimization of the quadratic form $\langle Qf, f\rangle$ under the constraints $z_j = T_f(b_j, a_j)$. This is the problem which was solved in [64] and [67]. It appears as a particular case of the more general procedure presented here. We give the result for completeness. In such a case, the optimization problem (8.10) is replaced with

$$\min_{f,\Lambda} \ \left(\langle Qf, f\rangle - \Lambda \cdot [Z - T_f(\cdot,\cdot)]\right), \tag{8.14}$$

where $\Lambda = \{\lambda_1, \ldots, \lambda_n\} \in \mathbb{C}^n$ is a set of Lagrange multipliers, and "$\cdot$" stands for the Euclidian product in $\mathbb{C}^n$. The solution of this problem is still of the form (8.11) where the "dual wavelets" are given in (8.12), and the λ_j's are the Lagrange multipliers of the problem, now given by

$$\Lambda = \tilde{\Sigma}^{-1} Z . \tag{8.15}$$

The advantages of the general smoothing spline approach were explained in the introduction. We shall not reproduce this discussion here.

Remark 8.3 Notice that the reconstructed signal appears as a linear function of the observations. Nevertheless, our whole analysis is nonlinear because of the procedure used to choose the sample points in the time-frequency/time-scale plane.

8.4 Reconstruction from the Extrema of the Dyadic Wavelet Transform

The purpose of this section is to show how the smoothing spline methodology can be used to derive in a natural way the algorithm introduced by Mallat and Zhong [214] to reconstruct a signal from the extrema of its dyadic wavelet transform. We borrow the details of the discussion from [64], whose motivation was to simplify and shed light on the original procedure of Mallat and Zhong.

Let us consider for example the transient signal plotted in the top part of Figure 8.2. The support of the transient is small compared to the length of the signal. A closer look at the plot shows that this transient is in fact an exponentially damped sine wave. The transient signal plotted in Figure 8.3 is of the same type, except for the fact that there are many fewer oscillations under the damping envelope.

In both cases, the modulus of the wavelet transform shows energy concentrations forming ridges in the vertical direction. These vertical wrinkles can be defined mathematically as the sets of local extrema in the time variable b of the function $b \hookrightarrow T_f(b, a)$ when the scale variable a is held fixed. Notice that these extrema are different from the extrema encountered in the analysis of the frequency modulated signals, in the sense that the general direction of the ridges they produce is essentially vertical instead of horizontal. Despite this important difference, there is a common feature which we emphasize: the location of these extrema indicate the regions of high energy concentration in the time-scale plane. As a consequence, it is likely that the knowledge of the transform of the signal on or near these ridges will be enough for a faithful reconstruction of the signals.

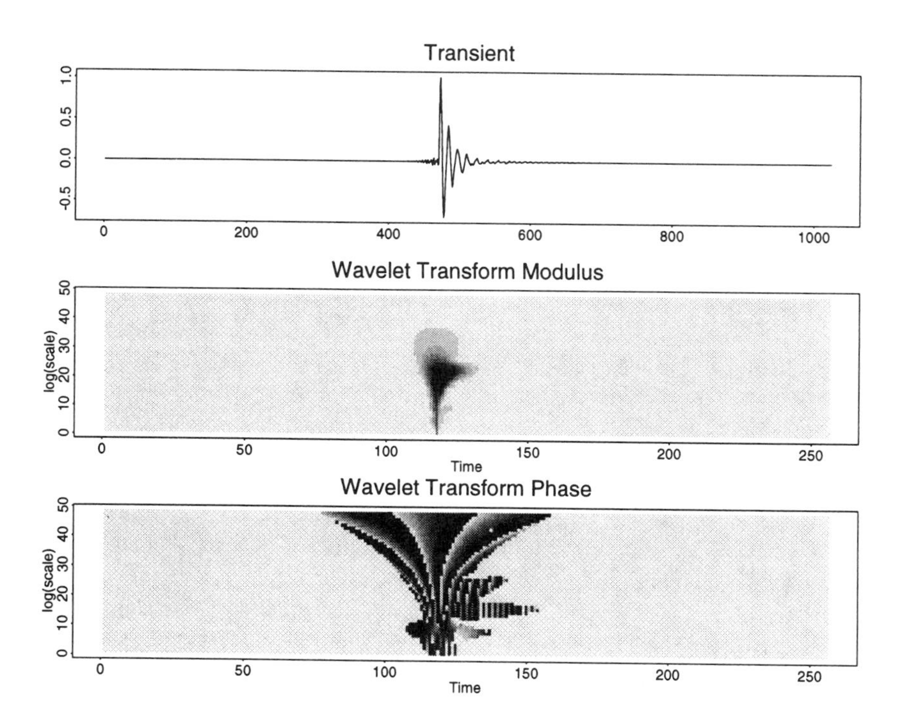

Figure 8.2. Plots of a pure transient (top), of the modulus of its wavelet transform (middle), and of the phase of this wavelet transform (bottom).

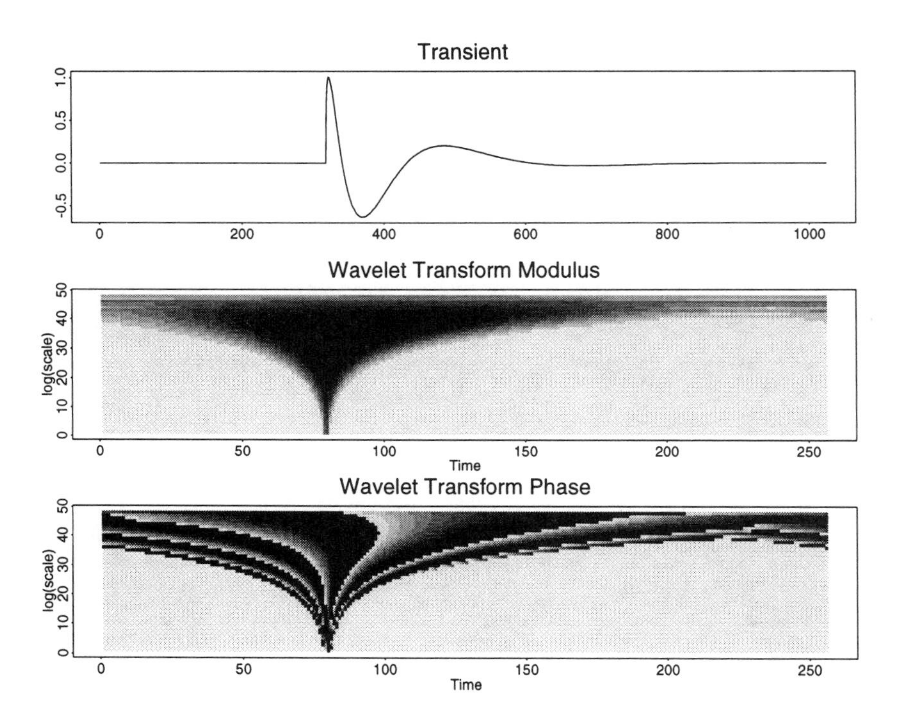

Figure 8.3. Plots of a pure transient D (top), of the modulus of its wavelet transform (middle), and of the phase of this wavelet transform (bottom).

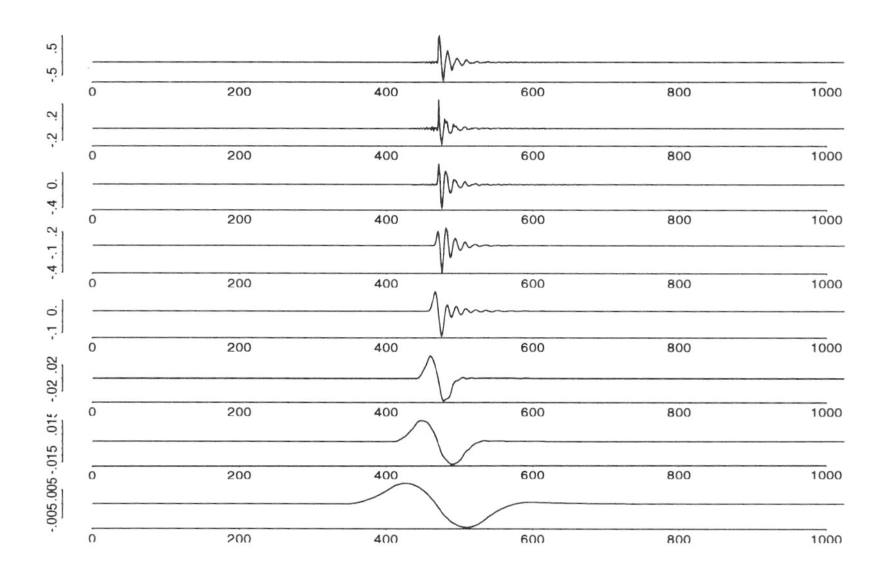

Figure 8.4. Dyadic wavelet transform of transient C.

The following is a reasonable proposal for the choice for the sample points (b_j, a_j) of the time-scale plane used for the reconstruction of the signal. We restrict the scales to the dyadic scales 2^n for integers n, and for each such scale we choose the values of the time variable b which are locations of local maxima of the function $b \hookrightarrow T_f(b, a)$. In other words, we choose to reconstruct the signal from the observations of the wavelet transform at the locations of the maxima of the modulus of the dyadic wavelet transform. These points of the time-scale plane were singled out in Section 5.2 because the analyzing wavelet was assumed to be the derivative of a smoothing spline. We were interested in these points because they were providing a map of the singularities of the signal. We are now interested in their providing a map of the energy distribution over the time-scale plane.

The reconstruction algorithm of Mallat and Zhong mentioned in Section 5.2 is a synthesis algorithm of the "smoothing spline" type as described in the previous section. Indeed, Mallat and Zhong assume that the signal is observed without error (i.e., $\epsilon_j \equiv 0$ or, equivalently, $\Sigma^{(1)} = 0$) and that the wavelet transform is computed without error (i.e., the variance of the noise component ϵ'_j is zero, $\sigma'^2 = 0$.) Since they assume $\Sigma = 0$, the minimization problem (8.10) has to be interpreted as the minimization of the

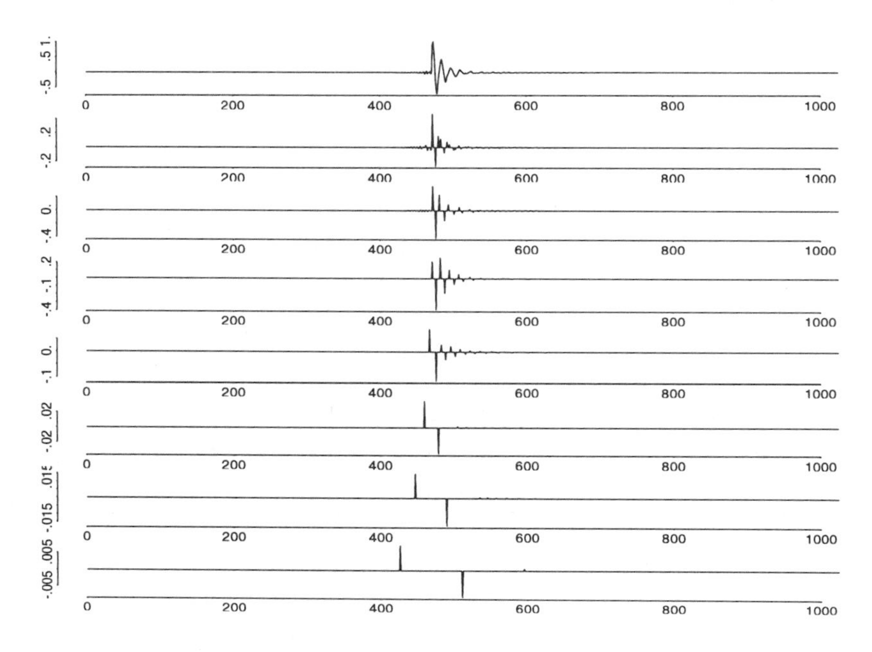

Figure 8.5. Local extrema of the dyadic wavelet transform of transient C.

cost $\langle Qf, f\rangle$ under the linear constraints

$$z_j = T_f(b_j, a_j), \qquad\qquad j = 1, \cdots, n \;, \tag{8.16}$$

and the associated coefficients λ_j take the form of Lagrange multipliers introduced to enforce the constraint (8.16). They use the quadratic penalty

$$\langle Qf, f\rangle = \sum_{j=1}^{J} \left(\int |T_f(x, 2^j)|^2 \, dx \;\; + \;\; 2^{2j} \int |\frac{d}{dx} T_f(x, 2^j)|^2 \, dx \right) \; . \tag{8.17}$$

The details concerning the computation of the quadratic form Q both in the time domain and in the Fourier domain can be found in [64].

Figure 8.6 gives examples of reconstruction using the algorithm presented in this note.

Unfortunately, not all the reconstructions are as precise. The reader can find in [64] examples showing that the algorithm has a few flaws, in particular for the reconstruction of very sharp variations, because of the presence of the smoothness penalty.

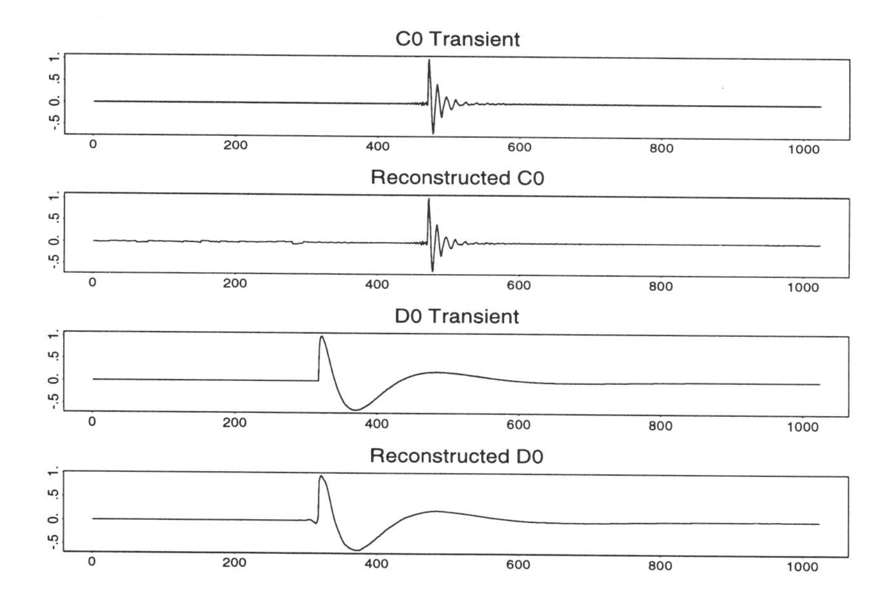

Figure 8.6. Reconstruction of transients: The plot on the top gives the original signal C, and the second plot from the top gives the reconstruction obtained by the method described earlier. The other two plots give similar results in the case of the transient D.

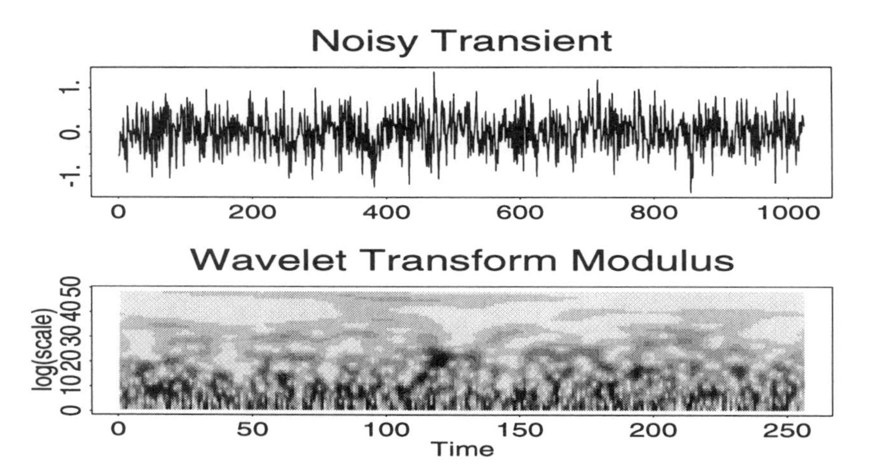

Figure 8.7. Plots of the first transient embedded in noise (top), of the modulus of its wavelet transform (middle), and of the phase of this wavelet transform (bottom).

In their original work [214], Mallat and Zhong designed an ad-hoc procedure to control these oscillations. We shall refrain from modifying the smoothing spline approach to do the same thing.

Remark 8.4 One of the enlightening features of our reconstruction procedure is its very intuitive interpretation. The reconstruction is obtained as the sum of pseudo-wavelets sitting at the extremum locations and with a specific amplitude given by the corresponding Lagrange multiplier. More precisely, each extremum (b_j, a_j) contributes to the reconstruction in the following way. At the very location of the extremum, i.e., at the point (b_j, a_j) of the time-scale plane, a pseudo-wavelet $\tilde{\psi}_j(x) = Q^{-1}\psi((x - b_j)/a_j)/a_j$ is set. This pseudo-wavelet has the scale a_j of the extremum and it is multiplied by the corresponding Lagrange multiplier λ_j. We call the functions $\tilde{\psi}_j$ pseudo-wavelets because their shapes are very reminiscent of the shapes of the original wavelets ψ_j. The terminology "dual wavelet" or "biorthogonal wavelet" could also be used as long as the fact that the duality or biorthogonality is with respect to the quadratic form defined by Q is understood.

Noisy Signals and Transient Detection

Let us consider now the case of noisy signals. Clearly, in the situations where the signal to noise ratio is significantly small, the procedure described

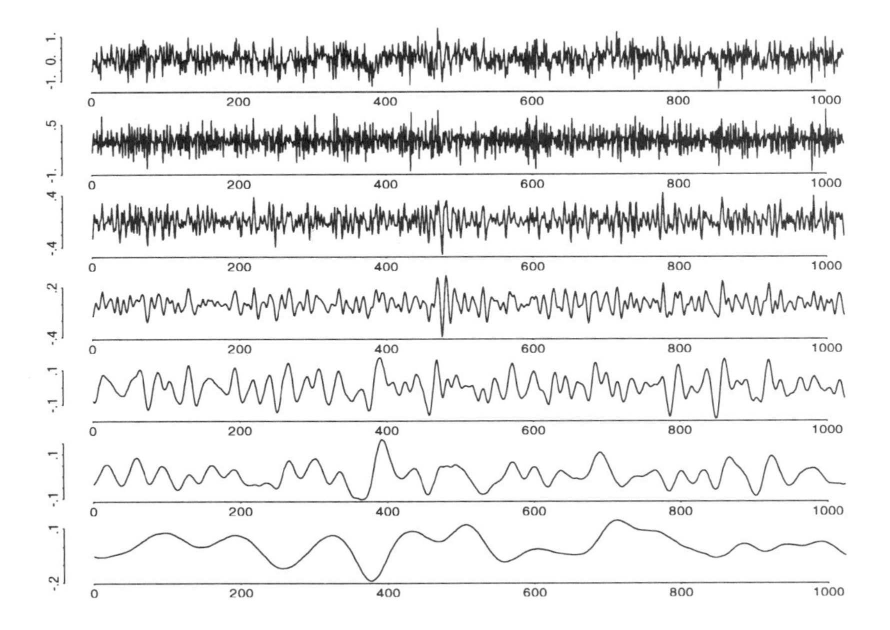

Figure 8.8. Dyadic wavelet transform of noisy transient C.

above has to be modified.

Consider as an example the pure transient of Figure 8.2 embedded in an additive mean zero Gaussian white noise. As Figure 8.7 shows, very few of the features of the modulus of the wavelet transform can still be seen. The only noticeable feature is the presence of a maximum of the modulus near the location of the transient. The same remark applies to the case of dyadic wavelet transform and the corresponding extrema (Figures 8.8 and 8.9). In particular, one has to notice the presence of a much larger number of extrema (here 968 extrema) than in Figure 8.5.

Let us now consider the problem of transient detection [65]. The goal is to test for the presence of transient in noise by using the wavelet transform maxima information. Clearly, an additional "trimming" of the maxima has to be done. We now describe such a trimming procedure.

The first step is a "learning step." The signal (or part of it) is used for statistical analysis of the noise, after subtraction of a local mean if necessary. Then a certain number of realizations of the noise are done (either by direct simulations, if some parametric characterization of the noise is known, or by bootstrapping) and the dyadic wavelet transform maxima are computed for each realization, yielding a histogram of the values of the maxima, scale by scale. Then, for a given significance level α,

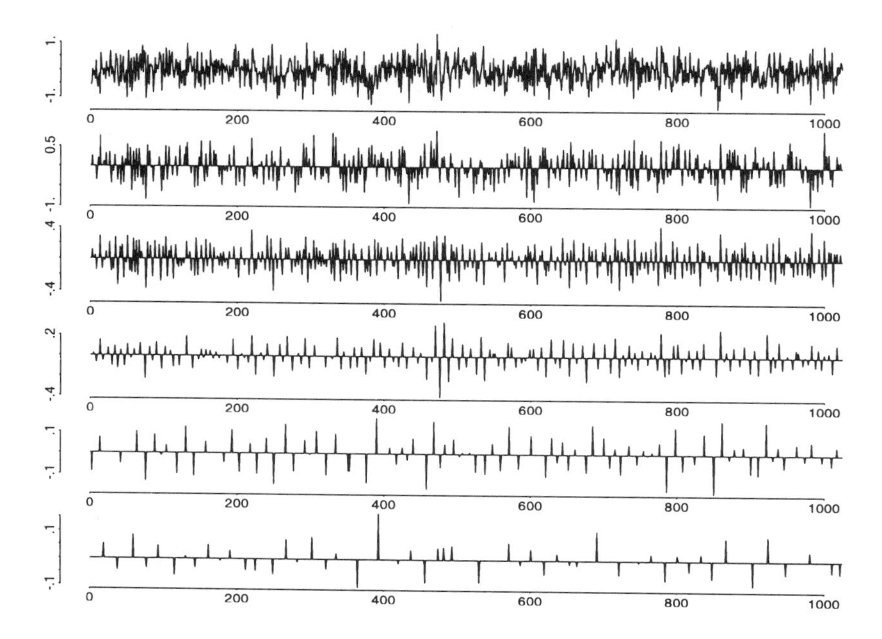

Figure 8.9. Local extrema of the dyadic wavelet transform of noisy transient C.

a corresponding threshold $h_\alpha(j)$ is associated with each scale of the dyadic wavelet transform, and only the extrema (at scale j) larger than $h_\alpha(j)$ are kept for the reconstruction.

As an example, we display in Figure 8.10 the thresholded extrema of the dyadic wavelet transform of noisy C, and in Figure 8.11 the reconstructed (denoised) C. The noise was Gaussian white noise, and the input signal to noise ratio was equal to -19 dB. Since the noise was assumed Gaussian, a direct Monte-Carlo simulation was used. The trimming (with a significance level of 95%) reduced the number of extrema from 968 to 5. As can be seen, the location of the transient is perfectly identified, as well as its frequency. The shape is not reconstructed perfectly, and we notice the presence of a spurious transient in the reconstruction (which would not be there for a smaller significance level).

Remark 8.5 *Possible Extensions:* The synthesis results presented here use the form of the reconstruction algorithm which assumes that the values z_j of the transform are exactly known at the points (b_j, a_j) of the time-scale plane. We explained in Remark 8.2 how one can use the full generality of the smoothing spline approach to account for the fact that these observations are noisy. Moreover, it should be possible to use Stein's idea to shrink the values of the transform in order to obtain better reconstructions. Even

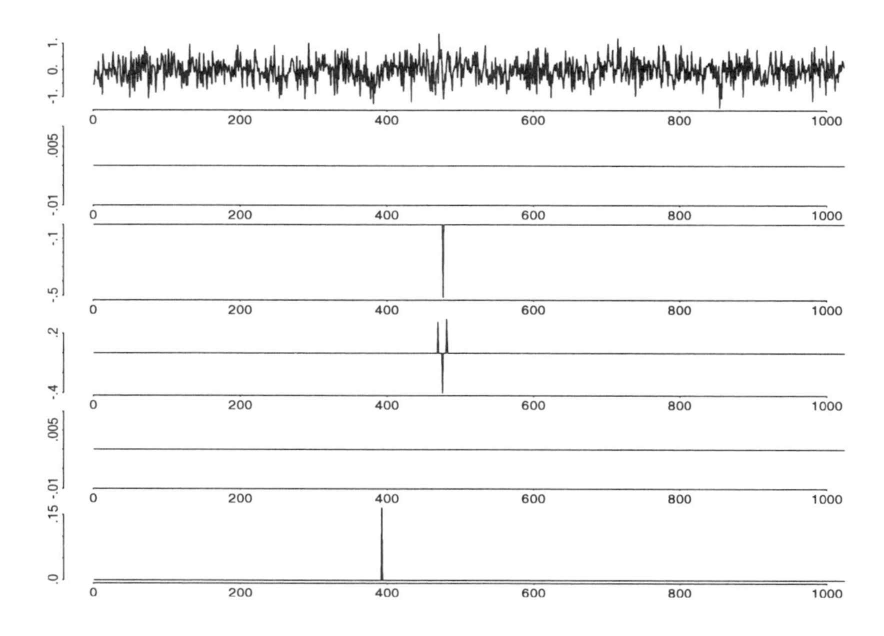

Figure 8.10. Local extrema of the dyadic wavelet transform of noisy transient C after trimming.

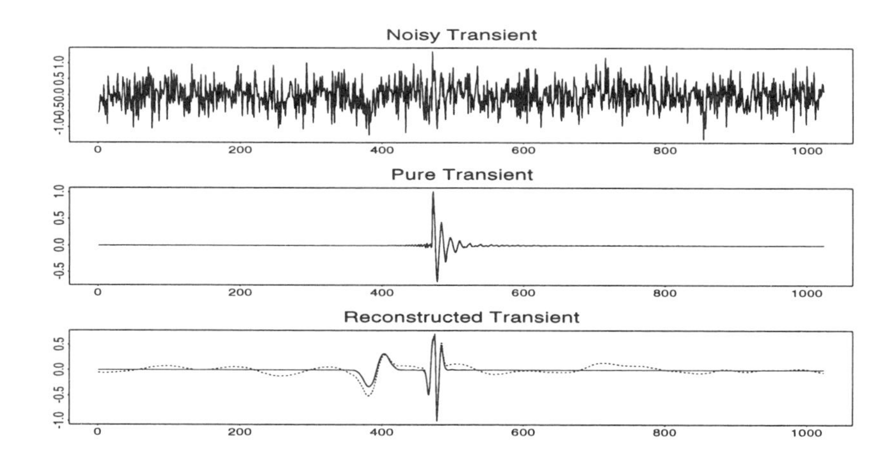

Figure 8.11. Reconstruction of transient C from trimmed local extrema (significance level: 95%): noisy C (top), pure C (middle), reconstructed C and reconstructed C + low pass component (dashed curve) at the bottom.

though this proposal seems very attractive, it needs to take into account the correlation structure of the random noise, and we do not know of any attempt in this direction.

8.5 Reconstruction from Ridge Skeletons

The purpose of this section is to show how the general smoothing spline approach presented in the first part of this chapter can be applied to the case of the reconstruction of a signal of finite energy from observations of a time-scale/time-frequency transform along the ridges of such a transform.

We are concerned with the implementation of the folk belief that *a signal can be characterized by the values of the restriction of its wavelet transform to its ridges.* Illustrations can be found in [102, 272], where it is shown that in the case of signals of the form $f(x) = A(x)\cos\phi(x)$ the restriction of the wavelet transform to its ridge (the so-called *skeleton* of the wavelet transform) essentially behaves as $A(x)\exp[i\phi(x)]$. Similar remarks (for the Gabor transform) also yield the so-called *sinusoidal model* for speech (see, e.g., [222, 223, 252]) which achieves good quality reconstruction with high compression rate. Such an approach can be used in non-noisy situations, but it may fail dramatically in the presence of a significant noise component.

Let us first present the simplest possible way of reconstructing a signal from a ridge of a wavelet or Gabor transform. It follows from the approximations we described in the previous chapter that the restriction of the wavelet transform and/or the Gabor transform to a ridge may be considered as good approximations of the original signal. Then, once a ridge has been estimated, there is a very simple recipe to reconstruct the corresponding component: in the wavelet case for instance, just compute

$$f_{\text{rec}}(x) = T_f(x, a_r(x)) \tag{8.18}$$

for each ridge component and sum up the various terms so obtained to get a global reconstruction of the signal. This reconstruction is particularly efficient, as may be seen in the middle plot of Figure 8.12, in the case of the CGT of the speech signal we have described throughout this book (represented in the top plot). This example will be described in more detail later. However, let us stress that it requires the knowledge of the whole wavelet (or Gabor) transform on the ridge, i.e., an important amount of data, especially when there are many ridges in the transform. In order to reduce the volume of data, alternatives have to be looked for.

Throughout this section, we shall develop a smoothing spline-based

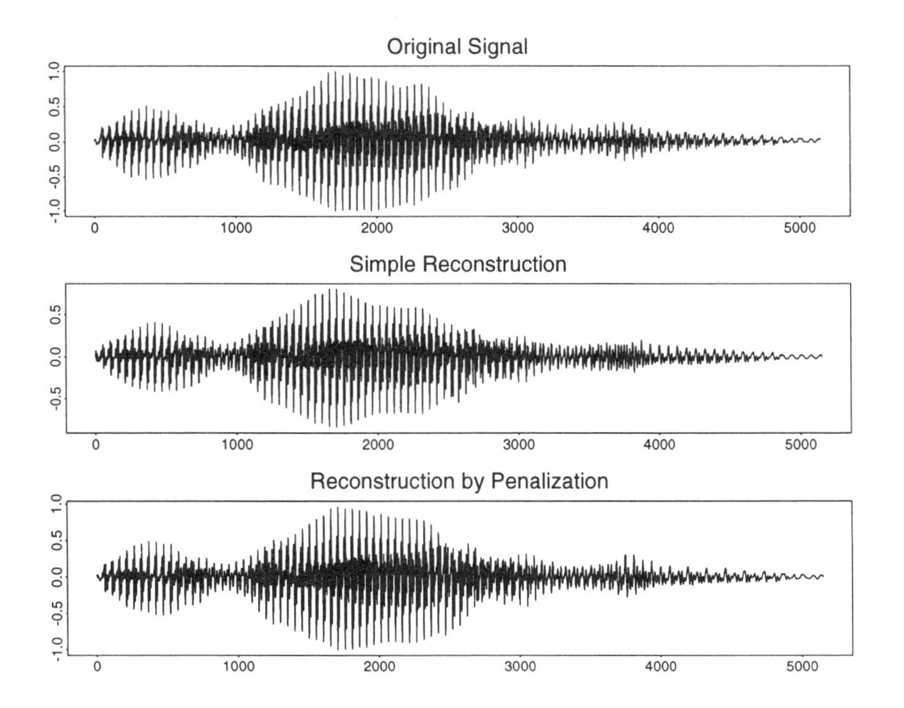

Figure 8.12. Speech Signal /How are you ?/ (top), reconstructions from the ridges of its CGT: simple reconstruction (medium), and reconstruction by penalization (bottom).

method to do signal reconstruction from ridges. Our approach will be the following. Assume that we are given a time-frequency representation $M(b,\omega)$ and a set of sample points (b_j,φ_j) of the time-frequency plane together with the values $z_j = M(b_j,\varphi_j)$ of the transform at these points. In the notation of the first section, this implies that we restrict ourselves to the simple case $\Sigma = 0$. We use the result of Section 8.3 for the problem of estimating a signal from such data.

We consider the cases of the wavelet and the Gabor transforms separately, even though their treatment is very similar.

8.5.1 The Case of the Wavelet Transform

We begin with the case of wavelet transform. For the sake of simplicity we shall restrict the discussion to the case of a single ridge, postponing the multiridge problem to Remark 8.8 at the end of the section.

We assume that the ridge can be parameterized by a continuous function:

$$[b_{\min}, b_{\max}] \ni b \hookrightarrow a_r(b) \ .$$

Such a ridge function is usually difficult to determine. A good starting point could be the result of a ridge estimation procedure such as those presented in Sections 7.3.1 and 7.2.4. In any case, one will presumably have to construct a smooth ridge function from the knowledge of a few points. Indeed, in practical applications one only knows sample points $(b_1,a_1),\ldots,(b_n,a_n)$, and the smooth function $b \hookrightarrow a_r(b)$, which we use in lieu of the true (unknown) ridge function, is merely a guess which we construct from the sample points. In the implementation proposed in the Swave package, we use a smoothing spline, but any other kind of nonlinear regression curve would do as well for the purpose of smoothing out the sample points of the ridge. From now on, $a_r(b)$ is a smooth ridge function which is constructed from sample data points.

Statement of the Problem

In order to obtain a result in the spirit of the smoothing spline approach, we assume that the values z_j of the wavelet transform of the unknown signal of finite energy f_0 are observed at sample points (b_j,a_j) which are tentatively regarded as representative of the ridge of the of the wavelet transform of the (unknown) signal f_0. For the purpose of the present discussion we assume that there is no observation noise in the sense that the observations are perfect and produce the *exact* values of the transform. We shall also regard values z_j as exact values, to be imposed as constraints. Then in the notation of Section 8.3, we have $\Sigma = 0$.

The set of sample points (b_j, a_j) together with the values z_j constitutes what we call the *skeleton of the wavelet transform* of the signal to be reconstructed.

As we already explained, we use a smooth function $b \hookrightarrow a_r(b)$ which *interpolates* the sample points, and we look at the graph of this function as our best guess for the ridge of the modulus of the wavelet transform of f_0.

We look for a signal f of finite energy whose wavelet transform $T_f(b,a)$ satisfies

$$T_f(b_j, a_j) = z_j, \qquad j = 1, \ldots, n \,, \tag{8.19}$$

and has the graph of the function $b \hookrightarrow a_r(b)$ as ridge. We are about to show how to construct such a signal in the framework of the approach developed in Section 8.3. We also show with examples that it is a very good approximation of the original signal f_0.

The Penalization Approach: Choice of the Quadratic Form

As we said before, the choice of the quadratic form Q in Equation (8.10) is application dependent. In the present case, the role of the quadratic form is to enforce localization of the wavelet transform near the ridge, i.e., its decay away from the ridge. A natural candidate is the $L^2(\mathbb{R})$ norm

$$F_1(f) = \frac{1}{c_\psi} \int db \int \frac{da}{|a|} |T_f(b,a)|^2 \,, \tag{8.20}$$

or its "discrete" alternative:

$$F_1(f) = \frac{1}{c_\psi} \sum_j \int \frac{da}{|a|} |T_f(b_j, a)|^2 \,. \tag{8.21}$$

In the present setting, we look for a signal f whose wavelet transform has prescribed values at the sample points of the ridge, i.e., which satisfies the constraints (8.16) while the L^2-norm is kept to a minimum. Notice that when the integration in (8.20) is performed on the whole half-plane, $F_1(f)$ is nothing but $||f||^2$.

Solving such a constrained optimization problem is easy [64] (the reconstruction coincides with that given in [150]). Since the cost function $F_1(f)$ is a quadratic form in the unknown function f, the solution is easily computed by means of Lagrange multipliers.

A solution can be constructed as a linear combination of the wavelets ψ_j defined in (8.9) at the sample points of the ridge, the coefficients being given by the solution of a $n \times n$ linear system.

This solution may not be completely satisfactory, especially when the number of sample points is small. In particular, it ignores the empirical fact that (in most of the practical applications and especially for the frequency modulated signals we are mostly interested in) the restriction of the modulus $|T_f(b,a)|$ to the ridge, i.e., the function $b \hookrightarrow |T_f(b,a_r(b))|$, is smooth and slowly varying. In order to force the solution of the constrained optimization problem to satisfy this requirement, it is desirable to add an extra term to the cost function $F_1(f)$. We describe below the choice proposed in [67] by the authors. It amounts to adding the contribution

$$\tilde{F}_2(f) = \epsilon \int_{b_{\min}}^{b_{\max}} \left| \frac{d}{db} |T_f(b,a_r(b))| \right|^2 db, \qquad (8.22)$$

and considering the minimization of the cost function

$$\tilde{F}(f) = F_1(f) + \epsilon \tilde{F}_2(f), \qquad (8.23)$$

where the free smoothing parameter $\epsilon > 0$ can be chosen to control the relative importance of the two contributions to the penalty. Unfortunately, $\tilde{F}_2(f)$ is not quadratic in f. This makes the problem more difficult to solve and goes beyond the scheme described at the beginning of the present chapter. Furthermore, the solution is not given by a linear combination of *reconstruction* wavelets placed at each of the sample points of the ridge. In order to remedy this problem we replace the contribution $\tilde{F}_2(f)$ by a quadratic form which gives a good approximation to $\tilde{F}_2(f)$ when the signal belongs to one of the classes of signals likely to be characterized by their skeletons on their ridges (see, for example, [102, 272]).

Let us denote by $\Omega_f(b,a)$ the phase of the wavelet transform $T_f(b,a)$, i.e., the argument of the complex number $T_f(b,a)$. We have

$$T_f(b,a) = M_f(b,a)e^{i\Omega(b,a)}.$$

For the purpose of the present discussion we assume that a smooth version of the phase exists whenever $M(b,a) \neq 0$ (of course, we ignore the 2π jumps, i.e., we perform *phase unwrapping*), and that such a choice was made for $\Omega_f(b,a)$. Notice that with these notations we have

$$\begin{aligned} \frac{d}{db}T_f(b,a_r(b)) &= \left(\frac{d}{db}M_f(b,a_r(b))\right) e^{i\Omega(b,a_r(b))} \\ &\quad + i\left(\frac{d}{db}\Omega_f(b,a_r(b))\right) T_f(b,a_r(b)), \end{aligned}$$

so that

$$|\frac{d}{db}T_f(b,a_r(b))|^2 = |\frac{d}{db}M_f(b,a_r(b))|^2 + \left(\frac{d}{db}\Omega_f(b,a_r(b))\right)^2 M_f(b,a_r(b))^2.$$

For the examples we have in mind we expect that, at least along the ridge, the wavelet transform of the signal can be approximated by the wavelet transform of a function of the form $A(x)\cos\phi(x)$. This essentially means that the signal itself can be modeled as a superposition of such components, and that the wavelet "separates the components" in the following sense: the distance between two close components has to be larger than the wavelet's bandwidth at the considered scale (if not, we face the problem of interferences of components we talked about in the first two chapters). In this case $\Omega(b, a_r(b)) \approx \phi_0 + \phi(b)$ for some constant phase ϕ_0, so that

$$\frac{d}{db}\Omega(b, a_r(b)) \approx \phi'(b) = \frac{\omega_0}{a_r(b)} .$$

As a consequence, we may replace the natural cost function $\tilde{F}_2(f)$ defined in Equation (8.22) by the approximation

$$F_2(f) = \int_{b_{\min}}^{b_{\max}} \left(|\frac{d}{db}T_f(b, a_r(b))|^2 - \frac{\omega_0^2}{a_r(b)^2}|T_f(b, a_r(b))|^2 \right) db \tag{8.24}$$

and the global cost function $\tilde{F}$ defined in Equation (8.23) by

$$F(f) = F_1(f) + \epsilon F_2(f) . \tag{8.25}$$

But the interpretation remains the same: one considers a smoothness penalty in the spirit of the H^1-norm of the restriction of the modulus of the wavelet transform to the ridge. Consequently, our reconstruction is given by the solution of the minimization of $F(f)$ subject to the linear constraints (8.16) which we rewrite in the form $L_j(f) = z_j$, where the linear functionals L_j are defined by

$$L_j(f) = \frac{1}{a_j} \int f(x) \overline{\psi\left(\frac{x - b_j}{a_j}\right)} dx . \tag{8.26}$$

Solution of the Optimization Problem

Since the penalization problem we are considering is slightly different from the general smoothing spline problem discussed at the beginning of this chapter, we give the details of its solution. A simple computation shows that

$$F_2(f) = \int\int Q_2(x, y) f(x) f(y)\, dx dy ,$$

where the kernel $Q_2(x,y)$ is defined by

$$
\begin{aligned}
Q_2(x,y) = \int \frac{db}{a_r(b)^4} \Bigg(& \overline{\psi\left(\frac{x-b}{a_r(b)}\right)} \psi\left(\frac{y-b}{a_r(b)}\right) [a_r'(b)^2 - \omega_0^2] \\
& + \overline{\psi'\left(\frac{x-b}{a_r(b)}\right)} \psi'\left(\frac{y-b}{a_r(b)}\right) \left[\frac{(x-b)(y-b)}{a_r(b)^2} + 1 + \frac{x-2b+y}{a_r(b)}\right] \\
& + \overline{\psi\left(\frac{x-b}{a_r(b)}\right)} \psi'\left(\frac{y-b}{a_r(b)}\right) a_r'(b) \left[1 + \frac{y-b}{a_r(b)}\right] \\
& + \overline{\psi'\left(\frac{x-b}{a_r(b)}\right)} \psi\left(\frac{y-b}{a_r(b)}\right) a_r'(b) \left[1 + \frac{x-b}{a_r(b)}\right] \Bigg) .
\end{aligned}
\tag{8.27}
$$

Using the definition of the total cost function $F(f)$ and the conservation of energy (to compute the first term $F_1(f)$), we see that

$$
F(f) = \int\int Q(x,y) f(x) f(y) \, dx dy \tag{8.28}
$$

is a quadratic functional given by the kernel

$$
Q(x,y) = \delta(x-y) + \epsilon Q_2(x,y). \tag{8.29}
$$

Remark 8.6 Notice that the kernel $Q(x,y)$ becomes a finite matrix for the purpose of practical applications. Notice further that formulas (8.29) and (8.27) give a practical way to compute the entries of the matrix Q.

The optimization problem can be reformulated as a minimization problem in the real domain rather than the complex domain by noticing the two obvious facts:

1. The n complex constraints (8.19) can be replaced by the $2n$ real constraints

$$
R_j(f) = r_j, \qquad j = 1, \cdots, 2n, \tag{8.30}
$$

   where

$$
\begin{aligned}
R_j(f) &= \frac{1}{a_j} \int f(x) \Re\psi\left(\frac{x-b_j}{a_j}\right) dx \\
R_{n+j}(f) &= -\frac{1}{a_j} \int f(x) \Im\psi\left(\frac{x-b_j}{a_j}\right) dx ,
\end{aligned}
\tag{8.31}
$$

   for $j = 1, \ldots, n$, $r_j = \Re z_j$, and $r_{n+j} = \Im z_j$.

2. The kernel $Q(x,y)$ defining the quadratic functional $F(f)$ can be replaced by its real part $\Re Q(x,y)$. Indeed, the kernel $Q(x,y)$ is sesquilinear and we are only computing $F(f)$ for real signals.

Consequently, there exist real numbers $\lambda_1, \ldots, \lambda_n, \lambda_{n+1}, \ldots, \lambda_{2n}$ (the Lagrange multipliers of the problem) such that the solution $\hat{f}$ of the optimization problem

$$\hat{f} = \arg\min F(f) \tag{8.32}$$

is given by the formula

$$\hat{f} = \sum_{j=1}^{2n} \lambda_j \tilde{\psi}_j \ , \tag{8.33}$$

where we have set

$$\begin{aligned} \tilde{\psi}_j &= \Re Q^{-1}\psi_j, \quad j = 1, \ldots n, \\ \tilde{\psi}_j &= \Im Q^{-1}\psi_j, \quad j = n+1, \ldots 2n, \end{aligned} \tag{8.34}$$

where the functions ψ_j are defined in equation (8.9).

The Lagrange multipliers are determined by requiring that the constraints (8.30) be satisfied. In other word, by demanding that the wavelet transform of the function $\hat{f}$ given in (8.33) be equal to the z_j's at the sample points (b_j, a_j) of the time-scale plane. This gives a system of $(2n) \times (2n)$ linear equations from which the Lagrange multipliers λ_j's can be computed.

Remark 8.7 It is instructive to consider the following particular case: Assume that we consider for b_j all the possible values of b, and we set $\epsilon = 0$. Then $F = F_1$, Q is a multiple of the identity, and it is easy to see that (at least formally) the solution of the constrained optimization problem yields a reconstructed signal of the form

$$\hat{f}(x) = \int \overline{T_f(b, a_r(b))\psi\left(\frac{x-b}{a_r(b)}\right)} \frac{db}{a_r(b)}.$$

This is a superposition of wavelets located on the ridge. We recover here a reconstruction formula close to those used in references [102, 222]. However, when the sampling of the ridge becomes coarser, this reconstruction procedure is no longer satisfactory. This lack of smoothness justifies (once more) the introduction of the second term F_2 of the cost function.

Remark 8.8 So far we have considered only the simple ridge case. The situations where the wavelet transform of the signal presents several ridges may be handled similarly as long as the ridges are well separated. The first term in the penalty function remains unchanged, and a term $F_2(f)$ for each ridge should be present. For example, if there are L ridges, denoted by a_ℓ, $\ell = 1, \dots L$, one has to minimize a cost function of the form

$$F(f) = F_1(f) + \epsilon \sum_\ell \int_{b_{\min}}^{b_{\max}} \left(|\frac{d}{db} T_f(b, a_\ell(b))|^2 - \frac{\omega_0^2}{a_\ell(b)^2} |T_f(b, a_\ell(b))|^2 \right) db$$

with the constraints

$$T_f(b_{\ell,j}, a_{\ell,j}) = z_{\ell,j},$$

where ℓ labels the ridges, and j labels the samples on the ridges. However, if the ridges are "well separated" in the sense that they are localized in different regions of the time-scale plane (see Remark 8.11 for a more precise definition of well-separated components), interactions between ridges may be neglected and all the components of the signal may be reconstructed separately along the lines of the synthesis procedure given earlier (notice that this is not always the case, as the example of the HOWAREYOU signal shows). The final reconstruction can then be obtained as the superposition of the reconstructions obtained from the individual ridges. This point is described with more details in the next section, devoted to the Gabor case.

The Reconstruction Algorithm

We summarize the results of the discussion of this section in an algorithmic walk through our solution to the reconstruction problem.

1. Determine a finite set of sample points $(b_1, a_1), \dots, (b_n, a_n)$ on a ridge.

2. Construct a smooth estimate $b \hookrightarrow a_r(b)$ of the ridge from the knowledge of the sample points.

3. Compute the matrix $Q(x, y)$ of the smoothness penalty along the ridge estimate.

4. Compute the reconstruction wavelets $\tilde{\psi}_j$ in (8.34) at the ridge sample points of the time-scale plane.

5. Compute the coefficients λ_j.

6. Use Equation (8.33) to get the estimated signal.

Notice that steps 3 and 4 are the most time-consuming steps of this scheme, whereas all other steps only involve simple operations.

Supplementary Remarks Concerning Noisy Signals

In most practical applications, the input signal is corrupted by noise and the identification and reconstruction problems are significantly more difficult. We have already explained how the presence of an additive noise can be handled by the ridge detection algorithms we presented in this paper. But even when sample points of a ridge estimate have been reliably determined, there are still several possible variations on the theme presented in this section when it comes to the reconstruction problem. We mention a few of them for the record.

Remark 8.9 In practical applications, the exact values the wavelet transform at the sample points of the ridge are not known exactly. The z_j's are merely noisy perturbations of the true values. The reconstruction problem can be solved in a very similar way, as described in Section 8.3. See, for example, [64] and [289] for the modifications which are necessary in order to get the solution of this seemingly more general problem.

Remark 8.10 We mentioned at the beginning of the chapter that there is an instructive Bayesian interpretation of the penalization problem. The details of this interpretation are given in [289] in the case of the smoothing splines. The argument can easily be adapted to the present situation.

Examples

As a first example, we consider the bat sonar signal we have been studying since the beginning. We can see later in Figures 8.15 and 8.16 that the reconstructions of the main component, and even of the first harmonic (there are actually two ridges), from sample ridges are of extremely good quality and respect the main features of the signal. These figures are included not only to show that a very good approximation of the signal can be recovered with a small number of sample points, but also as an illustration of Remark 8.8. The examples show that such an approach allows us to "disentangle" the two components of the signal. More comments on those examples are given later.

8.5.2 The Case of the Gabor Transform

Let us now turn to the case of Gabor transform. The discussion of this case closely follows the discussion of the wavelet case, so we shall not go into great detail.

Rather than repeating the same argument all over again, we try to illustrate the new difficulties occurring because of the presence of multiple ridges and the singularities and/or instabilities due to the possible confluences of the various ridge components.

As before, we assume that the ridges can be parameterized by continuous functions:

$$[b_{\ell,\min}, b_{\ell,\max}] \ni b \hookrightarrow \omega_\ell(b) \ ,$$

where $\ell \in \{1, \cdots, L\}$ labels the different ridges. These ridges are usually constructed as smooth functions resulting from fitting procedures (spline smoothing is an example we are using in practical applications) from the sample points obtained from ridge estimation algorithms such as the algorithms presented in the previous chapter.

Statement of the Problem

We assume that the values of the Gabor transform of an unknown signal of finite energy are known at sample points $(b_{\ell,j}, \omega_{\ell,j})$ which are regarded as representative of the ridges of the modulus of the Gabor transform of the unknown signal f_0. As before, we use the notation $z_{\ell,j}$ for the value of the Gabor transform of f_0 at the point $(b_{\ell,j}, \omega_{\ell,j})$. The set of sample points $(b_{\ell,j}, \omega_{\ell,j})$ together with the values $z_{\ell,j}$ constitute what we call the *skeleton* of the Gabor transform of the signal to be reconstructed.

As we have explained, we use smooth functions $b \hookrightarrow \omega_\ell(b)$ which fit the sample points and we look at the graphs of these functions as our best guesses for the ridges of the modulus of the Gabor transform of f_0.

The reconstruction problem is to find a signal f of finite energy whose Gabor transform $G_f(b, \omega)$ satisfies

$$G_f(b_{\ell,j}, \omega_{\ell,j}) = z_{\ell,j}, \qquad \ell = 1, \cdots L, \ \ j = 1, \cdots, n_\ell, \tag{8.35}$$

and has the union R of the graphs of the functions $b \hookrightarrow \omega_\ell(b)$ as set of ridges. Recall that for the purpose of the present discussion this last statement means that for each b, the points $(b, \omega_\ell(b))$ of the time-frequency plane are the local maxima of the function $\omega \hookrightarrow |G_f(b, \omega)|$. We are about to show how to construct such a signal. We will also show that it is a very good approximation of the original signal f_0.

The Penalization Approach

As before, the reconstruction problem is solved via a variational problem. The argument given for the (continuous) wavelet transform of frequency modulated signals can be reproduced here with only minor changes. This leads to the minimization of the cost function

$$\mathcal{G}(f) = \mathcal{G}_1(f) + \epsilon \sum_{\ell=1}^{L} \int \left[\left| \frac{d}{db} G_f(b, \omega_\ell(b)) \right|^2 - \omega_\ell(b)^2 |G_f(b, \omega_\ell(b))|^2 \right] db, \tag{8.36}$$

where, owing to the preceding discussion,

$$\begin{cases} \mathcal{G}_1(f) &= \int \left(\int |G_f(b,\omega)|^2 d\omega\right) db \qquad \text{or} \\ \mathcal{G}_1(f) &= \sum_{\ell,j} \int |G_f(b_{\ell,j},\omega)|^2 d\omega \end{cases} \tag{8.37}$$

The first term, together with the constraints (8.35), forces the localization of the energy distribution of the Gabor transform on the ridges as given by the graphs of the functions $b \hookrightarrow \omega_\ell(b)$. The second term is a quadratic form which provides a good approximation of the H^1-norm of the restriction of the Gabor transform to the ridges. As explained in the preceding discussion, this part of the cost function is designed to ensure that the tops of the ridges remain smooth and slowly varying. The free smoothing parameter $\epsilon > 0$ is to be chosen to control the relative importance of the two contributions to the penalty.

The reconstruction procedure is then given by the solution of the minimization of $\mathcal{G}(f)$ subject to the linear constraints (8.35), which we rewrite in the form $L_{\ell,j}(f) = z_{\ell,j}$, where the linear functionals $L_{\ell,j}$ are defined by

$$L_{\ell,j}(f) = \int f(x) e^{-i\omega_{\ell,j}(x-b_{\ell,j})} \overline{g(x-b_{\ell,j})}\, dx \; . \tag{8.38}$$

Solution of the Optimization Problem

A simple computation shows that the second term in the right-hand side of Equation (8.36) reads

$$\mathcal{G}_2(f) = \int\int \mathcal{G}_2(x,y) f(x) f(y)\, dxdy,$$

where the kernel $\mathcal{G}_2(x,y)$ is defined by the formula

$$\begin{aligned} &\mathcal{G}_2(x,y) \\ &\quad = \sum_\ell \Bigg(\int \Big(g'(x-b)g'(y-b) + g(x-b)g(y-b)\big[(x-b)(y-b)\omega_\ell'(b)^2 \\ &\quad - (x+y-2b)\omega_\ell(b)\omega_\ell''(b)\big] \Big) \cos\left(\omega_\ell(b)(x-y)\right)\, db \\ &\quad + \int \Big(g'(x-b)g(y-b)\big[\omega_\ell(b) - (y-b)\omega_\ell'(b)\big] \\ &\quad - g(x-b)g'(y-b)\big[\omega_\ell(b) - (x-b)\omega_\ell'(b)\big] \Big) \sin\left(\omega_\ell(b)(x-y)\right) \Bigg)\, db. \end{aligned} \tag{8.39}$$

The total cost function $\mathcal{G}(f)$ is then expressed as the sum of two terms,

$$\mathcal{G}(f) = \mathcal{G}_1(f) + \mathcal{G}_2(f) = \int\int \mathcal{G}(x,y) f(x) f(y)\, dxdy,$$

and is a quadratic functional given by the kernel

$$\mathcal{G}(x,y) = \mathcal{G}_1(x-y) + \epsilon\mathcal{G}_2(x,y), \tag{8.40}$$

where we have set

$$\begin{cases} \mathcal{G}_1(x,y) &= \delta(x-y) \qquad \text{or} \\ \mathcal{G}_1(x,y) &= \dfrac{1}{||g||^2}\delta(x-y)\sum_{\ell,j}|g(x-b_{\ell,j})|^2. \end{cases} \tag{8.41}$$

Again, for the sake of the numerical implementation, the reconstruction is formulated as a minimization problem in the real domain rather than the complex domain as follows:

1. For each ridge ω_ℓ, the n_ℓ complex constraints (8.35) are replaced with the $2n_\ell$ real constraints

$$R_{\ell,j}(f) = r_{\ell,j}, \qquad \ell = 1,\ldots,L,\ j = 1,\cdots,2n_\ell, \tag{8.42}$$

   where

$$\begin{aligned} R_{\ell,j}(f) &= \int f(x)g(x-b_{\ell,j})\cos(\omega_{\ell,j}(x-b_{\ell,j}))\,dx \\ R_{\ell,n_\ell+j}(f) &= \int f(x)g(x-b_{\ell,j})\sin(\omega_{\ell,j}(x-b_{\ell,j}))\,dx \end{aligned} \tag{8.43}$$

   for $\ell = 1,\ldots,L,\ j = 1,\cdots,n_\ell$; $r_{\ell,j} = \Re z_{\ell,j}$ and $r_{\ell,n_\ell+j} = \Im z_{\ell,j}$.

2. The kernel $\mathcal{G}(x,y)$ defining the quadratic functional $F(f)$ can be replaced by its real part $\Re\mathcal{G}(x,y)$. Indeed, the kernel $\mathcal{G}(x,y)$ is sesquilinear and we are only computing $F(f)$ for real signals.

Consequently, there exist real numbers $\lambda_{\ell,j}, \ell = 1,\ldots,L,\ j = 1,\ldots,2n_\ell$, (Lagrange multipliers), such that the solution $\hat{f}$ of the optimization problem is given by the formula

$$\tilde{f}(x) = \sum_{\ell=1}^{L}\sum_{j=1}^{2n_\ell}\lambda_{\ell,j}\tilde{g}_{\ell,j}(x), \tag{8.44}$$

where the functions $\tilde{g}_{\ell,j}$ are defined by

$$\begin{cases} \tilde{g}_{\ell,j}(x) &= \Re\mathcal{G}^{-1}g_{(b_{\ell,j},\omega_{\ell,j})}, \qquad j = 1,\cdots,n_\ell, \\ \tilde{g}_{\ell,j}(x) &= \Im\mathcal{G}^{-1}g_{(b_{\ell,j},\omega_{\ell,j})}, \qquad j = n_\ell+1,\cdots,2n_\ell. \end{cases} \tag{8.45}$$

The Lagrange multipliers are determined by requiring that the constraints (8.42) be satisfied, in other words, by demanding that the wavelet transform of the function $\hat{f}$ given in (8.44) be equal to the $z_{\ell,j}$'s at the sample points $(b_{\ell,j},\omega_{\ell,j})$ of the time-scale plane. This gives a system of linear equations from which the Lagrange multipliers $\lambda_{\ell,j}$'s can be computed.

The Reconstruction Algorithm

We summarize the results of the discussion of this section in an algorithmic walk through our solution to the reconstruction problem.

1. Determine a finite set $\{R_\ell\}_{\ell=1,\cdots,L}$ of ridges and, on each of them, a set of sample points $(b_1, \omega_{\ell,1}), \cdots, (b_n, \omega_{\ell,n_\ell}))$ on the ridge.
2. Construct smooth estimates $b \hookrightarrow \omega_\ell(b)$ of the ridges from the knowledge of the sample points.
3. Compute the matrix $\mathcal{G}(x, y)$ of the smoothness penalty along the ridge estimate.
4. Compute of the reconstruction time-frequency atoms in (8.45), localized (in the time-frequency plane) at the ridge sample points.
5. Compute the coefficients $\lambda_{\ell,j}$.

The solution $\hat{f}$ of the reconstruction problem is then given by formula (8.44).

Supplementary Remarks on Numerical Implementation

Remark 8.11 *Component Separation* Because the computation time needed to solve a linear system grows as the cube of the number of equations, it is important to find ways to speed up the computations. In this respect, the following simple remark is very important. For the sake of the present discussion, let us say that two elementary ridges R_ℓ and $R_{\ell'}$ associated with two signals f and f' are well separated if for all b, $G_f(b, \omega_{\ell'}(b))$ and $G_{f'}(b, \omega_\ell(b))$ are below a certain threshold (fixed by the required accuracy of the calculations). Our interest in this notion is the following. When individual ridges are well separated, their contributions to the reconstructions are non-overlapping, and consequently, they can be separated and computed by solving systems of smaller orders. This simple remark can significantly reduce the computing time of the reconstructions. The same holds true in the wavelet case (see Remark 8.8).

Remark 8.12 In practical applications the exact values the Gabor transform at the sample points of the ridges are not known exactly. The $z_{\ell,j}$'s are merely noisy perturbations of the true values. The reconstruction problem can be solved in a very similar way. See, for example, [64] and [289] for the modifications which are necessary in order to get the solution of this seemingly more general problem.

8.5.3 Examples

We now give examples of reconstructions from ridges.

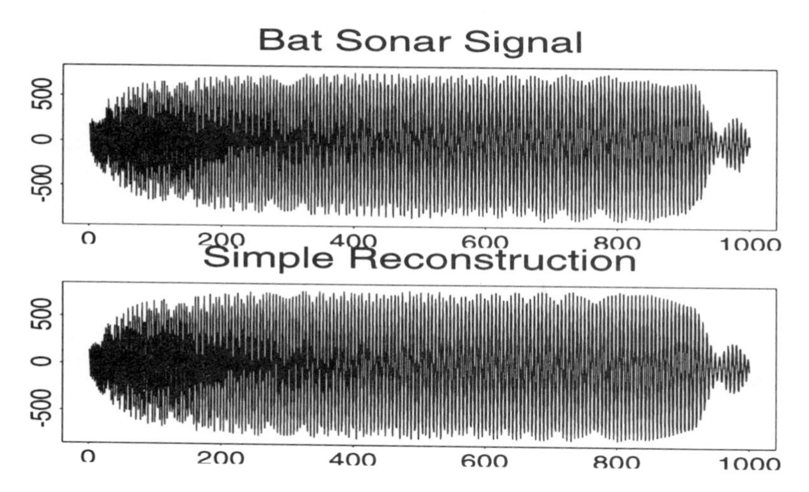

Figure 8.13. Bat sonar signal (top) and its simple reconstruction from the ridges of its CWT (bottom).

The Simple Reconstruction Formula

Let us start with our example on the bat signal. Figure 8.13 shows how faithful the simple reconstruction can be.

The /How are you ?/ speech signal is more complex. Let us explain how we obtained the reconstruction in Figure 8.12. We already described in the previous chapter the results of the crazy climbers and chaining algorithms on the CGT of this signal. We now have 27 ridges at hand, and since they are well separated, we just use for each of them the simple reconstruction formula to reconstruct the corresponding components. These 27 components are displayed in Figure 8.14. They are then simply summed up to get the reconstructed signal. It is remarkable that such a simple procedure is able to respect the phase coherence between all the components, and provide such a good reconstruction.

Reconstruction by penalization

As a simple example, we come back to the case of the bat sonar signal. The S commands used to generate this example are given in Example 8.6. The signal and its (main) ridge, estimated by the crazy climber method, are displayed in Figure 8.15 (after chaining, three ridges were found, only one of which is significan). The signal reconstructed by the penalization method is displayed at the bottom of the figure and is clearly seen to be of good quality.

However, it is possible to go further, and estimate another component

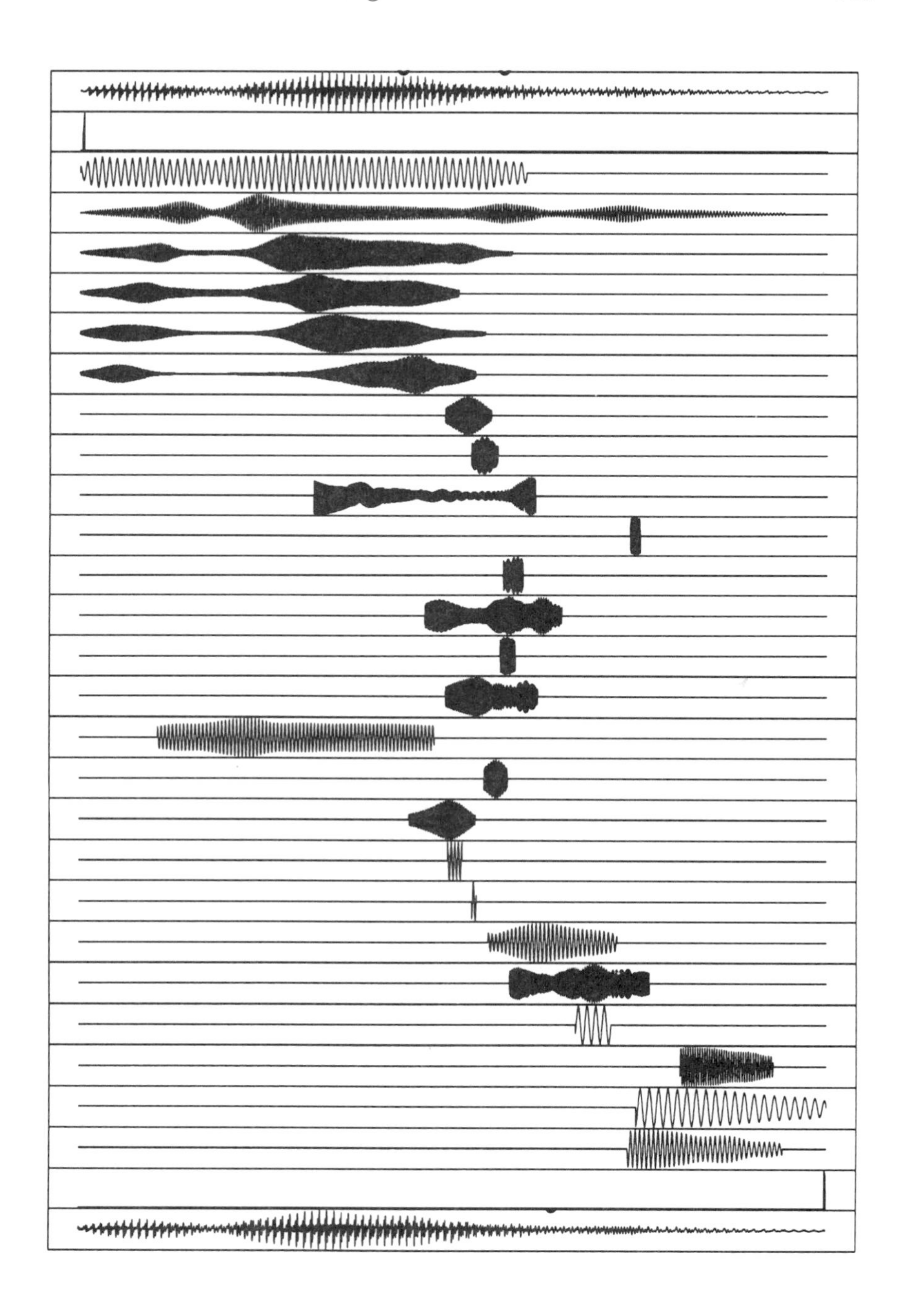

Figure 8.14. Speech signal /How are you ?/ (top), simple reconstructions from the ridges of its CGT (bottom), and all the components reconstructed independently.

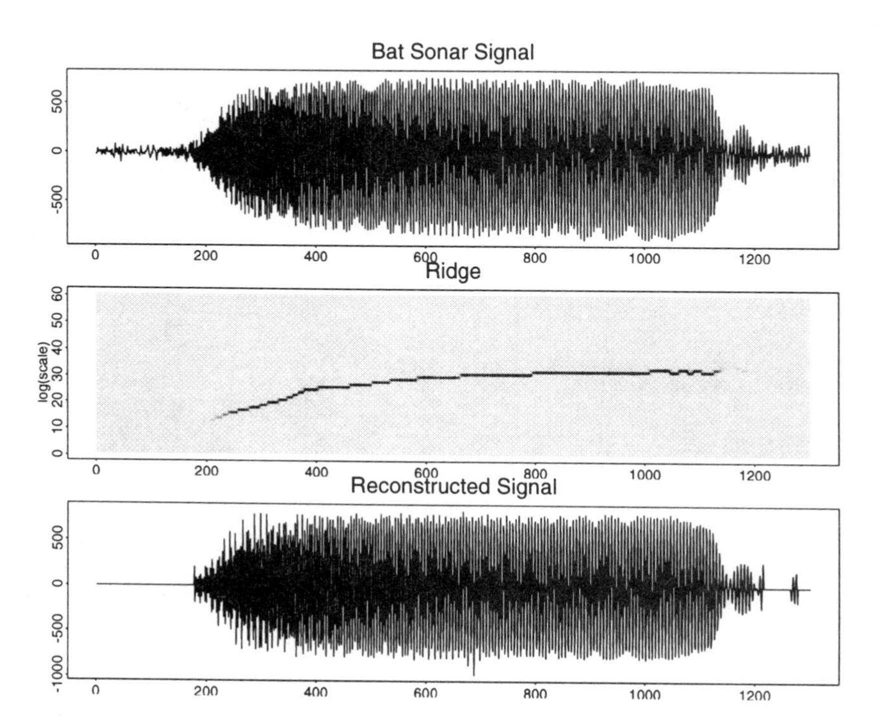

Figure 8.15. CWT of bat sonar signal, and reconstruction of the fundamental component by the penalization approach.

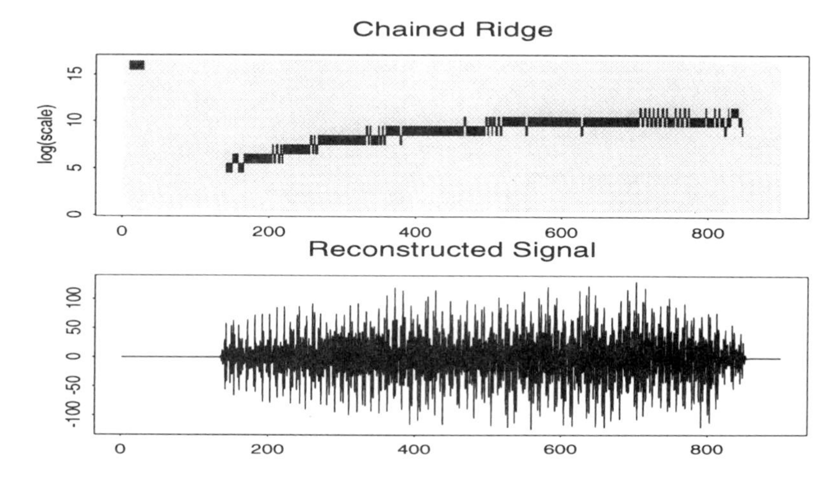

Figure 8.16. Reconstruction of the first harmonic component of bat sonar signal by the penalization approach.

of the signal, namely, its first harmonic. To do so, it is necessary to focus on the region in the time-scale plane where this component is located. This is done by considering the wavelet transform on the first octave only (see Example 8.7). The chained crazy climbers found two ridges, only one of which is significant, and reconstruction from the remaining ridge yields the first harmonic of the signal, as shown in Figure 8.16.

Another example may be seen at the bottom of Figure 8.12, where the speech signal /How are you ?/ has been reconstructed from Gabor transform ridges using the penalization approach. The S commands used for reconstruction are given in Example 8.8. The reconstruction is extremely faithful, even more precise than the simple one represented in the middle plot.

Application to Denoising

The ridge estimation and ridge reconstruction methods we have described so far have immediate applications to signal denoising. Indeed, they apply to signals which are localized in neighborhoods of curves in the time-frequency or time-scale plane. On the opposite, a random signal is very likely to be "completely delocalized" in time-frequency. Therefore, in the neighborhood of ridges, the contribution of noise is expected to be much smaller than that of the signal, and reconstruction from ridge yields *de facto* a denoised version of the signal.

This fact is illustrated in Figure 8.17, which represents the analysis of a

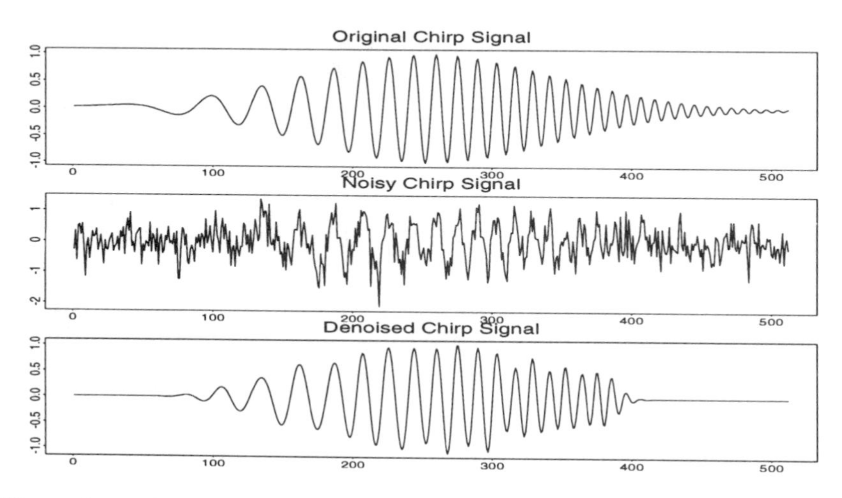

Figure 8.17. Reconstruction and denoising of noisy chirp by the penalization approach.

chirp buried into white noise, with input signal to noise equal to 0 dB (i.e., with equal variances for noise and signal). Even though the reconstruction is not perfect, it can be regarded as satisfactory given the importance of the noise. The S commands used to generate this example may be found in Example 8.9.

8.6 Examples and S-Commands

A series of functions have been developed for handling signals of the kind we have been considering since the beginning of this chapter. We describe here a series of examples of how to use these functions.

8.6.1 Transient Detection from Dyadic Wavelet Transform Extrema

The Swave package provides functions for dyadic wavelet transform (namely, the function `mw`) presented in Chapter 5 and a corresponding function for reconstruction from dyadic wavelet transform extrema. This function is named `mrecons`. It takes one argument, namely, a set of dyadic wavelet transform local extrema in the format described in Chapter 5, i.e., as an output of the function `ext`, and returns three different reconstructions: reconstruction from extrema f, f plus the mean of the signal, and f plus the large scale approximation (provided by the scaling function).

Swave also provides functions for the corresponding transient detection problem described in Section 8.4, in particular for local extrema trimming. The S functions `mntrim` and `mbtrim` perform such a trimming. Both functions require two arguments: a dyadic wavelet transform local maxima structure as an output of the function `ext`, and a confidence level α (set to 5% by default). `mntrim` trims the extrema under the hypothesis that the noise is white and Gaussian and uses direct simulations. `mtrim` does not suppose that the noise is Gaussian. It uses bootstrapping to perform simulations.

Let us describe an example. We consider a noisy transient C + white noise, with input signal to noise ratio set to -19 dB. For the sake of comparison, we describe the reconstruction of the transient C. The procedure is as in Example 8.1. The corresponding results are in Figures 8.4, 8.5, 8.8, and 8.9.

Example 8.1 "Pure" Transient Analysis/Synthesis

Compute the dyadic wavelet transform and the corresponding local extrema:

```
> dwC0 <- mw(C0,5)
> extC0<- ext(dwC0)
```

Reconstruction from the extrema:

```
> recC0 <- mrecons(dwC0)
```

The transients reconstructed using this procedure (as well as another example) are displayed in Figure 8.6. We now turn to the noisy case.

Example 8.2 "Noisy" Transient Analysis

Compute the dyadic wavelet transform and the corresponding local extrema:

```
> dwC4 <- mw(C4,5)
> extC4<- ext(dwC4)
```

We end up with a fairly large number of extrema, which is now reduced by the trimming procedure before reconstruction.

Example 8.3 : "Noisy" Transient Detection/Synthesis

Trim the extrema:

```
> trC4<- mntrim(extC4)
```

Reconstruction from the trimmed extrema:

```
> recC4 <- mrecons(trC4)
```

8.6.2 Reconstructions of Frequency Modulated Signals

Let us turn to the ridge finding methods described in the previous chapter and the corresponding reconstruction methods. The Swave package implements several reconstruction procedures.

Simple Reconstruction

The simplest such reconstruction is provided by the restriction of the transform to its ridge. The Splus function `sridrec` implements such a simple reconstruction. As an illustration, consider again the bat sonar signal.

Example 8.4 Simple Reconstruction of the Bat Signal
Reconstruct the signal from the ridge (see Example 7.3) and display the result:

```
> srecbat <- sridec(cwtbat[201:1200,],rbat$ridge)
> npl(2)
> tsplot(bat[201:1200])
> tsplot(srecbat)
```

In the case of ridges obtained via the crazy climber method (i.e., the function `crc`)), the simple reconstruction may be performed directly from the output of `crc` by using the function `scrcrec`).

Example 8.5 Simple Reconstruction of a Speech Signal
Compute the Gabor transform and estimate the ridges. In order to get precise ridge estimates, we use windows with good frequency localization (scale=60) and quite a large number of climbers:

```
> tsplot(HOWAREYOU)
> cgtHOW <- cgt(HOWAREYOU,70,0.01,60)
> clHOW <- crc(Mod(cgtHO),nbclimb=2000)
```

Simple reconstruction:

```
> srecHOW <- scrcrec(HOWAREYOU,cgtHOW,clHOW,ptile=0.0001,
+          plot=2)
```

Reconstruction by Penalization

For the case or ridges obtained via crazy climbers method (i.e., the function `crc`)), Swave implements two shortcuts for reconstruction by penalization, namely `crcrec`) for CWT and `gcrcrec`) for CGT. Examples follow.

Example 8.6 Bat Sonar Signal: Fundamental Component
Computation of the CWT:

```
> npl(3)
> tsplot(bat)
s > cwtbat<- cwt(bat,3,20,plot=F)
```

Crazy climber ridge detection and reconstruction from the main ridge:

```
> crcbat<- crc(Mod(cwtbat),nclimb=50)
> recbat <- crcrec(bat,cwtbat,crcbat,3,20,5,epsilon=0,plot=F,
+         ptile=0.01)
> tsplot(recbat$rec)
```

Example 8.7 Bat Sonar Signal: First Harmonic Component
Compute the CWT on the first octave, and estimate the ridge:

```
> cwtbat0 <- cwt(bat,1,16)
> crcbat0 <- crc(Mod(cwtbat0[301:1200,]),nbclimb=200)
```

Reconstruction:

```
> recbat0 <- crcrec(bat[301:1200], cwtbat0[301:1200, ],
+         crcbat0, 1, 16, 5,epsilon = 0, plot = F, minnbnodes =
5,
+         para = 5, ptile = 0.0005)
```

Display (two ridges are found; only the second one is significant):

```
> npl(2)
> image(recbat0$ordered)
> tsplot(recbat0$comp[2,])
```

Example 8.8 Speech Reconstruction (continued)
From the crazy climber ridges of Example 8.5, we use the penalization approach to reconstruct speech signal:

```
> crecHOW <- crcgrec(HOWAREYOU,cgtHOW,clHOW,70,0.01,60,20,
+        ptile=0.0001,epsilon=0,para=5)
> npl(3)
```

```
> tsplot(HOWAREYOU)
> title(''Original Signal'')
> tsplot(srecHOW$rec)
> title(''Simple Reconstruction'')
> tsplot(crecHOW$rec)
> title(''Reconstruction by Penalization method'')
```

Denoising

Example 8.9 Denoising a Noisy Chirp
Generate noisy chirp and compute its wavelet transform. Variance of the noise is chosen so as to have 0 *dB input SNR:*

```
> x <- 1:512
> gchirp <- exp(-4*(x-256)*(x-256)/(256*256))*
+         sin(2*pi*(x+0.002*(x-256)*(x-256))/16)
> z <- gchirp + rnorm(512,sd = 0.4)
> cwtx <- cwt(z,5,10)
```

Run the crazy climber detection algorithm and reconstruct the signal:

```
> crcx <- crc(Mod(cwtx),nbclimb=100)
> crcrecx <- crcrec(x,cwtx,crcx,5,10,5,epsilon=0,plot=2,
+         para=5,minnbnodes=6,ptile=0.1)
> npl(3)
> tsplot(gchirp)
> tsplot(z)
> tsplot(crcrecx$rec)
```

Still Another Example

We now analyze a data file downloaded from the Web page of Victor Chen from the Navy Research Lab in Washington, DC. The original data file contains 15,000 samples of a simulated chirp embedded in noise. The SNR of this data is approximately −5 dB. The data are posted on the Web to test algorithms to detect and extract the signal embedded in noise. This challenge is exactly of the type we have met with the bat sonar signal.

Example 8.10 Denoise and Reconstruct Chen's Chirp
Read and compute the wavelet transform of Chen's chirp:

```
> ch <- scan("./signals/chirpm5db.dat")
> npl(2)
> tsplot(ch)
> cwtch <- cwt(ch,7,8,plot=F)
```

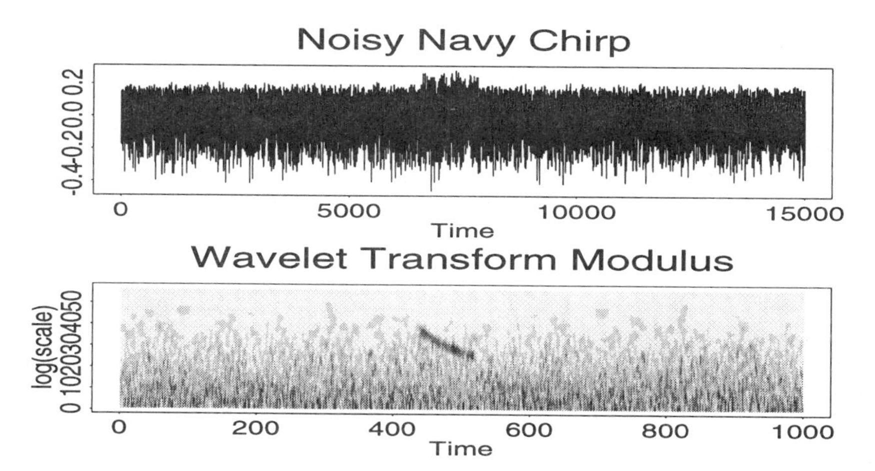

Figure 8.18. Original noisy chirp signal and the modulus of its wavelet transform.

```
> image(Mod(cwtch[15*(1:1000),]))
```

We can see from Figure 8.18 that the chirp is in the middle of the signal. So, in order to save memory allocation and to increase computing speed, we extract a small part of the signal containing the chirp to be detected. We shall work with the signal comprising the samples ranging from 5501 to 9596 from now.

Example 8.11 Ridge Estimation with the Crazy Climbers Algorithm
First, we estimate the "typical CWT" of the noise in a "noise alone region." Then we normalize the CWT of the interesting part of the signal and we run "crazy climbers."

```
> shortch <- ch[5501:9596]
> cwtshortch <- cwt(shortch, 7, 8,plot=F)
> shortn <- ch[1:4096]
> cwtshortn <- Mod(cwt(shortn,7,8,plot=F))
> wspecn <- tfmean(cwtshortn*cwtshortn,plot=F)
> ncwtsn <- t(t(cwtshortch[,1:56])/sqrt(wspecn))
> crshort <- crc(Mod(ncwtsn), nbclimb = 1000)
```

The ridge detected by the crazy climber algorithm is shown at the top of Figure 8.19. We run two synthesis algorithms to reconstruct the original chirp from the values of the wavelet transform on this ridge. We use a compression factor of 5 and as a consequence, the reconstructions are produced from 74 samples of the wavelet transform. The first reconstruction

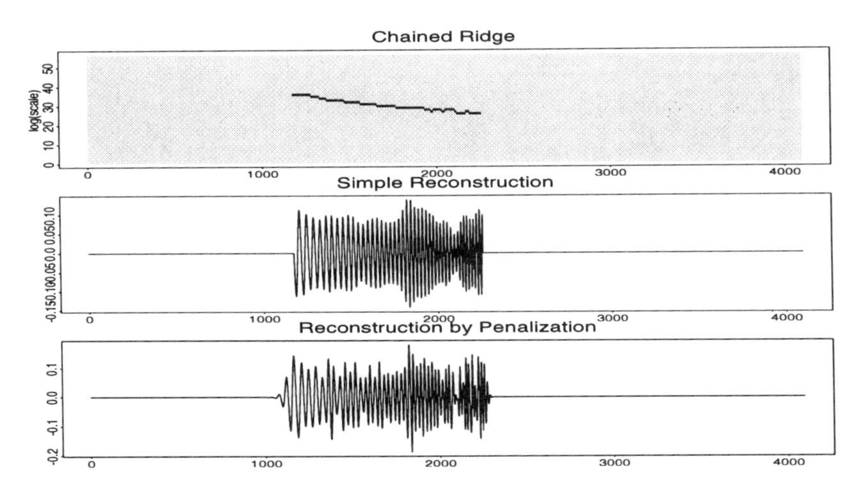

Figure 8.19. Ridge detected by the crazy climber algorithm and results of the two reconstructed procedures from the values of the wavelet transform on the ridge (with and without the smoothness penalty term).

procedure merely gives the restriction of the wavelet transform to the ridge, while the second one is the result of the optimization described in the text without the smoothness penalty on the top of the ridge. The lengths of the estimated signals end up being equal to 1394 in both cases. It is difficult to compare these results with the results of [72] since the reconstruction reported in [72] was done at a much higher SNR. This higher SNR is the reason why our reconstruction includes part of the noise component into the result.

Two reconstructions:

```
> srecshort <- simplecrcrec(shortch,cwtshortch,crshort,
+         ptile=0.2)
> recshort <- crcrec(shortch,cwtshortch,crshort,7,8,5,epsilon=0,
+         para=5,pile=0.2)
```

The results are shown in the middle and the bottom of Figure 8.19, respectively.

8.7 Notes and Complements

There are many excellent textbooks on nonparametric regression. See, for example, the books by Härdle [159], Eubank [122], or Bosq [55], and the S-related monographs by Venables and Ripley [Venables95]. Our discussion of the spline smoothing approach is borrowed from Grace Wahba's monograph [289]. A general introduction to signal detection and estimation can be found in Poor's book [244]. The shrinkage estimation strategy of Donoho and Johnstone was applied to the periodogram for the estimation of spectral densities in Gao's Ph.D. thesis [140]. The WaveShrink spectral estimator is implemented in the S+wavelets package [Bruce94]. There has been a huge number of publications on the applications of wavelet analysis to statistics. As we already did in the Notes and Complements to Chapter 5, we refer the interested reader to [Antoniadis94] for a recent overview of some of these efforts. First, most of them relied on expansions in a specific wavelet basis for the purpose of density estimation, denoising or regression. See, for example, [114] and the references therein, or even [17] for density estimation, and the series of works by Donoho and collaborators [110, 111, 112, 113, 115] and the articles by Antoniadis and collaborators [19, 13, 14, 15] for applications to denoising and regression. More effort is now devoted to investigations of the possible roles of these expansions in related problems such as change point detection or model design [18, 20, 16].

The usefulness of the local extrema of the wavelet transform can be traced back goes back to early works of Marr [218] and Canny [62], where it appears in disguised form. It was adapted to the (dyadic) wavelet framework by S. Mallat and Zhong in [214] (see also [212] for another application). The versatility of the Mallat and Zhong reconstruction algorithm [214] from the extrema of the dyadic transform was exploited by Carmona in [65] to detect and reconstruct transients as explained in the text and also by Carmona and Hudgins in [66] to analyze intrinsic signatures of EEG evoked response potentials.

The characterization of frequency modulated signals by the ridges of a time-frequency representation is an old folk tale in signal analysis (see, for example, [Flandrin93] or [Torrésani96] for extensive reviews). Surprisingly, the introduction of effective algorithms for signal reconstruction from estimated ridges is much more a recent phenomenon. Algorithms based on the simple reconstruction methods discussed in the text can be found in the work of Delprat *et al.* [102, 155] or Tchamitchian and Torrésani [272]. The algorithms based on the penalization techniques are due to the authors [67, 68, 69] .

Part IV

The Swave Library

Chapter 9

Downloading and Installing the Swave Package

This short chapter is devoted to the necessary instructions to download the C and S codes which are needed to install the Swave package on your system. This set of programs is made available free or charge for non-commercial use.

The installation procedures described here have been tested on several UNIX platforms, including SUN's and SGI's. Throughout the text of the first part of the book we used ¿ for the prompt of Splus. This convention is still in force here. We shall also use $ for the prompt of the Unix system on which the system is to be installed. As of the publication of the book, the installation of Swave on PCs was not tested. We are in the process of doing and the relevant information will be made available on the book's Web page at

```
http://chelsea.princeton.edu/~rcarmona/TFbook/
```

This page will be used to post corrections, updates, etc.

9.1 Downloading Swave

The package can be retrieved from the Web with your favorite browser from the URL given for the book Web page. The files can also be downloaded directly by anonymous ftp. The address of the server is

```
chelsea.princeton.edu
```

The following sequence should do the trick (at least from a UNIX system on the Internet).
Connect to the server by typing

```
$ ftp chelsea.princeton.edu
```

Once you are connected to the server, at the prompt

```
login:
```

type

```
anonymous
```

and when a password is requested

```
psswd:
```

type your e-mail address. Then type the following sequence of commands:

```
binary
cd pub
cd prog
get Swave.tar.gz
```

9.2 Installing Swave on a Unix Platform

The file which you downloaded is a compressed binary archive. In order to use it, type the following command on your system:

```
$ gunzip Swave.tar.gz
$ tar -xfv Swave.tar
```

A directory Swave will be created on your system. It contains all the files necessary to the use of the Swave package. In particular, this directory contains a subdirectory `.Data`, a file named `Makefile`, and a file named `README`. The purpose of the subdirectory `.Data` is to avoid cluttering the subdirectory `.Data` of your home directory. All the functions, the data sets, and the S objects which are included in the package, as well as those you are about to create, will be located in this local subdirectory. The Makefile is used to compile the C-code on your system. But the most important file is the `README` file. You should read it first. You will find there a detailed list of all the steps you will have to take to compile the C code and to create the S library which you want to attach to be able to use the time-frequency functions described in the book.

9.3 Troubleshooting

The data sets, the C codes, and the S functions have been tested on several platforms. It is difficult at this stage to anticipate the problems which will hinder the use of the package Swave. Also, because this package is placed in the public domain free of charge, the authors cannot be held responsible for any error or for the consequences of its use. Furthermore, the authors will not offer technical support for the package. They nevertheless encourage those experiencing problems with the downloading or the installation or even the use of the package to send an e-mail message at

`rcarmona@princeton.edu`

describing the problem. They will try to respond with suggestions and they will share the names and addresses of users on similar platforms. Also, they will use the Web page of the book to make available an updated list of reported bugs and proposed fixes.

Chapter 10

The Swave S Functions

This chapter is the user guide of the Swave package. It gives, one page at a time, the technical description of the functions that make up the library. When you are using Swave from Splus, these descriptions are available through the on-line help. In order to use the help facility, simply type:

> `help` *function-filename*

and the help file of the function *filename* will be displayed. For example, you would type:

```
> help cwt
```

to get information on the function `cwt`. The functions are organized in alphabetical order.

`cfamily`	Ridge Chaining Procedure	`cfamily`

DESCRIPTION: Chains the ridge estimates produced by the function `crc`.
USAGE: `cfamily(ccridge, bstep=1, nbchain=100, ptile=0.05)`
REQUIRED ARGUMENTS:

ccridge: Unchained ridge set as the output of the function `crc`

OPTIONAL ARGUMENTS:

bstep: Maximal length for a gap in a ridge.
nbchain: Maximal number of chains produced by the function.
ptile: Relative threshold for the ridges.

VALUE: Returns the results of the chaining algorithm

ordered map: Image containing the ridges (displayed with different colors)
chain: 2D array containing the chained ridges, according to the chain data structure:

- chain[,1]: first point of the ridge
- chain[,2]: length of the chain
- chain[,3:(chain[,2]+2)]: values of the ridge

nbchain: Number of chains produced by the algorithm

DETAILS: `crc` returns a measure in time-frequency (or time-scale) space. `cfamily` turns it into a series of one-dimensional objects (ridges). The measure is first thresholded, with a relative threshold value set to the input parameter ptile.
During the chaining procedure, gaps within a given ridge are allowed and filled in. The maximal length of such gaps is the input parameter bstep.
SEE ALSO: `crc` for the ridge estimation, and `crcrec`, `gcrcrec`, and `scrcrec` for corresponding reconstruction functions.
EXAMPLES: tmp <– cfamily(ccridge).
See Example 7.7 and 7.8.
REFERENCES: See discussion in Chapter 7, more particularly Section 7.4.

`cgt` Continuous Gabor Transform `cgt`

DESCRIPTION: Computes the continuous Gabor transform with Gaussian window.

USAGE: `cgt(input, nvoice, freqstep=(1/nvoice),scale=1, plot=TRUE)`

REQUIRED ARGUMENTS:

input: Input signal (possibly complex-valued).

nvoice: Number of frequencies for which Gabor transform is to be computed.

OPTIONAL ARGUMENTS:

freqstep: Sampling rate for the frequency axis.

scale: Size parameter for the window.

plot: Logical variable set to TRUE to display the modulus of the continuous gabor transform on the graphic device.

VALUE: Continuous (complex) Gabor transform (2D array).

DETAILS: The output contains the (complex) values of the Gabor transform of the input signal. The format of the output is a 2D array (signal_size $\times$ nb_scales).

SEE ALSO: `cwt, cwtp, DOG` for continuous wavelet transforms. `cwtsquiz` for synchrosqueezed wavelet transform.

WARNING: freqstep must be less than 1/nvoice to avoid aliasing. freqstep=1/nvoice corresponds to the Nyquist limit.

EXAMPLES: tmp <- cgt(input, nvoice,freqstep,scale)

See, e.g., Examples 3.2, 3.3, 3.4 in Chapter 3.

REFERENCES: See Chapter 3.

`cleanph`	Threshold Phase based on Modulus	`cleanph`

DESCRIPTION: Sets to zero the phase of time-frequency transform when modulus is below a certain value.

USAGE: `cleanph(tfrep, thresh=0.01, plot=TRUE)`

REQUIRED ARGUMENTS:

tfrep: Continuous time-frequency transform (2D array).

OPTIONAL ARGUMENTS:

thresh: (Relative) hreshold.

plot: If set to TRUE, displays the maxima of cwt on the graphic device.

VALUE: Thresholded phase (2D array).

EXAMPLES: See Example 3.2.

corona	Ridge Estimation by Corona Method	corona

DESCRIPTION: Estimate a (single) ridge from a time-frequency representation, using the corona method.

USAGE: `corona(tfrep, guess, tfspec=numeric(dim(modulus)[2]), subrate=1, temprate=3, mu=1, lambda=2*mu, iteration=1000000, seed=-7, stagnant=20000,costsub=1, plot=T)`

REQUIRED ARGUMENTS:

tfrep: Time-frequency representation (real valued).

guess: Initial guess for the algorithm.

OPTIONAL ARGUMENTS:

tfspec: Estimate for the contribution of the noise to modulus.

subrate: Subsampling rate for ridge estimation.

temprate: Initial value of temperature parameter.

lambda: Coefficient of the ridge's derivative in cost function.

mu: Coefficient of the ridge's second derivative in cost function.

iteration: Maximal number of moves.

seed: Initialization of random number generator.

stagnant: Maximum number of stationary iterations before stopping.

costsub: Subsampling of cost function in output.

plot: When set(default), some results will be shown on the display.

VALUE: Returns the estimated ridge and the cost function.

ridge: 1D array (of same length as the signal) containing the ridge.

cost: 1D array containing the cost function.

DETAILS: To accelerate convergence, it is useful to pre-process modulus before running annealing method. Such a pre-processing (smoothing and subsampling of modulus) is implemented in `corona`. The parameter subrate specifies the subsampling rate.

SEE ALSO: `icm`,`coronoid`,`snake`, `snakoid`.

WARNING: The returned cost may be a large array, which is time consuming. The argument costsub allows subsampling the cost function.

EXAMPLES: See Examples 7.3 and 7.4 in Chapter 7.

REFERENCES: See Chapter 7, and more particularly Section 7.3.1.

`coronoid` Ridge Estimation by Modified Corona Method `coronoid`

DESCRIPTION: Estimate a ridge using the modified corona method (modified cost function).

USAGE: `coronoid(modulus, guess, tfspec=numeric (dim(modulus)[2]), subrate=1, temprate=3, mu=1, lambda=2*mu, iteration=1000000, seed=-7, stagnant=20000, costsub=1, plot=T)`

REQUIRED ARGUMENTS:

modulus Time-frequency representation (real valued).

guess Initial guess for the algorithm.

OPTIONAL ARGUMENTS:

tfspec: Estimate for the contribution of the noise to modulus.

subrate: Subsampling rate for ridge estimation.

temprate: Initial value of temperature parameter.

lambda: Coefficient of the ridge's second derivative in cost function.

mu: Coefficient of the ridge's derivative in cost function.

iteration: Maximal number of moves.

seed: Initialization of random number generator.

stagnant: Maximum number of stationary iterations before stopping.

costsub: Subsampling of cost function in output.

plot: When set(default), some results will be shown on the display.

VALUE: Returns the estimated ridge and the cost function.

ridge: 1D array (of same length as the signal) containing the ridge.

cost: 1D array containing the cost function.

DETAILS: To accelerate convergence, it is useful to pre-process modulus before running annealing method. Such a pre-processing (smoothing and subsampling of modulus) is implemented in `coronoid`. The parameter subrate specifies the subsampling rate.

SEE ALSO: `corona, icm, snake, snakoid.`

WARNING: The returned cost may be a large array. The argument costsub allows subsampling the cost function.

EXAMPLES: See Example 7.3 in Chapter 7.

REFERENCES: See in Chapter 7, and more particularly Section 7.3.1.

crc Ridge Extraction by Crazy Climbers crc

DESCRIPTION: Uses the "crazy climber algorithm" to detect ridges in the modulus of a continuous wavelet or a Gabor transform.

USAGE: `crc(tfrep, tfspec=numeric(dim(modulus)[2]), bstep=3, iteration=10000, rate=0.001, seed=-7, nbclimb=1, flag.int=TRUE, chain=TRUE, flag.temp=FALSE)`

REQUIRED ARGUMENTS: tfrep: modulus of the (wavelet or Gabor) transform.

OPTIONAL ARGUMENTS:

tfspec: Numeric vector which gives, for each value of the scale or frequency, the expected size of the noise contribution.

bstep: Stepsize for random walk of the climbers.

iteration: Number of iterations.

rate: Initial value of the temperature.

seed: Initial value of the random number generator.

nbclimb: Number of crazy climbers.

flag.int: If set to T, the weighted occupation measure is computed.

chain: If set to T, chaining of the ridges is done.

flag.temp: If set to T: constant temperature.

VALUE: Returns a 2D array called beemap containing the (weighted or unweighted) occupation measure (integrated with respect to time).

SEE ALSO: `corona`, `icm`, `coronoid`, `snake`, `snakoid` for ridge estimation, `cfamily` for chaining and `crcrec`,`gcrcrec`,`scrcrec` for reconstruction.

EXAMPLES: See Examples 7.7 and 7.8 in Chapter 7.

REFERENCES: See Section 7.4 in Chapter 7.

`crcrec` Crazy Climbers Reconstruction by Penalization `crcrec`

DESCRIPTION: Reconstructs a real-valued signal from the output of `crc` (wavelet case) by minimizing an appropriate quadratic form.

USAGE: `crcrec(siginput, inputwt, beemap, noct, nvoice, compr, minnbnodes=2, w0=2*pi, bstep=5, ptile=.01, epsilon=0, fast=F, para=5, real=F, plot=2)`

REQUIRED ARGUMENTS:

- siginput: Original signal.
- inputwt: Wavelet transform.
- beemap: Occupation measure, output of `crc`.
- noct: Number of octaves.
- nvoice: Number of voices per octave.
- compr: Compression rate for sampling the ridges.

OPTIONAL ARGUMENTS:

- minnbnodes: Minimal number of points per ridge.
- w0: Center frequency of the wavelet.
- bstep: Size (in the time direction) of the steps for chaining.
- ptile: Relative threshold of occupation measure.
- epsilon: Constant in front of the smoothness term in penalty function.
- fast: If set to T, uses trapezoidal rule to evaluate Q_2.
- para: Scale parameter for extrapolating the ridges.
- real: If set to TRUE, uses only real constraints.
- plot: 1: Displays signal, components, and reconstruction one after another. 2: Displays signal, components, and reconstruction.

VALUE: Returns a structure containing the following elements:

- rec: Reconstructed signal.
- ordered: Image of the ridges (with different colors).
- comp: 2D array containing the signals reconstructed from ridges.

DETAILS: When ptile is high, boundary effects may appear. para controls extrapolation of the ridge.

SEE ALSO: `crc, cfamily, crcgrec, scrcrec.`

EXAMPLES: See Examples 8.6, 8.7, and 8.9 in Chapter 8.

REFERENCES: See the discussion in Section 8.5 in Chapter 8.

`cwt`	Continuous Wavelet Transform	`cwt`

DESCRIPTION: Computes the continuous wavelet transform for the (complex-valued) Morlet wavelet.

USAGE: `cwt(input, noctave, nvoice=1, w0=2*pi, open=T, twoD=T, plot=T)`

REQUIRED ARGUMENTS:

input: Input signal (possibly complex-valued).

noctave: Number of powers of 2 for the scale variable.

OPTIONAL ARGUMENTS:

nvoice Number of scales in each octave (i.e., between two consecutive powers of 2).

w0: Central frequency of the wavelet.

open: Logical variable set to T to load the C object codes.

twoD: Logical variable set to T to organize the output as a 2D array (signal_size × nb_scales); otherwise, the output is a 3D array (signal_size × noctave × nvoice).

plot: If set to T, display the modulus of the continuous wavelet transform on the graphic device.

VALUE: Continuous (complex) wavelet transform.

DETAILS: The output contains the (complex) values of the wavelet transform of the input signal. The format of the output can be

- 2D array (signal_size × nb_scales)
- 3D array (signal_size × noctave × nvoice)

SEE ALSO: `cwtp`, `cwtTh`, `DOG`, `gabor`.

WARNING: Since Morlet's wavelet is not, strictly speaking, a wavelet (it is not of vanishing integral), artifacts may occur for certain signals.

EXAMPLES: `tmp <- cwt(input, noctave, nvoice)`

See the examples in Section 4.6.

REFERENCES: See Chapter 4.

`cwtp` Continuous Wavelet Transform with Phase Derivative `cwtp`

DESCRIPTION: Computes the continuous wavelet transform with (complex-valued) Morlet wavelet and its phase derivative.

USAGE: `cwtp(input, noctave, nvoice=1, w0=2*pi, open=T, twoD=T,plot=T)`

REQUIRED ARGUMENTS:

- input: Input signal (possibly complex-valued).
- noctave: Number of powers of 2 for the scale variable.

OPTIONAL ARGUMENTS:

- nvoice: Number of scales in each octave (i.e., between two consecutive powers of 2).
- w0: Central frequency of the wavelet.
- open: Logical variable set to T to load the C object codes.
- twoD: Logical variable set to T to organize the output as a 2D array (signal_size $\times$ nb_scales), otherwise, the output is a 3D array (signal_size $\times$ noctave $\times$ nvoice).
- plot: If set to T, display the modulus of the continuous wavelet transform on the graphic device.

VALUE: List containing the continuous (complex) wavelet transform and the phase derivative.

- wt: Array of complex numbers for the values of the continuous wavelet transform.
- f: Array of the same dimensions containing the values of the derivative of the phase of the continuous wavelet transform.

SEE ALSO: `cgt`, `cwt`, `cwtTh`, `DOG` for wavelet transform, and `gabor` for continuous Gabor transform.

EXAMPLES: `tmp <- cwtp(input, noctave, nvoice)`

`cwtsquiz`	Squeezed Continuous Wavelet Transform	`cwtsquiz`

DESCRIPTION: Computes the synchrosqueezed continuous wavelet transform with the (complex-valued) Morlet wavelet.

USAGE: `cwtsquiz(input, noctave, nvoice=1, w0=2*pi, twoD=T, plot=T)`

REQUIRED ARGUMENTS:

input: Input signal (possibly complex-valued).

noctave: Number of powers of 2 for the scale variable.

OPTIONAL ARGUMENTS:

nvoice: Number of scales in each octave (i.e., between two consecutive powers of 2).

w0: Central frequency of the wavelet.

twoD: Logical variable set to T to organize. the output as a 2D array (signal_size × nb_scales), otherwise, the output is a 3D array (signal_size × noctave × nvoice).

plot: Logical variable set to T to T to display the modulus of the squeezed wavelet transform on the graphic device.

VALUE: Synchrosqueezed continuous (complex) wavelet transform.

DETAILS: The output contains the (complex) values of the squeezed wavelet transform of the input signal. The format of the output can be

- 2D array (signal_size × nb_scales)
- 3D array (signal_size × noctave × nvoice)

SEE ALSO: `cwt`, `cwtp`, `DOG`, `cgt`.

EXAMPLES: `tmp <- cwtsquiz(input, noctave, nvoice)`

See Example 7.9 in Chapter 7.

REFERENCES: Reassignment methods and synchrosqueezed wavelet transform are described in Section 7.5.

DOG Continuous Wavelet Transform with Derivative of Gaussian DOG

DESCRIPTION: Computes the continuous wavelet transform with (complex-valued) derivative of Gaussian wavelets.

USAGE: `DOG(input, noctave, nvoice=1, moments, twoD=T, plot=T)`

REQUIRED ARGUMENTS:

input: Input signal (possibly complex-valued).

noctave: Number of powers of 2 for the scale variable.

moments: Number of vanishing moments of the wavelet (order of the derivative).

OPTIONAL ARGUMENTS:

nvoice: Number of scales in each octave (i.e. between two consecutive powers of 2).

twoD: Logical variable set to T to organize the output as a 2D array (signal_size $\times$ nb_scales), otherwise, the output is a 3D array (signal_size $\times$ noctave $\times$ nvoice).

plot: If set to T, display the modulus of the continuous wavelet transform on the graphic device.

VALUE: Continuous (complex) wavelet transform.

DETAILS: The output contains the (complex) values of the wavelet transform of the input signal. The format of the output can be

- 2D array (signal_size $\times$ nb_scales)
- 3D array (signal_size $\times$ noctave $\times$ nvoice)

SEE ALSO: `cwt`, `cwtp`, `cwtsquiz`, `cgt`.

EXAMPLES: `tmp <- DOG(input, noctave, nvoice, moments)`
See the examples in Chapter 6.

REFERENCES: See the description in Section 4.3.

`dwinverse`	Inverse Dyadic Wavelet Transform	`dwinverse`

DESCRIPTION: Invert the dyadic wavelet transform.
USAGE: `dwinverse(wt, filtername="Gaussian1")`
REQUIRED ARGUMENTS:

wt: Dyadic wavelet transform.

OPTIONAL ARGUMENTS:

filtername: Filters used. ("Gaussian1" stands for the filters corresponding to those of Mallat and Zhong's wavelet; "Haar" stands for the filters of Haar basis.)

VALUE: Reconstructed signal.
SEE ALSO: `mw`, `ext`, `mrecons`.
EXAMPLES: `recsig <- dwinverse(wt)`
REFERENCES: See Section 5.2.

`ext`	Extrema of Dyadic Wavelet Transform	`ext`

DESCRIPTION: Compute the local extrema of the dyadic wavelet transform modulus.

USAGE: `ext(wt, scale=F, plot=T)`

REQUIRED ARGUMENTS:

wt: Dyadic wavelet transform.

OPTIONAL ARGUMENTS:

scale: Flag indicating if the extrema at each resolution will be plotted at the same scale.

plot: If set to T, displays the transform on the graphics device.

VALUE: Structure containing:

original: Original signal.

extrema: Extrema representation.

Sf: Coarse resolution of signal.

maxresoln: Number of decomposition scales.

np: Size of signal.

SEE ALSO: `mw`, `mrecons`.

EXAMPLES: `extrema <- ext(dwt)`

See Example 5.3 in Chapter 5.

REFERENCES: See Section 5.2, and more particularly Section 5.2.3.

`gabor`	Generate Gabor function	`gabor`

DESCRIPTION: Generates a Gabor for given location and frequency.
USAGE: `gabor(sigsize, location, frequency, scale)`
REQUIRED ARGUMENTS:

sigsize: Length of the Gabor function.

location: Position of the Gabor function.

frequency: Frequency of the Gabor function.

scale: Size parameter for the Gabor function.

VALUE: Complex 1D array of size sigsize.
SEE ALSO: `morlet`
EXAMPLES: `tmp <- gabor(size, position, frequency, scale)`
See Example 3.1 in Chapter 3.
REFERENCES: See Chapter 3.

`gcrcrec` Crazy Climbers Reconstruction by Penalization `gcrcrec`

DESCRIPTION: Reconstructs a real-valued signal from ridges found by crazy climbers on a Gabor transform.

USAGE: `gcrcrec(siginput, inputgt, beemap, nvoice, freqstep, scale, compr, bstep=5, ptile=.01, epsilon=0, fast=T, para=0, minnbnodes=3, hflag=F, real=F, plot=2)`

REQUIRED ARGUMENTS:

- siginput: Original signal.
- inputgt: Gabor transform.
- beemap: Occupation measure, output of `crc`.
- nvoice: Number of frequencies.
- freqstep: Sampling step for frequency axis.
- scale: Size of windows.
- compr: Compression rate to be applied to the ridges.

OPTIONAL ARGUMENTS:

- bstep: Size (in the time direction) of the steps for chaining.
- ptile: Threshold of ridge.
- epsilon: Constant in front of the smoothness term in penalty function.
- fast: If set to T, uses trapezoidal rule to evaluate Q_2.
- para: Scale parameter for extrapolating the ridges.
- minnbnodes: Minimal number of points per ridge.
- hflag: If set to F, uses the identity as first term in the kernel. If not, uses Q_1 instead.
- real: If set to T, uses only real constraints.
- plot: 1: Displays signal, components, and reconstruction one after another. 2: Displays signal, components, and reconstruction.

VALUE: Returns a structure containing the following elements:

- rec: Reconstructed signal.
- ordered: Image of the ridges (with different colors).
- comp: 2D array containing the signals reconstructed from ridges.

DETAILS: When ptile is high, boundary effects may appear. para controls extrapolation of the ridge.

SEE ALSO: `crc`, `cfamily`, `crcrec`, `scrcrec`.

EXAMPLES: See Example 8.8 in Chapter 8.
REFERENCES: See the discussion in Section 8.5 in Chapter 8.

gregrec	Reconstruction from a Ridge	gregrec

DESCRIPTION: Reconstructs signal from a "regularly sampled" ridge, in the Gabor case.

USAGE: `gregrec(siginput, gtinput, phi, nbnodes, nvoice, freqstep, scale, epsilon=0, fast=F, plot=F, para=5, hflag=F, real=F, check=F)`

REQUIRED ARGUMENTS:

siginput: Input signal.

gtinput: Gabor transform, output of `cgt`.

phi: Unsampled ridge.

nbnodes: Number of nodes used for the reconstruction.

nvoice: Number of different scales per octave.

freqstep: Sampling rate for the frequency axis.

scale: Size parameter for the Gabor function.

OPTIONAL ARGUMENTS:

epsilon: Coefficient of the Q_2 term in reconstruction kernel.

fast: If set to T, the kernel is computed using trapezoidal rule.

plot: If set to T, displays original and reconstructed signals.

para: Scale parameter for extrapolating the ridges.

hflag: If set to T, uses Q_1 as first term in the kernel.

real: If set to T, uses only real constraints on the transform.

check: If set to T, computes `cwt` of reconstructed signal.

VALUE: Returns a list containing:

sol: Reconstruction from a ridge.

A: <gaborlets, dualgaborlets> matrix.

lam: Coefficients of dual wavelets in reconstructed signal.

dualwave: Array containing the dual wavelets.

gaborets: Array containing the wavelets on sampled ridge.

solskel: Gabor transform of sol, restricted to the ridge.

inputskel: Gabor transform of signal, restricted to the ridge.

Q2: Second part of the reconstruction kernel.

SEE ALSO: `regrec`
EXAMPLES: `recon <- gregrec(siginput, gtinput, phi, nbnodes, nvoice, freqstep, scale)`
REFERENCES: See Chapter 8.

gridrec	Reconstruction from a Ridge	gridrec

DESCRIPTION: Reconstructs signal from sample of a ridge, in the Gabor case.

USAGE: `gridrec(gtinput, node, phinode, nvoice, freqstep, scale, Qinv, epsilon, np, real=F, check=F)`

REQUIRED ARGUMENTS:

- gtinput: Gabor transform, output of `cgt`.
- node: Time coordinates of the ridge samples.
- phinode: Frequency coordinates of the ridge samples.
- nvoice: Number of different frequencies.
- freqstep: Sampling rate for the frequency axis.
- scale: Scale of the window.
- Qinv: Inverse of the matrix Q of the quadratic form.
- epsilon: Coefficient of the Q_2 term in reconstruction kernel.
- np: Number of samples of the reconstructed signal.

OPTIONAL ARGUMENTS:

- real: If set to T, uses only constraints on the real part of the transform.
- check: If set to T, computes `cgt` of reconstructed signal.

VALUE: Returns a list containing the reconstructed signal and the chained ridges.

- sol: Reconstruction from a ridge.
- A: <gaborlets,dualgaborlets> matrix.
- lam: Coefficients of dual gaborlets in reconstructed signal.
- dualwave: Array containing the dual gaborlets.
- gaborets: Array of gaborlets located on the ridge samples.
- solskel: Gabor transform of sol, restricted to the ridge.
- inputskel: Gabor transform of signal, restricted to the ridge.

SEE ALSO: `sridrec, gregrec, regrec, regrec2`.

EXAMPLES: `recon <- gridrec(gtinput, node, phinode, nvoice, freqstep, scale, Qinv, epsilon, np)`

REFERENCES: See Chapter 8.

`hurst.est`	Estimate Hurst Exponent	`hurst.est`

DESCRIPTION: Estimates Hurst exponent from a wavelet transform.
USAGE: `Hurst.est(wspec, range, nvoice, plot=T)`
REQUIRED ARGUMENTS:

wspec: Wavelet spectrum (output of `tfmean`)

range: Range of scales from which to estimate the exponent.

nvoice: Number of scales per octave of the wavelet transform.

OPTIONAL ARGUMENTS:

plot: If set to T, displays regression line on current plot.

VALUE: Complex 1D array of size sigsize.
SEE ALSO: `tfmean`, `wspec.pl`.
EXAMPLES: expo <– Hurst.est(wspec, nscale).
REFERENCES: See Section 6.3.2 in Chapter 6.

`icm` Ridge Estimation by ICM Method `icm`

DESCRIPTION: Estimate a (single) ridge from a time-frequency representation, using the ICM minimization method.

USAGE: `icm(modulus, guess, tfspec=numeric(dim(modulus)[2]), subrate=1, mu=1, lambda=2*mu, iteration=100)`

REQUIRED ARGUMENTS:

modulus: Time-frequency representation (real valued).

guess: Initial guess for the algorithm.

OPTIONAL ARGUMENTS:

tfspec: Estimate for the contribution of the noise to modulus.

subrate: Subsampling rate for ridge estimation.

lambda: Coefficient of the ridge's derivative in cost function.

mu: Coefficient of the ridge's second derivative in cost function.

iteration: Maximal number of moves.

VALUE: Returns the estimated ridge and the cost function.

ridge: 1D array (of same length as the signal) containing the ridge.

cost: 1D array containing the cost function.

DETAILS: To accelerate convergence, it is useful to pre-process modulus before running annealing method. Such a pre-processing (smoothing and subsampling of modulus) is implemented in `icm`. The parameter subrate specifies the subsampling rate.

SEE ALSO: `corona`, `coronoid`, and `snake`, `snakoid`.

EXAMPLES: `tmp <- icm(modulus, guess)`

REFERENCES: See the discussion in Chapter 7, and more particularly Section 7.3.2.

`mbtrim`	Trim Dyadic Wavelet Transform Extrema	`mbtrim`

DESCRIPTION: Trimming of dyadic wavelet transform local extrema, using bootstrapping.

USAGE: `mbtrim(extrema, scale=F)`

REQUIRED ARGUMENTS:

extrema: Dyadic wavelet transform extrema (output of `ext`).

OPTIONAL ARGUMENTS:

scale: When set, the wavelet transform at each scale will be plotted with the same scale.

VALUE: Structure containing:

original: Original signal.

extrema: Trimmed extrema representation.

Sf: Coarse resolution of signal.

maxresoln: Number of decomposition scales.

np: Size of signal.

DETAILS: The distribution of extrema of dyadic wavelet transform at each scale is generated by bootstrap method, and the 95% critical value is used for thresholding the extrema of the signal.

SEE ALSO: `mntrim, mrecons, ext`.

EXAMPLES: `trext <- mbtrim(extrema)`

REFERENCES: See Chapter 8.

mntrim	Trim Dyadic Wavelet Transform Extrema	mntrim

DESCRIPTION: Trimming of dyadic wavelet transform local extrema, assuming normal distribution.

USAGE: `mntrim(extrema, scale=F)`

REQUIRED ARGUMENTS:

extrema: Dyadic wavelet transform extrema (output of `ext`).

OPTIONAL ARGUMENTS:

scale: When set, the wavelet transform at each scale will be plotted with the same scale.

VALUE: Structure containing:

original: Original signal.

extrema: Trimmed extrema representation.

Sf: Coarse resolution of signal.

maxresoln: Number of decomposition scales.

np: Size of signal.

DETAILS: The distribution of extrema of dyadic wavelet transform at each scale is generated by simulation, assuming a normal distribution, and the 95% critical value is used for thresholding the extrema of the signal.

SEE ALSO: `mbtrim`, `mrecons`, `ext`.

EXAMPLES: `trext <- mntrim(extrema)`

REFERENCES: See Chapter 8.

`morlet`	Morlet Wavelets	`morlet`

DESCRIPTION: Computes a Morlet wavelet at the point of the time-scale plane given in the input.

USAGE: `morlet(sigsize,location,scale,w0=2*pi)`

REQUIRED ARGUMENTS:

sigsize: Length of the output.

location: Time location of the wavelet.

scale: Scale of the wavelet.

OPTIONAL ARGUMENTS:

w0: Central frequency of the wavelet.

VALUE: Returns the values of the complex Morlet wavelet at the point of the time-scale plane given in the input.

DETAILS: The details of this construction (including the definition formulas) are given in the text.

SEE ALSO: `gabor`

EXAMPLES:
```
tmp <- morlet(sigsize,location,scale)
tsplot(Re(tmp))
tsplot(Mod(tmp))
```

REFERENCES: See Chapter 4.

`mrecons`	Reconstruct from Wavelet Transform Extrema.	`mrecons`

DESCRIPTION: Reconstruct from dyadic wavelet transform modulus extrema. The reconstructed signal preserves locations and values at extrema.

USAGE: `mrecons(extrema, filtername=''Gaussian1", readflag= T)`

REQUIRED ARGUMENTS:

extrema: The extrema representation.

OPTIONAL ARGUMENTS:

filtername: Filter used for dyadic wavelet transform.

readflag: If set to T, read reconstruction kernel from pre-computed file.

VALUE: Structure containing

f: The reconstructed signal.

g: Reconstructed signal plus mean of original signal.

h: Reconstructed signal plus coarse scale component of original signal.

DETAILS: The reconstruction involves only the wavelet coefficients, without taking care of the coarse scale component. The latter may be added a posteriori.

SEE ALSO: `mw`, `ext`.

EXAMPLES: `recsig <- mrecons(extrema)`

See Examples 8.1, 8.2, and 8.3 in Chapter 8.

REFERENCES: See Section 5.2 in Chapter 5, and Section 8.4 in Chapter 8.

`mw`	Dyadic Wavelet Transform	`mw`

DESCRIPTION: Dyadic wavelet transform, with Mallat's wavelet. The reconstructed signal preserves locations and values at extrema.

USAGE: `mw(inputdata, maxresoln, filtername="Gaussian1", scale=F, plot=T)`

REQUIRED ARGUMENTS:

inputdata: Either a text file or an S object containing data.

maxresoln: Number of decomposition scales.

OPTIONAL ARGUMENTS:

filename: Name of filter.

scale: When set, the wavelet transform at each scale is plotted with the same scale.

plot: Indicate if the wavelet transform at each scale will be plotted.

VALUE: Structure containing:

original: Original signal.

Wf: Dyadic wavelet transform of signal.

Sf: Multiresolution of signal.

maxresoln: Number of decomposition scales.

np: Size of signal.

DETAILS: The decomposition goes from resolution 1 to the given maximum resolution.

SEE ALSO: `dwinverse`, `mrecons`, `ext`.

EXAMPLES: `dwt <- mw(signal)`

See Example 5.3 in Chapter 5, and Examples 8.1 and 8.2 in Chapter 8.

REFERENCES: See Chapter 8.

`scrcrec` Simple Reconstruction from Crazy Climbers Ridges `scrcrec`

DESCRIPTION: Reconstructs signal from ridges obtained by `crc`, using the restriction of the transform to the ridge.

USAGE: `scrcrec(siginput, tfinput, beemap, bstep=5, ptile=.01, plot=2)`

REQUIRED ARGUMENTS:

siginput: Input signal.

tfinput: Time-frequency representation (output of `cwt` or `cgt`).

beemap: Output of crazy climber algorithm.

OPTIONAL ARGUMENTS:

bstep: Used for the chaining (see `cfamily`).

ptile: Threshold on the measure beemap (see `cfamily`).

plot: 1: Displays signal, components, and reconstruction one after another. 2: Displays signal, components, and reconstruction. Else, no plot.

VALUE: Returns a list containing the reconstructed signal and the chained ridges.

rec: Reconstructed signal.

ordered: Image of the ridges (with different colors).

comp: 2D array containing the signals reconstructed from ridges.

SEE ALSO: `crc,cfamily` for crazy climbers method, `crcrec, crcgrec` for reconstruction methods.

EXAMPLES: `recon <- scrcrec(siginput, tfinput, beemap)`
See Examples 8.4 and 8.5 in Chapter 8.

REFERENCES: See Chapter 8.

`regrec`	Reconstruction from a Ridge	`regrec`

DESCRIPTION: Reconstructs signal from a "regularly sampled" ridge, in the wavelet case.

USAGE: `regrec(siginput, cwtinput, phi, compr, noct, nvoice, epsilon=0, w0=2*pi, fast=F, plot=F, para=5, hflag=F, check=F, minnbnodes=2, real=F)`

REQUIRED ARGUMENTS:

siginput: Input signal.

cwtinput: Wavelet transform, output of `cwt`.

phi: Unsampled ridge.

compr: Subsampling rate for the wavelet coefficients (at scale 1).

noct: Number of octaves (powers of 2).

nvoice: Number of different scales per octave.

OPTIONAL ARGUMENTS:

epsilon: Coefficient of the Q_2 term in reconstruction kernel.

w0: Central frequency of Morlet wavelet.

fast: If set to T, the kernel is computed using trapezoidal rule.

plot: If set to T, displays original and reconstructed signals.

para: Scale parameter for extrapolating the ridges.

hflag: If set to T, uses Q_1 as first term in the kernel.

check: If set to T, computes `cwt` of reconstructed signal.

minnbnodes: Minimum number of nodes for the reconstruction.

real: If set to T, uses only real constraints on the transform.

VALUE: Returns a list containing:

sol: Reconstruction from a ridge.

A: <wavelets,dualwavelets> matrix.

lam: coefficients of dual wavelets in reconstructed signal.

dualwave: Array containing the dual wavelets.

morvelets: Array containing the wavelets on sampled ridge.

solskel: Wavelet transform of sol, restricted to the ridge.

inputskel: Wavelet transform of signal, restricted to the ridge.

Q2: Second part of the reconstruction kernel.

nbnodes: Number of nodes used for the reconstruction.

SEE ALSO: `regrec2, ridrec, gregrec, gridrec.`

EXAMPLES: `recon <- regrec(siginput, cwtinput, phi, compr, noct, nvoice)`

REFERENCES: See Chapter 8.

regrec2	Reconstruction from a Ridge	regrec2

DESCRIPTION: Reconstructs signal from a "regularly sampled" ridge, in the wavelet case, from a pre-computed kernel.

USAGE: `regrec(siginput, cwtinput, phi, nbnodes, noct, nvoice, Q2, epsilon=0, w0=2*pi, plot=F)`

REQUIRED ARGUMENTS:

siginput: Input signal.

cwtinput: Wavelet transform, output of `cwt`.

phi: Unsampled ridge.

nbnode: Number of samples on the ridge.

noct: Number of octaves (powers of 2).

nvoice: Number of different scales per octave.

Q2: Second term of the reconstruction kernel.

OPTIONAL ARGUMENTS:

epsilon: Coefficient of the Q_2 term in reconstruction kernel.

w0: Central frequency of Morlet wavelet.

plot: If set to T, displays original and reconstructed signals.

VALUE: Returns a list containing:

sol: Reconstruction from a ridge.

A: <wavelets,dualwavelets> matrix.

lam: Coefficients of dual wavelets in reconstructed signal.

dualwave: Array containing the dual wavelets.

morvelets: Array containing the wavelets on sampled ridge.

solskel: Wavelet transform of sol, restricted to the ridge.

inputskel: Wavelet transform of signal, restricted to the ridge.

DETAILS: The computation of the kernel may be time consuming. This function avoids recomputing it if it was computed already.

SEE ALSO: `regrec`, `gregrec`, `ridrec`, `sridrec`.

EXAMPLES: `recon <- regrec2(siginput, cwtinput, phi, noct, nvoice, Q2)`

REFERENCES: See Chapter 8.

`ridrec`	Reconstruction from a Ridge	`ridrec`

DESCRIPTION: Reconstructs signal from sample of a ridge, in the wavelet case.

USAGE: `ridrec(cwtinput, node, phinode, noct, nvoice, Qinv, epsilon, np, w0=2*pi, check=F, real=F)`

REQUIRED ARGUMENTS:

- cwtinput: Wavelet transform, output of `cwt`.
- node: Time coordinates of the ridge samples.
- phinode: Scale coordinates of the ridge samples.
- noct: Number of octaves (powers of 2).
- nvoice: Number of different scales per octave.
- Qinv: Inverse of the matrix Q of the quadratic form.
- epsilon: Coefficient of the Q_2 term in reconstruction kernel.
- np: Number of samples of the reconstructed signal.

OPTIONAL ARGUMENTS:

- w0: Central frequency of Morlet wavelet.
- check: If set to T, computes `cwt` of reconstructed signal.
- real: If set to T, uses only constraints on the real part of the transform.

VALUE: Returns a list containing the reconstructed signal and the chained ridges.

- sol: Reconstruction from a ridge.
- A: <wavelets,dualwavelets> matrix.
- lam: Coefficients of dual wavelets in reconstructed signal.
- dualwave: Array containing the dual wavelets.
- morvelets: Array of morlet wavelets located on the ridge samples.
- solskel: Wavelet transform of sol, restricted to the ridge.
- inputskel: Wavelet transform of signal, restricted to the ridge.

SEE ALSO: `sridrec, regrec, regrec2`.

EXAMPLES: `recon <- ridrec(cwtinput, node, phinode, noct, nvoice, Qinv, epsilon)`

REFERENCES: See Chapter 8.

`skeleton`	Reconstruction from Dual Wavelets	`skeleton`

DESCRIPTION: Computes the reconstructed signal from the ridge, given the inverse of the matrix Q.

USAGE: `skeleton(cwtinput, Qinv, morvelets, bridge, aridge, N)`

REQUIRED ARGUMENTS:

cwtinput: Continuous wavelet transform (as the output of cwt).

Qinv: Inverse of the reconstruction kernel (2D array).

morvelets: Array of Morlet wavelets located at the ridge samples.

bridge: Time coordinates of the ridge samples.

aridge: Scale coordinates of the ridge samples.

N: Size of reconstructed signal.

VALUE: Returns a list of the elements of the reconstruction of a signal from sample points of a ridge.

sol: Reconstruction from a ridge.

A: Matrix of the inner products.

lam: Coefficients of dual wavelets in reconstructed signal. They are the Lagrange multipliers λ's of the text.

dualwave: Array containing the dual wavelets.

DETAILS: The details of this reconstruction are given in the text, in Chapter 8.

SEE ALSO: `skeleton2`, `zeroskeleton`, `zeroskeleton2`.

EXAMPLES: `tmp <- skeleton(cwtinput, Qinv, morvelets, bridge, aridge, N)`

REFERENCES: See Chapter 8.

`skeleton2`	Reconstruction from Dual Wavelets	`skeleton2`

DESCRIPTION: Computes the reconstructed signal from the ridge in the case of real constraints.

USAGE: `skeleton2(cwtinput, Qinv, morvelets, bridge, aridge, N)`

REQUIRED ARGUMENTS:

- cwtinput: Continuous wavelet transform (as the output of cwt).
- Qinv: inverse of the reconstruction kernel (2D array).
- morvelets: Array of Morlet wavelets located at the ridge samples.
- bridge: Time coordinates of the ridge samples.
- aridge: Scale coordinates of the ridge samples.
- N: Size of reconstructed signal.

VALUE: Returns a list of the elements of the reconstruction of a signal from sample points of a ridge.

- sol: Reconstruction from a ridge.
- A: Matrix of the inner products.
- lam: Coefficients of dual wavelets in reconstructed signal. They are the Lagrange multipliers λ's of the text.
- dualwave: Array containing the dual wavelets.

DETAILS: The details of this reconstruction are given in the text in Chapter 8.

EXAMPLES: `tmp <- skeleton2(cwtinput, Qinv, morvelets, bridge, aridge, N)`

REFERENCES: See Chapter 8.

snake	Ridge Estimation by Snake Method	snake

DESCRIPTION: Estimate a ridge from a time-frequency representation, using the snake method.

USAGE:
```
snake(tfrep, guessA, guessB, snakesize=length(guessB),
tfspec=numeric(dim(modulus)[2]), subrate=1,
temprate=3, muA=1, muB=muA, lambdaB=2*muB,
lambdaA=2*muA, iteration=1000000, seed=-7,
costsub=1, stagnant=20000)
```

REQUIRED ARGUMENTS:

tfrep: Time-frequency representation (real valued).

guessA: Initial guess for the algorithm (frequency variable).

guessB: Initial guess for the algorithm (time variable).

OPTIONAL ARGUMENTS:

snakesize: The length of the initial guess of time variable.

tfspec: Estimate for the contribution of the noise to modulus.

subrate: Subsampling rate for ridge estimation.

temprate: Initial value of temperature parameter.

lambdaA: Coefficient of the ridge's second derivative in cost function (frequency component).

muA: Coefficient of the ridge's derivative in cost function (frequency component).

lambdaB: Coefficient of the ridge's second derivative in cost function (time component).

muB: Coefficient of the ridge's derivative in cost function (time component).

iteration: Maximal number of moves.

seed: Initialization of random number generator.

stagnant: Maximum number of stationary iterations before stopping.

costsub: Subsampling of cost function in output.

stagnant: Maximum number of steps without move (for the stopping criterion).

plot: When set (by default), certain results will be displayed.

VALUE: Returns a structure containing:

ridge: 1D array (of same length as the signal) containing the ridge.

cost: 1D array containing the cost function.

SEE ALSO: `corona, coronoid, icm, snakoid.`
EXAMPLES: `sn <- snake(Mod(tfrep), guessA, guessB)`
See Examples 7.5 and 7.6 in Chapter 7.
REFERENCES: See Section 7.2.4 in Chapter 7.

snakoid	Modified Snake Method	snakoid

DESCRIPTION: Estimate a ridge from a time-frequency representation, using the modified snake method (modified cost function).

USAGE: `snakoid(modulus, guessA, guessB, snakesize =length(guessB), tfspec=numeric(dim(modulus)[2]), subrate=1, temprate=3, muA=1, muB=muA, lambdaB=2*muB, lambdaA=2*muA, iteration=1000000, seed=-7, costsub=1, stagnant=20000)`

REQUIRED ARGUMENTS:

- modulus: Time-frequency representation (real-valued).
- guessA: Initial guess for the algorithm (frequency variable).
- guessB: Initial guess for the algorithm (time variable).

OPTIONAL ARGUMENTS:

- tfspec: Estimate for the contribution of the noise to modulus.
- subrate: Subsampling rate for ridge estimation.
- temprate: Initial value of temperature parameter.
- lambdaA: Coefficient of the ridge's second derivative in cost function (frequency component).
- muA: Coefficient of the ridge's derivative in cost function (frequency component).
- lambdaB: Coefficient of the ridge's second derivative in cost function (time component).
- muB: Coefficient of the ridge's derivative in cost function (time component).
- iteration: Maximal number of moves.
- seed: Initialization of random number generator.
- stagnant: Maximum number of stationary iterations before stopping.
- costsub: Subsampling of cost function in output.

VALUE: Returns a structure containing:

- ridge: 1D array (of same length as the signal) containing the ridge.
- cost: 1D array containing the cost function.

SEE ALSO: `corona, coronoid, icm, snake.`

EXAMPLES: `sn <- snakoid(Mod(tfrep), guessA, guessB)`

REFERENCES: See Section 7.2.4 in Chapter 7.

`sridrec`	Simple Reconstruction from Ridge	`sridrec`

DESCRIPTION: Simple reconstruction of a real-valued signal from a ridge, by restriction of the transform to the ridge.
USAGE: `sridrec(tfinput, ridge)`
REQUIRED ARGUMENTS:

tfinput: Time-frequency representation.

ridge: Ridge (1D array).

VALUE: (real) Reconstructed signal (1D array).
SEE ALSO: `ridrec, gridrec.`
EXAMPLES: `tmp <- sridrec(tfinput, ridge)`
See Example 8.4.
REFERENCES: See Chapter 8.

`tfgmax`	Time-Frequency Transform Global Maxima	`tfgmax`

DESCRIPTION: Computes the maxima (for each fixed value of the time variable) of the modulus of a continuous wavelet transform.
USAGE: `tfgmax(input)`
REQUIRED ARGUMENTS:

input: Wavelet transform (as the output of the function "cwt").

OPTIONAL ARGUMENTS:
VALUE:

output: Values of the maxima (1D array).

pos: Positions of the maxima (1D array).

SEE ALSO: `tflmax`
EXAMPLES: `gmax <- tfgmax(input)`
REFERENCES: See Chapter 7.

tflmax	Time-Frequency Transform Local Maxima	tflmax

DESCRIPTION: Computes the local maxima (for each fixed value of the time variable) of the modulus of a time-frequency transform.
USAGE: `tflmax(input, plot=T)`
REQUIRED ARGUMENTS:

input: Time-frequency transform (real 2D array).

OPTIONAL ARGUMENTS:

plot: If set to T, displays the local maxima on the graphic device.

VALUE: Values of the maxima (2D array).
SEE ALSO: `tfgmax`
EXAMPLES: `lmax <- tlgmax(input)`
REFERENCES: See Chapter 7.

`tfmean`	Average Frequency by Frequency	`tfmean`

DESCRIPTION: Compute the mean of time-frequency representation frequency by frequency.

USAGE: `tfmean(input, plot=T)`

REQUIRED ARGUMENTS:

input: Time-frequency transform (output of `cwt` or `cgt`).

OPTIONAL ARGUMENTS:

plot: If set to T, displays the values of the energy as a function of the scale (or frequency).

VALUE: 1D array containing the noise estimate.

SEE ALSO: `tfpct,tfvar`.

EXAMPLES: See Examples 6.3 and 6.5.

REFERENCES: See Chapter 8.

`tfpct`	Percentile Frequency by Frequency	`tfpct`

DESCRIPTION: Compute a percentile of time-frequency representation frequency by frequency.
USAGE: `tfpct(input, percent=.8, plot=T)`
REQUIRED ARGUMENTS:

input: Time-frequency transform (output of `cwt` or `cgt`).

OPTIONAL ARGUMENTS:

percent: Percentile to be retained.

plot: If set to T, displays the values of the energy as a function of the scale (or frequency).

VALUE: 1D array containing the noise estimate.
SEE ALSO: `tfmean`, `tfvar`.

`tfvar`	Variance Frequency by Frequency	`tfvar`

DESCRIPTION: Compute the variance of time-frequency representation frequency by frequency.

USAGE: `tfvar(input, plot=T)`

REQUIRED ARGUMENTS:

input: Time-frequency transform (output of `cwt` or `cgt`).

OPTIONAL ARGUMENTS:

plot: If set to T, displays the values of the energy as a function of the scale (or frequency).

VALUE: 1D array containing the noise estimate.

SEE ALSO: `tfmean`, `tfpct`.

`vgt`	Gabor Transform on One Voice	`vgt`

DESCRIPTION: Compute Gabor transform for fixed frequency.
USAGE: `vgt(input, frequency, scale, plot = F)`
REQUIRED ARGUMENTS:

input: Input signal (1D array).

frequency: Frequency at which the Gabor transform is to be computed.

scale: Size parameter for the window.

OPTIONAL ARGUMENTS:

plot: If set to T, plots the real part of cgt on the graphic device.

VALUE: 1D (complex) array containing Gabor transform at specified frequency.
SEE ALSO: `vwt`, `vDOG`.
EXAMPLES: `tmp <- vgt(inputsig, frequency, scale)`

`vDOG`	DOG Wavelet Transform on One Voice	`vDOG`

DESCRIPTION: Compute DOG wavelet transform at one scale.
USAGE: `vDOG(input, scale, moments)`
REQUIRED ARGUMENTS:

input: Input signal (1D array).

scale: Scale at which the wavelet transform is to be computed.

moments: Number of vanishing moments.

VALUE: 1D (complex) array containing wavelet transform at one scale.
SEE ALSO: `vgt`, `vwt`.
EXAMPLES: `tmp <- vDOG(inputsig, scale, moments)`

vwt	Voice Wavelet Transform	vwt

DESCRIPTION: Compute Morlet's wavelet transform at one scale.
USAGE: `vwt(input,scale,w0=2*pi)`
REQUIRED ARGUMENTS:

input: Input signal (1D array).

scale: Scale at which the wavelet transform is to be computed.

OPTIONAL ARGUMENTS:

w0: Center frequency of the wavelet.

VALUE: 1D (complex) array containing wavelet transform at one scale.
SEE ALSO: `vgt`, `vDOG`.
EXAMPLES: `tmp <- vwt(inputsig, scale)`

`zeroskeleton` Reconstruction from Dual Wavelets `zeroskeleton`

DESCRIPTION: Computes the the reconstructed signal from the ridge when the epsilon parameter is set to zero.

USAGE: `zeroskeleton(cwtinput, Qinv, morvelets, bridge, aridge, N)`

REQUIRED ARGUMENTS:

- ccwtinput: Continuous wavelet transform (as the output of cwt).
- Qinv: Inverse of the reconstruction kernel (2D array).
- morvelets: Array of Morlet wavelets located at the ridge samples.
- bridge: Time coordinates of the ridge samples.
- aridge: Scale coordinates of the ridge samples.
- N: Size of reconstructed signal.

VALUE: Returns a list of the elements of the reconstruction of a signal from sample points of a ridge.

- sol: Reconstruction from a ridge.
- A: Matrix of the inner products.
- lam: Coefficients of dual wavelets in reconstructed signal. They are the Lagrange multipliers λ's of the text.
- dualwave: Array containing the dual wavelets.

DETAILS: The details of this reconstruction are the same as for the function skeleton. They can be found in the text.

SEE ALSO: `skeleton`, `skeleton2`, `zeroskeleton2`.

EXAMPLES: `tmp <- zeroskeleton(cwtinput, Qinv, morvelets, bridge, aridge, N)`

REFERENCES: See Chapter 8.

`zeroskeleton2` Reconstruction from Dual Wavelets `zeroskeleton2`

DESCRIPTION: Computes the the reconstructed signal from the ridge when the epsilon parameter is set to zero, in the case of real constraints.

USAGE: `zeroskeleton2(cwtinput, Qinv, morvelets, bridge, aridge, N)`

REQUIRED ARGUMENTS:

- cwtinput: Continuous wavelet transform (output of `cwt`).
- Qinv: Inverse of the reconstruction kernel (2D array).
- morvelets: Array of Morlet wavelets located at the ridge samples.
- bridge: Time coordinates of the ridge samples.
- aridge: Scale coordinates of the ridge samples.
- N: Size of reconstructed signal.

VALUE: Returns a list of the elements of the reconstruction of a signal from sample points of a ridge.

- sol: Reconstruction from a ridge.
- A: Matrix of the inner products.
- lam: Coefficients of dual wavelets in reconstructed signal. They are the Lagrange multipliers λ's of the text.
- dualwave: Array containing the dual wavelets.

DETAILS: The details of this reconstruction are the same as for the function skeleton. They can be found in the text.

SEE ALSO: `skeleton`, `skeleton2`, `zeroskeleton`.

EXAMPLES: `tmp <- zeroskeleton2(cwtinput, Qinv, morvelets, bridge, aridge,N)`

REFERENCES: See Chapter 8.

Chapter 11

The Swave S Utilities

This is a list of functions which either are intended to make life easier, or are called by the main functions of Swave. The description follows.

`adjust.length`	Zero Padding	`adjust.length`

DESCRIPTION: Add zeroes to the end of the data if necessary so that its length is a power of 2. It returns the data with zeroes added if necessary and the length of the adjusted data.

USAGE: `adjust.length(inputdata)`

REQUIRED ARGUMENTS:

inputdata: Either a text file or an S object containing data.

VALUE: Zero-padded 1D array.

EXAMPLES: `tmp<- adjust.length(inputdata)`

REFERENCES: See Chapter 5.

`check.maxresoln`	Verify Maximum Resolution	`check.maxresoln`

DESCRIPTION: Stop when $2^{\text{maxresoln}}$ is larger than the signal size.
USAGE: `check.maxresoln( maxresoln, np )`
REQUIRED ARGUMENTS:

maxresoln: Number of decomposition scales.

np: Signal size.

SEE ALSO: `mw, mrecons.`
REFERENCES: See Chapter 5.

`crfview`	Display Chained Ridges	`crfview`

DESCRIPTION: Displays a family of chained ridges, output of `cfamily`.
USAGE: `crfview(beemap, twod=T)`
REQUIRED ARGUMENTS:

beemap: Family of chained ridges, output of `cfamily`.

OPTIONAL ARGUMENTS:

twod: If set to T, displays the ridges as an image. If set to F, displays as a series of curves.

SEE ALSO: `crc,cfamily` for crazy climbers and corresponding chaining algorithms.
EXAMPLES: `crfview(beemap)`
REFERENCES: See Section 7.2.4 in Chapter 7.

`cwtimage`	Continuous Wavelet Transform Display	`cwtimage`

DESCRIPTION: Converts the output (modulus or argument) of cwtpolar to a 2D array and displays on the graphic device.
USAGE: `cwtimage(input)`
REQUIRED ARGUMENTS:

input: 3D array containing a continuous wavelet transform (output of cwtpolar).

VALUE: 2D array continuous (complex) wavelet transform.
DETAILS: The output contains the (complex) values of the wavelet transform of the input signal. The format of the output can be

- 2D array (signal_size $\times$ nb_scales).
- 3D array (signal_size $\times$ noctave $\times$ nvoice).

SEE ALSO: `cwtpolar, cwt, DOG`.
EXAMPLES: `tmp <- cwt(input, noctave, nvoice)`
REFERENCES: See Chapter 4.

`cwtpolar`	Conversion to Polar Coordinates	`cwtpolar`

DESCRIPTION: Converts one of the possible outputs of the function `cwt` to modulus and phase.

USAGE: `cwtpolar(cwt, threshold = 0)`

REQUIRED ARGUMENTS:

cwt: 3D array containing the values of a continuous wavelet transform in the format (signal_size $\times$ noctave $\times$ nvoice) as in the output of the function `cwt` with the logical flag `twodimension` set to FALSE.

OPTIONAL ARGUMENTS:

threshold: Value of a level for the absolute value of the modulus below which the value of the argument of the output is set to $-\pi$.

VALUE: Modulus and argument of the values of the continuous wavelet transform.

output1: 3D array giving the values (in the same format as the input) of the modulus of the input.

output2: 3D array giving the values of the argument of the input.

DETAILS: The output contains the (complex) values of the wavelet transform of the input signal. The format of the output can be:

- 2D array (signal_size $\times$ nb_scales).
- 3D array (signal_size $\times$ noctave $\times$ nvoice).

SEE ALSO: `cwt, DOG, cwtimage`.

EXAMPLES: `tmp <- cwtpolar(cwt)`

`epl`	Plot Dyadic Wavelet Transform Extrema	`epl`

DESCRIPTION: Plot dyadic wavelet transform extrema (output of `ext`).
USAGE: `epl(dwext)`
REQUIRED ARGUMENTS:

dwext: Dyadic wavelet transform (output of `ext`).

SEE ALSO: `mw, ext, wpl`.
EXAMPLES: `epl(extrema)`
REFERENCES: See Chapter 8.

`fastgkernel`	Kernel for Gabor Ridges	`fastgkernel`

DESCRIPTION: Computes the cost from the sample of points on the estimated ridge and the matrix used in the reconstruction of the original signal, using simple trapezoidal rule for integrals.

USAGE: `fastgkernel(node, phinode, freqstep, scale, x.inc=1, x.min=node[1], x.max=node[length(node)], plot=F)`

REQUIRED ARGUMENTS:

node: Values of the variable b for the nodes of the ridge.

phinode: Values of the frequency variable ω for the nodes of the ridge.

freqstep: Sampling rate for the frequency axis.

scale: Size of the window.

OPTIONAL ARGUMENTS:

x.inc: Step unit for the computation of the kernel.

x.min: Minimal value of x for the computation of $\mathcal{G}_2$.

x.max: Maximal value of x for the computation of $\mathcal{G}_2$.

plot: If set to TRUE, displays the modulus of the matrix of $\mathcal{G}_2$.

VALUE: Matrix of the $\mathcal{G}_2$ kernel.

DETAILS: Uses trapezoidal rule (instead of Romberg's method) to evaluate the kernel.

SEE ALSO: `gkernel, kernel, fastkernel, rkernel, zerokernel.`

REFERENCES: See Chapter 8.

`fastkernel`	Kernel for Wavelet Ridges	`fastkernel`

DESCRIPTION: Computes the cost from the sample of points on the estimated ridge and the matrix used in the reconstruction of the original signal, using simple trapezoidal rule for integrals.

USAGE: `fastkernel(node, phinode, nvoice, x.inc=1, x.min=node[1], x.max=node[length(node)], w0=2*pi, plot=F)`

REQUIRED ARGUMENTS:

- node: Values of the variable b for the nodes of the ridge.
- phinode: Values of the scale variable a for the nodes of the ridge.
- nvoice: Number of scales within 1 octave.

OPTIONAL ARGUMENTS:

- x.inc: Step unit for the computation of the kernel.
- x.min: Minimal value of x for the computation of Q_2.
- x.max: Maximal value of x for the computation of Q_2.
- w0: Central frequency of the wavelet.
- plot: If set to TRUE, displays the modulus of the matrix of Q_2.

VALUE: Matrix of the Q_2 kernel.

DETAILS: Uses trapezoidal rule (instead of Romberg's method) to evaluate the kernel.

SEE ALSO: `kernel`, `rkernel`, `gkernel`, `zerokernel`.

REFERENCES: See Chapter 8.

gkernel	Kernel for Reconstruction from Gabor Ridges	gkernel

DESCRIPTION: Computes the cost from the sample of points on the estimated ridge and the matrix used in the reconstruction of the original signal.

USAGE: `kernel(node, phinode, freqstep, scale, x.inc=1, x.min=node[1], x.max=node[length(node)], plot=F)`

REQUIRED ARGUMENTS:

node: Values of the variable b for the nodes of the ridge.

phinode: Values of the scale variable a for the nodes of the ridge.

freqstep: Sampling rate for the frequency axis.

scale: Size of the window.

OPTIONAL ARGUMENTS:

x.inc: Step unit for the computation of the kernel.

x.min: Minimal value of x for the computation of Q_2.

x.max: Maximal value of x for the computation of Q_2.

plot: If set to TRUE, displays the modulus of the matrix of Q_2.

VALUE: Matrix of the Q_2 kernel.

SEE ALSO: `fastgkernel`, `kernel`, `rkernel`, `fastkernel`, `zerokernel`.

REFERENCES: See Chapter 8.

`gsampleone`	Sampled Identity	`gsampleone`

DESCRIPTION: Generate a "sampled identity" matrix.
USAGE: `gsampleone(node,scale,np)`
REQUIRED ARGUMENTS:

node: Location of the reconstruction Gabor functions.

scale: Scale of the Gabor functions.

np: Size of the reconstructed signal.

VALUE: Diagonal of the "sampled" Q_1 term (1D vector)
SEE ALSO: `kernel, gkernel`.
REFERENCES: See Chapter 8.

`gwave` Gabor Functions on a Ridge `gwave`

DESCRIPTION: Generation of Gabor functions located on the ridge.
USAGE: `gwave(bridge, omegaridge, nvoice, freqstep, scale, np, N)`
REQUIRED ARGUMENTS:

- bridge: Time coordinates of the ridge samples.
- omegaridge: Frequency coordinates of the ridge samples.
- voice: Number of different scales per octave.
- freqstep: Sampling rate for the frequency axis.
- scale: Scale of the window.
- np: Size of the reconstruction kernel.
- N: Number of complex constraints.

VALUE: Array of Gabor functions located on the ridge samples.
SEE ALSO: `gwave2`, `morwave`, `morwave2`.
EXAMPLES: `tmp <- gwave(bridge, omegaridge, nvoice, freqstep, scale, np, N)`
REFERENCES: See Chapter 8.

`gwave2`	Real Gabor Functions on a Ridge	`gwave2`

DESCRIPTION: Generation of the real parts of Gabor functions located on a ridge. (Modification of `gwave`.)
USAGE: `gwave2(bridge,omegaridge,nvoice,freqstep,scale,np,N)`
REQUIRED ARGUMENTS:

bridge: Time coordinates of the ridge samples.

omegaridge: Frequency coordinates of the ridge samples.

voice: Number of different scales per octave.

freqstep: Sampling rate for the frequency axis.

scale: Scale of the window.

np: Size of the reconstruction kernel.

N: Number of complex constraints.

VALUE: Array of real Gabor functions located on the ridge samples.
SEE ALSO: `gwave`, `morwave`, `morwave2`.
EXAMPLES: `tmp <- gwave2(bridge, omegaridge, nvoice, freqstep, scale, np, N)`
REFERENCES: See Chapter 8.

`kernel` Kernel for Reconstruction from Wavelet Ridges `kernel`

DESCRIPTION: Computes the cost from the sample of points on the estimated ridge and the matrix used in the reconstruction of the original signal.

USAGE: `kernel(node, phinode, nvoice, x.inc=1, x.min=node[1], x.max=node[length(node)], w0=2*pi, plot=F)`

REQUIRED ARGUMENTS:

node: Values of the variable b for the nodes of the ridge.

phinode: Values of the scale variable a for the nodes of the ridge.

nvoice: Number of scales within 1 octave.

OPTIONAL ARGUMENTS:

x.inc: Step unit for the computation of the kernel.

x.min: Minimal value of x for the computation of Q_2.

x.max: Maximal value of x for the computation of Q_2.

w0: Central frequency of the wavelet.

plot: If set to T, displays the modulus of the matrix of Q_2.

VALUE: Matrix of the Q_2 kernel

DETAILS: The kernel is evaluated using Romberg's method.

SEE ALSO: `gkernel`, `rkernel`, `zerokernel`.

REFERENCES: See Chapter 8.

`morwave`	Ridge Morvelets	`morwave`

DESCRIPTION: Generates the Morlet wavelets at the sample points of the ridge.
USAGE: `morwave(bridge, aridge, nvoice, np, N, w0=2*pi)`
REQUIRED ARGUMENTS:

bridge: Time coordinates of the ridge samples.

aridge: Scale coordinates of the ridge samples.

nvoice: Number of different scales per octave.

np: Number of samples in the input signal.

N: Size of reconstructed signal.

OPTIONAL ARGUMENTS:

w0: Central frequency of the wavelet.

VALUE: Returns the Morlet wavelets at the samples of the time-scale plane given in the input: complex array of Morlet wavelets located on the ridge samples.
SEE ALSO: `morwave2, gwave, gwave2`.
EXAMPLES: `tmp <- morwave(bridge, aridge, nvoice, np, N,)`
REFERENCES: See Chapter 8.

`morwave2`	Real Ridge Morvelets	`morwave2`

DESCRIPTION: Generates the real parts of the Morlet wavelets at the sample points of a ridge.
USAGE: `morwave2(bridge, aridge, nvoice, np, N, w0=2*pi)`
REQUIRED ARGUMENTS:

bridge: Time coordinates of the ridge samples.

aridge: Scale coordinates of the ridge samples.

nvoice: Number of different scales per octave.

np: Number of samples in the input signal.

N: Size of reconstructed signal.

OPTIONAL ARGUMENTS:

w0: Central frequency of the wavelet.

VALUE: Returns the real parts of the Morlet wavelets at the samples of the time-scale plane given in the input: array of Morlet wavelets located on the ridge samples.
SEE ALSO: `morwave`, `gwave`, `gwave2`.
EXAMPLES: `tmp <- morwave2(bridge, aridge, nvoice, np, N)`
REFERENCES: See Chapter 8.

`npl`	Prepare Graphics Environment	`npl`

DESCRIPTION: Splits the graphics device into prescribed number of windows.

USAGE: `npl(nbrow)`

REQUIRED ARGUMENTS:

nbrow: Number of plots.

EXAMPLES: `npl(n)`

`plot.result` Plot Dyadic Wavelet Transform Extrema `plot.result`

DESCRIPTION: Plot extrema or local extrema of dyadic wavelet transform.

USAGE: `plotwt(original, psi, phi, maxresoln, scale=F, yaxtype="s")`

REQUIRED ARGUMENTS:

original: Input signal.

psi: Dyadic wavelet transform.

phi: Scaling function transform at last resolution.

maxresoln: Number of decomposition scales.

OPTIONAL ARGUMENTS:

scale: When set, the wavelet transform at each scale is plotted with the same scale.

yaxtype: y axis type (see Splus manual).

SEE ALSO: `plotwt`, `epl`, `wpl`.

REFERENCES: See Chapters 5 and 8.

`plotwt`	Plot Dyadic Wavelet Transform	`plotwt`

DESCRIPTION: Plot dyadic wavelet transform.
USAGE: `plotwt(original, psi, phi, maxresoln, scale=F, yaxtype="s")`
REQUIRED ARGUMENTS:

original: Input signal.

psi: Dyadic wavelet transform.

phi: Scaling function transform at last resolution.

maxresoln: Number of decomposition scales.

OPTIONAL ARGUMENTS:

scale: When set, the wavelet transform at each scale is plotted with the same scale.

yaxtype: y axis type (see Splus manual).

SEE ALSO: `plot.result`, `epl`, `wpl`.
REFERENCES: See Chapter 5.

`RidgeSampling`	Sampling Gabor Ridge	`RidgeSampling`

DESCRIPTION: Given a ridge phi (for the Gabor transform), returns a (regularly) subsampled version of length nbnodes.
USAGE: `RidgeSampling(phi, nbnodes)`
REQUIRED ARGUMENTS:

phi: Ridge (1D array).

nbnodes: Number of samples.

OPTIONAL ARGUMENTS:
VALUE: Returns a list containing the discrete values of the ridge.

node: Time coordinates of the ridge samples.

phinode: Frequency coordinates of the ridge samples.

DETAILS: Gabor ridges are sampled uniformly.
SEE ALSO: `wRidgeSampling`
EXAMPLES: `tmp <- RidgeSampling(phi, nbnodes)`
REFERENCES: See Chapter 8.

`rkernel` Kernel for Reconstruction from Wavelet Ridges `rkernel`

DESCRIPTION: Computes the cost from the sample of points on the estimated ridge and the matrix used in the reconstruction of the original signal, in the case of real constraints. Modification of the function `kernel`.

USAGE: `rkernel(node, phinode, nvoice, x.inc=1, x.min=node[1], x.max=node[length(node)], w0=2*pi, plot=F)`

REQUIRED ARGUMENTS:

node: Values of the variable b for the nodes of the ridge.

phinode: Values of the scale variable a for the nodes of the ridge.

nvoice: Number of scales within 1 octave.

OPTIONAL ARGUMENTS:

x.inc: Step unit for the computation of the kernel.

x.min: Minimal value of x for the computation of Q_2.

x.max: Maximal value of x for the computation of Q_2.

w0: Central frequency of the wavelet.

plot: If set to TRUE, displays the modulus of the matrix of Q_2.

VALUE: Matrix of the Q_2 kernel.

DETAILS: Uses Romberg's method for computing the kernel.

SEE ALSO: `kernel`, `fastkernel`, `gkernel`, `zerokernel`.

REFERENCES: See Chapter 8.

`smoothts` Smoothing Time Series `smoothts`

DESCRIPTION: Smooth a time series by averaging window.
USAGE: `smoothts(ts, windowsize)`
REQUIRED ARGUMENTS:

ts: Time series.

windowsize: Length of smoothing window.

VALUE: Smoothed time series (1D array).
EXAMPLES: `sts <- smoothts(ts,size)`.

`smoothwt` Smoothing and Time-Frequency Representation `smoothwt`

DESCRIPTION: Smooth the wavelet (or Gabor) transform in the time direction.

USAGE: `smoothwt(modulus, subrate, flag=F)`

REQUIRED ARGUMENTS:

modulus: Time-frequency representation (real valued).

subrate: Length of smoothing window.

OPTIONAL ARGUMENTS:

flag: If set to T, subsample the representation.

VALUE: 2D array containing the smoothed transform.

SEE ALSO: `corona, coronoid, snake, snakoid`.

`snakeview`	Restriction to a Snake	`snakeview`

DESCRIPTION: Restrict time-frequency transform to a snake.
USAGE: `snakeview(modulus, snake)`
REQUIRED ARGUMENTS:

modulus: Time-frequency representation (real valued).

snake: Time and frequency components of a snake.

VALUE: 2D array containing the restriction of the transform modulus to the snake.
DETAILS: Recall that a snake is a (two components) S structure.
EXAMPLES: `tmp <- snakeview(modulus, snake)`
REFERENCES: See Chapters 7 and 8.

SVD	Singular Value Decomposition	SVD

DESCRIPTION: Computes singular value decomposition of a matrix.
USAGE: `SVD(a)`
REQUIRED ARGUMENTS:

a: Input matrix.

VALUE: A structure containing the 3 matrices of the singular value decomposition of the input.
DETAILS: S interface for Numerical Recipes singular value decomposition routine.
EXAMPLES: `tmp <- SVD(input)`
REFERENCES: See Chapter 6.

vecgabor	Gabor Functions on a Ridge	vecgabor

DESCRIPTION: Generate Gabor functions at specified positions on a ridge.

USAGE: `vecgabor(sigsize, nbnodes,location, frequency, scale)`

REQUIRED ARGUMENTS:

sigsize: Signal size.

nbnodes: Number of wavelets to be generated.

location: b coordinates of the ridge samples (1D array of length nbnodes).

frequency: Frequency coordinates of the ridge samples (1D array of length nbnodes).

scale: Size parameter for the Gabor functions.

VALUE: 2D (complex) array containing Gabor functions located at the specific points.

SEE ALSO: `vecmorlet`

EXAMPLES: `tmp <- vecgabor(sigsize, nbnodes, location, frequency, scale)`

`vecmorlet`	Morlet Wavelets on a Ridge	`vecmorlet`

DESCRIPTION: Generate Morlet wavelets at specified positions on a ridge.
USAGE: `vecmorlet(sigsize, nbnodes, bridge, aridge, w0=2*pi)`
REQUIRED ARGUMENTS:

sigsize: Signal size.

nbnodes: Number of wavelets to be generated.

bridge: b coordinates of the ridge samples (1D array of length nbnodes).

aridge: a coordinates of the ridge samples (1D array of length nbnodes).

OPTIONAL ARGUMENTS:

w0: Center frequency of the wavelet.

VALUE: 2D (complex) array containing wavelets located at the specific points.
SEE ALSO: `vecgabor`
EXAMPLES: `tmp <- vecmorlet(sigsize, nbnodes, bridge, aridge)`

`wpl`	Plot Dyadic Wavelet Transform.	`wpl`

DESCRIPTION: Plot dyadic wavelet transform (output of `mw`).
USAGE: `wpl(dwtrans)`
REQUIRED ARGUMENTS:

dwtrans: Dyadic wavelet transform (output of `mw`).

SEE ALSO: `mw, ext, epl.`
EXAMPLES: `wpl(dwtrans)`
REFERENCES: See Section 5.2 in Chapter 8.

wRidgeSampling	Sampling wavelet Ridge	wRidgeSampling

DESCRIPTION: Given a ridge phi (for the wavelet transform), returns an (appropriately) subsampled version with a given subsampling rate.
USAGE: `wRidgeSampling(phi, compr, nvoice)`
REQUIRED ARGUMENTS:

phi: Ridge (1D array).

compr: Subsampling rate for the ridge.

nvoice: Number of voices per octave.

VALUE: Returns a list containing the discrete values of the ridge.

node: Time coordinates of the ridge samples.

phinode: Scale coordinates of the ridge samples.

nbnode: Number of nodes of the ridge samples.

DETAILS: To account for the variable sizes of wavelets, the sampling rate of a wavelet ridge is not uniform and is proportional to the scale.
SEE ALSO: `RidgeSampling`
EXAMPLES: `tmp <- wRidgeSampling(phi, compr, nvoice)`
REFERENCES: See Chapter 8.

`zerokernel`	Reconstruction from Wavelet Ridges	`zerokernel`

DESCRIPTION: Generate a zero kernel for reconstruction from ridges.
USAGE: `zerokernel(x.inc=1, x.min, x.max)`
REQUIRED ARGUMENTS:

x.min: Minimal value of x for the computation of Q_2.

x.max: Maximal value of x for the computation of Q_2.

OPTIONAL ARGUMENTS:

x.inc: Step unit for the computation of the kernel.

VALUE: Matrix of the Q_2 kernel.
SEE ALSO: `kernel`, `fastkernel`, `gkernel`, `gkernel`.
REFERENCES: See Chapter 8.

Bibliographies

General References

[1] A. Abramovici, W. Althouse, R. Drever, Y. Gursel, S. Kawamura, F. Raab, D. Shoemaker, L. Seivers, R. Spero, K. Thorne, R. Vogt, R. Weiss, S. Withcomb, and M. Zuker (1992): LIGO: The Laser Interferometer Gravitational-Wave Observatory. *Science* **256**, 325–333.

[2] D.P. Abrams and M.A. Sozen (1979): Experimental study of a frame-wall interaction in reinforced concrete structures subjected to strong earthquake motions. Rep. No. UILU-ENG-79-2002, SRS 460, University of Illinois, Urbana, Ill.

[3] P. Abry and A. Aldroubi (1996): Designing multiresolution analysis type wavelets and their fast algorithms, *Int. J. of Fourier Anal. and Appl.* **2**, 135–159.

[4] P. Abry and P. Flandrin (1994): On the initialisation of the discrete wavelet transform algorithm. *IEEE Signal Processing Lett.* **1**, 32–34.

[5] P. Abry, P. Gonçalves, and P. Flandrin (1993): Wavelet-based spectral analysis of $1/f$ processes. *Proc. IEEE-ICASSP'93*, III.237–III.240.

[6] P. Abry, P. Gonçalves, and P. Flandrin (1994): Wavelets, spectrum analysis and $1/f$ processes. *in Wavelets and Statistics*, A. Antoniadis and G. Oppenheim, Eds., *Lecture Notes in Statistics*, **103**, 15–29.

[7] P. Abry and F. Sellan (1996): The wavelet-based synthesis for fractional Brownian motion: Remarks and Fast Implementation. *Appl. and Comp. Harmonic Anal.* **3**, 377–383.

[8] P. Abry and D. Veitch (1998): Wavelet analysis of long-range dependent traffic. *IEEE Trans. Inf. Theory* **44**, 2–15.

[9] A. Aldroubi and M. Unser (1993): Families of multiresolution and wavelet spaces with optimal properties. *Numer. Funct. Anal. Optimiz.* **14**, 417–446.

[10] J.P. d'Ales and A. Cohen (1997): Nonlinear approximation of random functions. *SIAM J. Appl. Math.* **57**, 518–540.

[11] S.T. Ali, J.P. Antoine, and J.P. Gazeau (1991): Square integrability of group representations on homogeneous spaces, I: reproducing triples and frames, *Ann. Inst. Henri Poincaré* **55**, 829–856; II, generalized square integrability and generalized coherent states, *Ann. Inst. Henri Poincaré* **55**, 860–890.

[12] S.T. Ali, J.P. Antoine, J.P. Gazeau, and U.A. Müller (1995): Coherent states and their generalizations: an overview. *Rev. Math. Phys.* **7**, 1013–1104.

[13] A. Antoniadis (1994): Smoothing noisy data with Coiflets, *Statistica Sinica* **4**, 651–678.

[14] A. Antoniadis (1994): Wavelet methods for smoothing noisy data, *in Wavelets, Images, and Surface Fitting*, AK Peters Ltd, 21–29.

[15] A. Antoniadis (1996): Smoothing noisy data with tapered coiflet series. *Scand. J. Stat.*, **23**, 313–330.

[16] A. Antoniadis (1997) Wavelets estimators for change-point regression models, *Proc. Amer. Math. Soc.* (to appear)

[17] A. Antoniadis and R. Carmona (1991): Multiresolution analyses and wavelets for density estimation. U.C. Irvine Dept. Math. Tech. Rep. (June 17, 1991)

[18] A. Antoniadis, I. Gijbels, and G. Grégoire (1997): Model selection using wavelet decomposition and applications. Discussion paper 9508, Institute of Statistics, Louvain-la-neuve *Biometrika* **84**, 751–763.

[19] A. Antoniadis, G. Grégoire, and I. McKeague (1994): Wavelet methods for curve estimation. *J. Amer. Statist. Assoc.*, **89**, 1340–1353.

[20] A. Antoniadis, G. Grégoire, and P. Vial (1997): Random design wavelet curve smoothing. *Statistics and Probability Letters* **35**, 225–232.

[21] A. Arneodo, G. Grasseau, and M. Holschneider (1988): On the wavelet transform of multifractals. *Phys. Rev. Lett.* **61**, 2281–2284.

[22] F. Auger, E. Chassande-Mottin, I. Daubechies, and P. Flandrin (1997): Differential reassignment. *IEEE Signal Proc. Letters*, **4**, 293–294.

[23] F. Auger and P. Flandrin (1993): Improving the readability of time-frequency and time-scale representations by the reassignment method, *IEEE Trans. Signal Proc.* **40**, 1068–1089.

[24] P. Auscher (1991): Wavelet bases of $L^2(\mathbb{R})$ with rational dilation factor. *Wavelets and their Applications*. M.B. Ruskai et al. Eds, Jones and Bartlett, Boston, pp. 439–451.

[25] P. Auscher (1994): Remarks on the local Fourier bases. In *Wavelets: Mathematics and Applications*, J.J. Benedetto and M.W. Frazier, Eds., CRC Press, 203–218.

[26] P. Auscher, G. Weiss, and M.V. Wickerhauser (1992): Local sine and cosine basis of Coifman and Meyer and the construction of wavelets. In *Wavelets: a Tutorial in Theory and Applications*, C.K. Chui Ed. Academic Press, 237–256.

[27] L. Auslander and R. Tolimieri (1987): Radar ambiguity function and group theory. *SIAM J. Math. Anal.* **1**, 847–897.

[28] R. Balian (1981): Un principe d'incertitude fort en théorie du signal. *C.R. Acad. Sci. Paris* **292**, 1357–1362.

[29] R.J. Barton and V. Poor (1988): Signal detection in fractional Gaussian noise. *IEEE Trans. Info. Th.* **34**, 943–959.

[30] E. Bacry, J.F. Muzy, and A. Arneodo (1993): Singularity spectrum of fractal signals from wavelet transform: exact results. *J. Stat. Phys.* **70**, 635–674.

[31] G. Battle (1987): A block spin construction of ondelettes I: Lemarié functions. *Comm. Math. Phys.* **110**, 601–615.

[32] G. Battle (1988): A block spin construction of ondelettes II: the QFT connection *Comm. Math. Phys.* **114**, 93–102.

[33] G. Battle (1988): Heisenberg proof of the Balian Low theorem. *Lett. Math. Phys.* **15**, 175–177.

[34] A. Benassi (1994): Locally self-similar Gaussian processes. *in Wavelets and Statistics*, A. Antoniadis and G. Oppenheim, Eds., *Lecture Notes in Statistics*, **103**, 43–54.

[35] A. Benassi, S. Jaffard, and D. Roux (1997): Elliptic Gaussian random processes. *Rev. Mat. Iberoam.* **13**, 19–90.

[36] J. Benedetto (1989): Gabor representations and wavelets. *Math. Soc. Contemporary Math.* **91**, 833–852.

[37] J. Benedetto, C. Heil, and D. Walnut: Remarks on the proof of the Balian-Low theorem. *Ann. Scuola Norm. Pisa.*

[38] J. Benedetto and D. Walnut (1994): Gabor frames for L^2 and related spaces. *Wavelets, Mathematics and Applications*, J.J. Benedetto and M.W. Frazier Eds. CRC Press, 97–162.

[39] G. Benke, M. Bozek-Kuzmicki, D. Collela, J.M. Jacyna, and J.J. Benedetto (1994): Wavelet-based analysis of EEG signals for detection and localization of epileptic seizures. *SPIE* **2491**, 760–769.

[40] J. Beran, S. Sherman, M. Taqqu, and W. Willinger (1995): Long range dependence in variable bit-rate video traffic. *IEEE Trans. Commun.* **43**, 1566–1579.

[41] M.O. Berger (1993): Towards dynamic adaptation of snake contours. In *International Conference on Image Analysis and Processing, Como (Italy).* IAPR, september 1991.

[42] Z. Berman and J.S. Baras (1993): Properties of multiscale maxima and zero-crossings representations. *IEEE Trans. Signal Processing* **41**, 3216–3231.

[43] D. Bernier and M. Taylor (1996): Wavelets and square integrable representations. *SIAM J. Math. Anal.* **27**, 594–608

[44] P. Bertrand and J. Bertrand (1985): Représentation temps-fréquence des signaux à large bande. *La Recherche Aérospatiale (in French)* **5**, 277–283

[45] J. Besag (1986): On the statistical analysis of dirty pictures. *J. Royal Statist. Soc. Ser. B vol* **48**, 259–302.

[46] G. Beylkin (1992): On the representation of operators in bases of compactly supported wavelets. *SIAM J. Num. Anal.* **29**, 1716–1740.

[47] G. Beylkin, R. Coifman, and V. Rokhlin (1991): Fast wavelet transforms and numerical algorithms I. *Comm. Pure Appl. Math.* **44**, 141–183.

[48] G. Beylkin and B. Torrésani (1996): Implementation of operators with filter banks, autocorrelation shell and Hardy wavelets. *Appl. and Comput. Harmonic Anal.* **3**, 164–185.

[49] R.E. Blahut (1987): *Principles and practice of information theory, Addison-Wesley.*

[50] R.E. Blahut, W. Miller Jr, and C.H. Wilcox, Eds. (1991): Radar and Sonar, Part I. *Springer Verlag*, New York, N.Y.

[51] L. Blanchet, T. Damour, and B.R. Iyer (1995): Gravitational waves from inspiralling compact binaries: energy loss and waveform to second-post-Newtonian order. *Phys. Rev.* **D 51**, 5360–5386.

[52] P. Bloomfield (1976): *Fourier Analysis of Time Series: An Introduction.* Wiley Interscience, New York.

[53] B. Boashash (1992): Estimating and interpreting the instantaneous frequency of a signal, Part I: Fundamentals. *Proc. IEEE* **80**, 520–538.

[54] B. Boashash (1992): Estimating and interpreting the instantaneous frequency of a signal, Part II: applications and algorithms. *Proc. IEEE* **80**, 540–568.

[55] D. Bosq (1996): Nonparametric statistics for stochastic processes: estimation and prediction. *Lecture Notes in Statistics*, **110**, Springer Verlag, New York.

[56] C. Bradaschia, R. Del Fabbro, A. Di Virgilio, A. Giazotto, H. Kautzky, V. Montelatici, D. Passuello, A. Brillet, O. Cregut, P. Hello, C. N. Man, P. T. Manh, A. Marraud, D. Shoemaker, J.Y. Vinet, F. Barone, L. Di Fiore, L. Milano, G. Russo, J.M. Aguirregabiria, H. Bel, J. P. Duruisseau, G. Le Denmat, Ph. Tourrenc, M. Capozzi, M. Longo, M. Lops, I. Pinto, G. Rotoli, T. Damour, S. Bonazzola, J. A. Marck, Y. Gourghoulon, L.E. Holloway, F. Fuligni, V. Iafolla, and G. Natale (1990): The VIRGO project: a wide band antenna for gravitational wave detection. *Nucl. Instrum. Methods Phys. Res.* **A289**, 518–525.

[57] L. Breiman, J.H. Friedman, R.O. Olshen, and C.J. Stone (1984): *Classification and Regression Trees.* Wadsworth Intern. Group.

[58] D.R. Brillinger (1981): *Time Series; Data Analysis and Theory,* Holden Day Inc.

[59] P.J. Burt and E.H. Adelson (1983): The Laplacian pyramid as a compact image code. *IEEE Trans. Comm.* **31**, 482–540.

[60] A. Calderón (1964): Intermediate spaces and interpolation, the complex method. *Studia Math.* **24**, 113

[61] S. Cambanis and C. Houdré (1995): On continuous wavelet transforms of second order random processes. *IEEE Trans. Inf. Theory* **41**, 628–642.

[62] J.A. Canny (1986): A computational approach to edge detection. *IEEE Trans. PAMI* **8**, 679–698.

[63] A. L. Carey (1976): Square-integrable representations of non-unimodular groups. *Bull. Austr. Math. Soc.* **15**, 1–12.

[64] R. Carmona (1994): Spline smoothing and extrema representation: variations on a reconstruction algorithm of Mallat and Zhong. *Wavelets and Statistics, Lecture Notes in Statistics.* **103**, A. Antoniadis and G. Oppenheim, Eds., Springer Verlag, New York, 83–94.

[65] R. Carmona (1993): Wavelet identification of transients in noisy signals. *Proc. SPIE 15–16 June 1993, San Diego, Mathematical Imaging: Wavelet Applications in Signal and Image Processing,* vol. **2034**, 392–400.

[66] R. Carmona and L. Hudgins (1994): Wavelet denoising of EEG signals and identification of evoked response potentials. *Proc. SPIE July 1994 San Diego, Signal Processing and Image Analysis,* vol **2303**, 91–104.

[67] R. Carmona, W.L. Hwang, and B. Torrésani (1997): Characterization of signals by the ridges of their wavelet transform, *IEEE Trans. Signal Processing,* **45**, 2586–2589.

[68] R. Carmona, W.L. Hwang, and B. Torrésani (1994): Identification of chirps with continuous wavelet transform. *Wavelets and Statistics,Lecture Notes in Statistics.* **103**, A. Antoniadis and Oppenheim Eds, Springer Verlag, New York, 95–108.

[69] R. Carmona, W.L. Hwang, and B. Torrésani (1998): Multiridge detection and time-frequency reconstruction. *IEEE Trans. Signal Processing,* to appear.

[70] H. Cecen (1973): Response of ten story reinforced concrete model frames to simulated earthquakes. Ph.D. Dissertation, University of Illinois, Urbana, Ill.

[71] S.S. Chen, D. Donoho, and M.A. Saunders (1996): Atomic decomposition by basis pursuit. (preprint)

[72] V. Chen and S. Qian (1996): CFAR detection and extraction of unknown signal in noise with time-frequency Gabor transform, *Proceedings SPIE*, **2762**, Wavelet Applications III, 285–293.

[73] C.K. Chui (1991): *An overview of wavelets. Approximation theory and Functional Analysis* Academic Press, 47–71.

[74] C.K. Chui and C. Li (1993): Non orthogonal wavelet packets *SIAM J. Math. Anal.* **24**, 712–738.

[75] C.K. Chui and E. Quak (1992): Wavelets on a bounded interval *Numerical Methods in Approximation Theory*, Birkhauser, 1–24.

[76] C.K. Chui and J.Z. Wang (1991): A cardinal spline approach to wavelets. *Proc. Am. Math. Soc.* **113**, 785–793.

[77] C.K. Chui and J.Z. Wang (1992): On compactly supported spline wavelets and a duality principle. *Trans. Am. Math. Soc.* **330**, 903–915.

[78] A. Cohen (1991): Ondelettes, analyses multirésolutions et filtres miroir en quadrature, *Annales de l'Inst. H. Poincaré, Anal. non-lin.* **7**, 439–459.

[79] A. Cohen and I. Daubechies (1993): Orthonormal bases of compactly supported wavelets III: better frequency localization, *SIAM J. Math. Anal.* **24**, 520–527.

[80] A. Cohen, I. Daubechies, and J.C. Feauveau (1992): Biorthogonal bases of compactly supported wavelets. *Comm. Pure and Appl. Math.* **45**, 485–560.

[81] A. Cohen, I. Daubechies, and P. Vial (1994): Wavelets on the interval and fast wavelet transforms *Appl. and Comp. Harmonic Anal.* **1**, 54–81.

[82] L. Cohen (1966): Generalized phase-space distribution functions. *J. Math. Phys.* **7**, 781–786.

[83] L. Cohen (1989): Time-frequency distributions: a review, *Proc. IEEE* **77**, 941–981.

[84] R.R. Coifman and D. Donoho (1995): Time invariant wavelet denoising. *Wavelets and Statistics, Lecture Notes in Statistics.* **103**, A. Antoniadis and G. Oppenheim, Eds., Springer Verlag, New York, 125–150.

[85] R.R. Coifman and Y.R. Meyer (1990): Orthonormal wavepacket bases. Preprint, Yale University.

[86] R.R. Coifman and Y.R. Meyer (1991): Remarques sur l'analyse de Fourier à fenêtre. *C.R. Acad. Sci. Paris Série I Math.*, **312**, 259–261.

[87] R.R. Coifman, Y.R. Meyer, and M.V. Wickerhauser (1992): Size properties of wavelet packets. *Wavelets and their Applications*, B. Ruskai et al. Eds, Jones and Bartlett, Boston, 153–178.

[88] R.R. Coifman, Y.R. Meyer, S. Quake, and M.V. Wickerhauser (1993): Signal processing and compression with wavelet packets. *Progress in Wavelet Analysis and Applications*, Proceedings of the International Conference *Wavelets and Applications*, Toulouse, Editions Frontière, 77–93.

[89] R.R. Coifman and N. Saito (1994): Construction of local orthonormal bases for classification and regression. *C. R. Acad. Sci. Paris ser. A* **319**, 191–196.

[90] R.R. Coifman and N. Saito (1996): Improved local discriminant bases using empirical probability density estimation. *Amer. Statist. Assoc.* (to appear)

[91] E. Copson (1965): *Asymptotic expansions.* Cambridge University Press Cambridge, U.K.

[92] I. Csiszar (1978): Information measures: a critical survey. *Trans. 7th Prague Conf. on Information Theory,* vol. **B**, 73–86, Academia Praha.

[93] I. Daubechies (1990): The wavelet transform, time-frequency localization and signal analysis. *IEEE Trans. Inf. Th.* **36**, 961–1005.

[94] I. Daubechies (1988): Orthonormal bases of compactly supported wavelets. *Comm. Pure Appl. Math.* **41**, 909–996.

[95] I. Daubechies (1993): Orthonormal bases of compactly supported wavelets II. Variations on a theme, *SIAM J. Math. Anal.* **24**, 499–519.

[96] I. Daubechies, A. Grossmann, and Y. Meyer (1986): Painless nonorthogonal expansions. *J. Math. Phys.*, **27**, 1271–1283.

[97] I. Daubechies, S. Jaffard, and J.L. Journé (1991): A simple Wilson basis with exponential decay. *SIAM J. Math. An.***22**, 554–572.

[98] I. Daubechies, H. Landau, and Z. Landau (1995): Gabor time-frequency lattices and the Wexler-Raz identity. *J. Fourier Anal. Applications* **4**, 437–478.

[99] I. Daubechies and S. Maes (1996): A non-linear squeezing of the continuous wavelet analysis based on auditory nerve models. *in Wavelets and Biology,* M. Unser and A. Aldroubi, Eds. CRC Press.

[100] J. Daugman (1986): Complete discrete 2D Gabor transforms. *IEEE Trans. Acoust. Speech Signal Processing* **36**, 1169–1179.

[101] N. Delprat (1994): *Extraction of frequency modulation laws in sound synthesis.* PhD Thesis, Marseille.

[102] N. Delprat, B. Escudié, P. Guillemain, R. Kronland-Martinet, Ph. Tchamitchian, and B. Torrésani (1992): Asymptotic wavelet and Gabor analysis: extraction of instantaneous frequencies. *IEEE Trans. Inf. Th.* **38**, 644–664.

[103] B. Delyon and A. Juditsky (1997): On the computation of wavelet coefficients. *Journal of Approximation Theory,* **88, # 1**, 47–79.

[104] M. Deriche and A.H. Tewfik (1993): Signal modeling with filtered discrete fractional noise processes. *IEEE Trans. Sig. Proc.* **41**, 2839–2849.

[105] M. Deriche and A.H. Tewfik (1993): Maximum likelihood estimation of the parameters of discrete fractionally difference Gaussian noise processes. *IEEE Trans. Sig. Proc.* **41**, 2977–2989.

[106] R.B. Dingle (1973): *Asymptotic expansions, their derivation and interpretation.* Academic Press.

[107] E. DiPasquale and A.S. Cakmak (1990): Detection of seismic structural damage using parameter based global damage indices. *Probab. Eng. Mechanics* **5**, 60–65.

[108] D. Dobrushin (1979): Gaussian and their subordinated self-similar random fields. *Ann. Prob.* **7**, 1–28.

[109] D. Donoho (1992): Interpolating wavelet transforms. Preprint, Stanford University.

[110] D.L. Donoho (1993): Wavelet shrinkage and W.V.D.: A ten minute tour. In *Wavelet Analysis and Applications* Y. Meyer and S. Roques, Eds., Editions Frontières.

[111] D. Donoho (1995): De-noising by soft thresholding. *IEEE Trans. Info. Theo.* **41**, 613–627.

[112] D. Donoho and I. Johnstone (1992): Minimax estimation by wavelet shrinkage. Preprint, Stanford University.

[113] D. Donoho and I. Johnstone (1995): Adapting to unknown smoothness by wavelet shrinkage. *J. Am. Stat. Assoc.* **90**, 1200–1224.

[114] D. Donoho, I. Johnstone, G. Kerkyacharian, and D. Picard (1993): Density estimation by wavelet thresholding. *Ann. Statist.* **24**, 508–539.

[115] D. Donoho, I. Johnstone, G. Kerkyacharian, and D. Picard (1995): Wavelet shrinkage: asymptotia ? *J. Royal Stat. Soc.* **57**, 301–337.

[116] R. Duffin and A. Schaeffer (1952): A class of non-harmonic Fourier series. *Trans. Am. Math. Soc.* **72**, 341–366.

[117] M. Duflo and C. C. Moore (1976): On the regular representation of a non-unimodular locally compact group. *J. Funct. An.* **21**, 209–243.

[118] M. Duval-Destin, M.A. Muschietti, and B. Torrésani (1993): Continuous wavelet decompositions: multiresolution and contrast analysis. *SIAM J. Math. An.* **24**, 739–755.

[119] H. Dym and H. P. McKean (1972): *Fourier Series and Integrals.* Probability and Mathematical Statistics **14**, Academic Press.

[120] B. Escudié, A. Grossmann, R. Kronland-Martinet, and B. Torrésani (1989): Analyse en ondelettes de signaux asymptotiques: emploi de la phase stationnaire. *Proceedings of the GRETSI Conference*, Juan-les-Pins, France.

[121] D. Esteban and C. Galand (1977): Applications of quadrature mirror filters to split band voice coding schemes. *Proc. ICASSP.*

[122] R.L.Eubank (1988): *Spline Smoothing and Nonparametric Regression.* M. Dekker, New York.

[123] H. Feichtinger (1994): Optimal iterative algorithms in Gabor analysis. *Proc. IEEE-SP Int. Symp. on Time-Frequency and Time-Scale Analysis*, Philadelphia.

[124] H. Feichtinger and K.H. Gröchenig (1992): Banach spaces related to integrable group representations and their atomic decompositions I. *J. Funct Anal.* **86**, 307–340.

[125] H. Feichtinger and K.H. Gröchenig (1992): Banach spaces related to integrable group representations and their atomic decompositions II. *Mh. Math.* **108**, 129–148.

[126] H. Feichtinger, K.H. Gröchenig, and D. Walnut (1992): Wilson bases in modulation spaces. *Math. Nachr.* **1551**, 7–17.

[127] A. Feldman, A.C. Gilbert, W. Willinger, and T.G. Kurtz (1997): Looking behind and beyond self-similarity: on scaling phenomena in measured WAN traffic. In *Proc. 35th Alberton Conf. on Communication, Control and Computing.* Alberton House, Monticello, IL 9/29–10/1 1997.

[128] P. Flandrin (1986): On the positivity of Wigner-Ville spectrum. *Signal Processing*, **11**, 187–189.

[129] P. Flandrin (1988): A time-frequency formulation of optimum detection. *IEEE Trans. ASSP*, **36**, 1377–1384.

[130] P. Flandrin (1989): On the spectrum of fractional Brownian motion. *IEEE Trans. Info. Theory*, **IT-35**, 197–199.

[131] P. Flandrin (1992): Wavelet analysis and synthesis of fractional Brownian motion. *IEEE Trans. Info. Theory*, **IT-38**, 910–917.

[132] P. Flandrin (1994): Time-scale analyses and self-similar stochastic processes. In *Wavelets and their Applications*, J.Byrnes et al. Eds., Kluwer Dordrecht, 121–142.

[133] P. Flandrin, E. Chassande-Mottin, and P. Abry (1996): Reassigned scalograms. *Proc. SPIE* vol. **2569**, 152–158.

[134] P. Flandrin and P. Gonçalves (1996): Geometry of affine time-frequency distributions. *Appl. and Comp. Harm. Anal.* **3**, 10–39.

[135] K. Flornes, A. Grossmann, M. Holschneider, and B. Torrésani (1994): Wavelets on Discrete Fields. *Appl. and Comp. Harmonic Anal.* **1** 137–146.

[136] M. Frazier, B. Jawerth, and G. Weiss (1991): *Littlewood-Paley Theory and the Classification of Function Spaces*, Regional Conference Series in Mathematics **79**, Providence, RI, American Mathematical Society.

[137] B. Friedlander and B. Porat (1992): Performance analysis of transient detectors based on a class of linear data transforms. *IEEE Trans. Info. Theory* **38**, 665–673.

[138] J.H. Friedman and W. Stuetzle (1981): Projection pursuit regression. *J. Amer. Statist. Assoc.* **76**, 817–823.

[139] D. Gabor (1946): Theory of communication. *J. Inst. Elec. Eng.* **903**, 429–.

[140] H. Gao (1993): *Wavelet Estimation of Spectral Densities in Time Series Analysis.* Ph.D. thesis, UC Berkeley.

[141] I.M. Gelfand and G.E. Shilov (1964): *Generalized Functions.* Academic Press, New York and London.

[142] S. Geman and D. Geman (1984): Stochastic relaxation, Gibbs distributions and Bayesian restoration of images. *IEEE Proc. Pattern Ana. Mach. Intell.* **6**, 721–741.

[143] B. Gidas (1985): Nonstationary Markov chains and convergence of the annealing algorithm. *J. Statist. Phys.* **39**, 73–131.

[144] R. J. Glauber (1963): The quantum theory of optical coherence. *Phys. Rev.* **130**, 2529–2539.

[145] R. Godement (1947): Sur les relations d'orthogonalit'e de V. Bargmann. *C.R. Acad. Sci. Paris*, **255**, 521–523; 657–659.

[146] C. Gonnet and B. Torrésani (1984): Local frequency analysis with two-dimensional wavelet analysis. *Signal Processing* **37**, 389–404.

[147] P. Goupillaud, A. Grossmann, and J. Morlet (1984): Cycle-octave and related transforms in seismic signal analysis. *Geoexploration* **23**, 85.

[148] K.H. Gröchenig (1992): Acceleration of the frame algorithm. *IEEE Trans. Signal Processing* **41**, 3331–3340.

[149] A. Grossmann (1988): Wavelet transform and edge detection. In *Stochastic Processes in Physics and Engineering*, S. Albeverio, Ph. Blanchard, L. Streit, and M. Hazewinkel, Eds., 149–157.

[150] A. Grossmann, R. Kronland-Martinet, and J. Morlet (1987): Reading and understanding the continuous wavelet transform. In *Wavelets, Time-Frequency Methods and Phase Space*, J.M. Combes, A. Grossmann and Ph. Tchamitchian, Eds., IPTI, Springer Verlag, 2–20.

[151] A. Grossmann, J. Morlet (1984): Decomposition of Hardy functions into square integrable wavelets of constant shape. *SIAM J. of Math. An.* **15**, 723–736.

[152] A. Grossmann, J. Morlet (1984): Decomposition of functions into wavelets of constant shape. In *Mathematics + Physics*, **1**, L. Streit Ed., World Scientific, 135–166.

[153] A. Grossmann, J. Morlet, and T. Paul (1985): Transforms associated with square integrable group representations I. *J. Math. Phys.* **27**, 2473–2479.

[154] A. Grossmann, J. Morlet, and T. Paul (1986): Transforms associated with square integrable group representations II. *Ann. Inst. H. Poincaré* **45**, 293–309.

[155] P. Guillemain (1995): *Estimation of Spectral Lines with the Help of the Wavelet Transform, Applications in NMR Spectroscopy*, Ph.D. Thesis, Marseille.

[156] P. Guillemain and R. Kronland-Martinet (1996): Characterization of acoustic signals through continuous linear time-frequency representations. *Proceedings of the IEEE* **84**, 561–585.

[157] L. Györfi, W. Härdle, P. Sarda, and P. Vieu (1989): *Nonparametric Curve Estimation from Time Series. Lecture Notes in Statistics* **60**. Springer Verlag.

[158] P. Hall, W. Qian, and D.M. Titterington (1992): Ridge Finding from Noisy Data. *J. Comput. and Graph. Statist.* **1**, 197–211.

[159] W. Härdle (1990): *Applied Nonparametric Regression.* Econometric Society Monographs, Cambridge Univ. Press.

[160] T.J. Healey and M.A. Sozen (1978): Experimental study if the dynamic response of a ten story reinforced concrete frame with a tall first story. Rep. No. UILU-ENG-78-2012, SRS 450, University of Illinois, Urbana, Ill.

[161] C. Heil and D. Walnut (1989): Discrete and continuous wavelet transforms. *SIAM reviews* **31**, 628–666.

[162] G. Hewer and W. Kuo (1993): Fractional Brownian motion, wavelets and infrared detector noise. Tech. Rep. NAWCWPNS 8103, China Lake, CA.

[163] G. Hewer and W. Kuo (1994): Wavelet transform of fixed pattern noise in focal plane arrays. Tech. Rep. NAWCWPNS 8185, China Lake, CA.

[164] F. Hlawatsch and G.F. Boudreaux-Bartels (1992): Linear and quadratic time-frequency signal representations. *IEEE Signal Processing Magazine* **9**, 21–37.

[165] F. Hlawatsch and R. Urbanke (1994): Bilinear time-frequency representations of signals: the shift-scale invariant case. *IEEE Trans. Sig. Proc.* **42**, 357–366.

[166] J.A. Hogan and J.D. Lakey (1995): Extensions of the Heisenberg group by dilations and frames. *Appl. Comp. Harm. Anal.* **2**, 174–199.

[167] M. Holschneider (1988): On the wavelet transformation of fractal objects. *J. Stat. Phys.* **50**, 953–993.

[168] M. Holschneider (1991): Inverse Radon transforms through inverse wavelet transforms. *Inverse Problems* **7**, 853–861.

[169] M. Holschneider (1995): Wavelet analysis over abelian groups. *Appl. and Comp. Harm. Anal.* **2**, 52–60.

[170] M. Holschneider, R. Kronland-Martinet, J. Morlet, and Ph. Tchamitchian (1987): A real-time algorithm for signal analysis with the help of wavelet transform. In *Wavelets, Time-Frequency Methods and Phase Space*, Combes, Grossmann, and Tchamitchian, Eds., IPTI, Springer Verlag, 286–297.

[171] M. Holschneider and Ph. Tchamitchian (1991): Pointwise analysis of Riemann-Weierstrass "nowhere differentiable" function. *Invent. Math.* **105**, 157–176.

[172] L. Hörmander (1983): *The Analysis of Partial Differential Operators.* Springer Verlag, New York.

[173] C. Houdré (1994): Wavelets, probability and statistics: some bridges. In *Wavelets: Mathematics and Applications*, J.J. Benedetto and M.W. Frazier, Eds., CRC Press, 365–398.

[174] L. Hudgins, M. Mayer, and K. Frehe (1992): Wavelet spectrum and turbulence. In *Wavelets and Applications*, Y. Meyer and S. Roques Eds., Editions Frontière, 491–498.

[175] R. Hummel and R. Moniot (1989): Reconstruction from zero-crossings in scale space. *IEEE Trans. Acoustics, Speech and Sig. Proc.* **40**, 2111–2310.

[176] W. L. Hwang (1998): Estimation of fractional Brownian motion embedded in a noisy environment using non-orthogonal wavelets. *IEEE Trans. Signal Processing* (to appear).

[177] J.M. Innocent and B. Torrésani (1996): Wavelet transforms and binary collapse detection. In *Mathematical Aspects of Gravitation*, Banach Center Publications **41**, part II, Warsaw, 179–208.

[178] J.M. Innocent and B. Torrésani (1996): A Multiresolution Strategy for Detecting Gravitational Waves Generated by Binary Collapses. Preprint, Marseille.

[179] J.M. Innocent and B. Torrésani (1996): Wavelet Transforms and Binary Coalescences Detection. *Appl. and Comp. Harmonic Anal.* **4**, 113–116.

[180] J.M. Innocent and J.Y. Vinet (1992): Time-Frequency Analysis of Gravitational Signals from Coalescing Binaries, VIRGO Technical Report.

[181] S. Jaffard (1989): Exposants de Hölder en des points donnés, et coefficients d'ondelettes. *Comptes Rendus Acad. Sci. Paris Ser. A* **308**, 79–81.

[182] S. Jaffard (1991): Pointwise smoothness, two-localization and wavelet coefficients. *Publicaciones Mathematiques* **35**, 155–168.

[183] S. Jaffard and Y. Meyer (1996): Wavelet Methods for Pointwise Regularity and Local Oscillations of Functions. *Memoirs of the AMS* **587**.

[184] S. Jaggi, W.C. Karl, S. Mallat, and A.S. Willsky (1997): High resolution pursuit for feature extraction. Preprint.

[185] B. Jawerth and W. Sweldens (1994): An overview of wavelet-based multiresolution analyses. *SIAM Reviews* **36**, 377–412.

[186] S. Kadambe and G.F. Boudreaux-Bartels (1992): A comparison of the existence of "cross terms" in the Wigner distribution and the squared modulus of the wavelet transform and the short time Fourier transform. *IEEE Trans. Sig. Proc.* **40**, 2498–2517.

[187] G. Kaiser (1996): Wavelet filtering with the Mellin transform. *Appl. Math. Lett.* **9**, 69–74.

[188] G. Kaiser and R. Streater (1992): Windowed Radon transforms, analytic signals and the wave equation. In *Wavelets: a Tutorial in Theory and Applications*, Academic Press, 399–441.

[189] L. Kaplan and C.C. Kuo (1993): Fractal estimation from noisy data via DFGN and the Haar basis. *IEEE Trans. Sig. Proc.* **41**, 3554–3562.

[190] L. Kaplan and C.C. Kuo (1993): Fast fractal feature extraction for texture segmentation using wavelets. *Mathematical imaging: wavelet applications in signal and image processing*, Proc. SPIE **2034**, 144–155.

[191] M. Kass, A. Witkin, and D.Terzopoulos (1988): Snakes: Active Contour Models, *Int. J. Computer Vision*, 321–331.

[192] M.S. Keshner (1982): $1/f$ Noise. *Proc. IEEE* **70**, 212–218.

[193] J. Klauder (1963): Continuous-representation theory. I: postulates of continuous representation theory. *J. Math. Phys.* **4**, 1055–1058. II: Generalized relation between quantum and classical dynamics. *J. Math. Phys.* **4**, 1059–1073.

[194] K. Kodera, R. Gendrin, and C. de Villedary (1978): Analysis of time-varying signals with small BT values. *IEEE Trans. ASSP* **26**, 64–.

[195] L.H. Koopmans (1995): *The Spectral Analysis of Time Series. Academic Press* New York.

[196] W. Kozek (1996): *Spectral Estimation in Non-Stationary Environments.* Ph.D. Thesis, Vienna.

[197] W. Kozek (1992): Time-frequency signal processing based on the Wigner-Weyl framework. *Signal Processing* **29**, 77–92.

[198] W. Kozek (1996): On the underspread/overspread classification of nonstationary random processes. *Proc. ICIAM-95, Special Issue ZAMM*, **3**, 63–66.

[199] R. Kronland-Martinet and A. Grossmann (1990): Application of time-frequency and time-scale methods to the analysis, synthesis and transformation of natural sounds. In *Representation of Musical Sounds*, G. dePoli, A. Picciali, and C. Roada, Eds., MIT Press, 47–85.

[200] R. Kronland-Martinet, J. Morlet, and A. Grossmann (1987): Analysis of sound patterns through wavelet transform. *Int. J. Pattern Recognition and Artificial Intelligence*, **1**, 273–302.

[201] P.J.M. van Laarhoven and E.H.L. Aarts (1987): *Simulated Annealing: Theory and Applications.* Reidel.

[202] W. Lawton (1991): Necessary and sufficient conditions for constructing orthonormal wavelet bases. *J. Math. Phys.* **31** , 1898–1901.

[203] N. Lee and S.C. Schwartz (1995): Linear time-frequency representations for transient signal detection and classification. Technical Report 23, Department of Electrical Engineering, Princeton University.

[204] W.E. Leland, M.S. Taqqu, W. Willinger, and D.V. Wilson (1994): On the self-similar nature of ethernet traffic (extended version). *IEEE/ACM Trans. Networking* **2**, 1–15.

[205] B. Logan (1977): Information in the zero crossings of band pass signals. *Bell Syst. Tech. J.* **56**, 510.

[206] F. Low (1985): Complete sets of wave packets. *A passion for physics, essays in honour of Geoffrey Chew*, C. DeTar et al., Eds., World Scientific, Singapore, 17–22.

[207] T. Lundhal, W.J. Ohley, S.M. Kay, and R. Siffert (1986): Fractional Brownian motion: maximum likelihood estimator and its application to image texture. *IEEE Trans. Medical Imag.* **MI-5**, 152–161.

[208] S. Maes (1994): *The Wavelet Transform in Signal Processing with Applications to the Speech Modulation Model Features.* Ph.D. Dissertation, UCL, Louvain-la-Neuve (Belgium).

[209] S. Maes and T. Hastie (1996): The maximum likelihood estimation based living cubic spline extractor and its applications to saliency grouping in the time-frequency plane. Technical report, IBM J. Watson Research Center, J2-N52, PO Box 794, Yorktown Heights, NY.

[210] S. Mallat (1989): Multiresolution approximation and wavelets. *Trans. Amer. Math. Soc.* **615**, 69–88.

[211] S. Mallat (1989): A theory for multiresolution signal decomposition: the wavelet representation. *IEEE Trans. PAMI* **11**, 674–693.

[212] S. Mallat and W.L. Hwang (1992): Singularities detection and processing with wavelets. *IEEE Trans. Info. Theory* **38**, 617–643.

[213] S. Mallat, G. Papanicolaou, and Z. Zhang (1995): Adaptive covariance estimation of locally stationary processes. *Ann. Stat.* **26**, 1–47.

[214] S. Mallat and S. Zhong (1991): Characterization of signals from multiscale edges. *IEEE Trans. Pattern Anal. Machine Intel.*, **14**, 710–732.

[215] S. Mallat and Z. Zhang (1993): Matching pursuits with time-frequency dictionaries. *IEEE Trans. Sign. Proc.* **41**, 3397–3415.

[216] E. Malvar (1990): Lapped transforms for efficient sub-band/transform coding *IEEE Trans. ASSP* **38**, 969–978.

[217] B.B. Mandelbrot and J. Van Ness (1968): Fractional Brownian motion, fractional noise and applications, *SIAM Reviews* **10**, 422–437.

[218] D. Marr (1982): *Vision*, Freeman.

[219] J.S. Marron, Ed. (1986): *Function Estimates. Contemporary Mathematics,* **59**. Amer. Math. Soc. Providence, RI.

[220] W. Martin and P. Flandrin (1985): Wigner-Ville spectral analysis of non-stationary processes. *IEEE Trans. Sig. Proc.* **33**, 1461–1470.

[221] E. Masry (1993): The wavelet transform of stochastic processes with stationary increments and its application to fractional Brownian motion. *IEEE Trans. Info. Theory* **IT-39**, 260–264.

[222] R.J. McAulay and T.F. Quatieri (1986): Speech analysis/synthesis based on a sinusoidal representation. *IEEE Trans. on Audio, Speech and Sign. Proc.* **34** 744–754.

[223] R.J. McAulay and T.F. Quatieri (1992): Low rate speech coding based on the sinusoidal model. In *Advances in Speech Signal Processing*, S. Furui and M. Mohan Sondui, Eds.

[224] Y. Meyer (1989): Wavelets and operators. In *Analysis at Urbana I*, London Mathematical Society Lecture Notes Series, **137**, Cambridge University Press, 257–365.

[225] Y. Meyer (1991): Orthonormal wavelets. *Frontiers in Pure and Applied Mathematics*, North Holland, 235–245 .

[226] Y. Meyer (1991): Ondelettes sur l'intervalle. *Revista Matematica IberoAmericana* **7**, 115–133.

[227] Y. Meyer (1994): The Riemann function $\sum_1^\infty n^{-2}\sin(n^2 x)$, chirps and microlocal analysis. Preprint, CEREMADE, Université de Paris-Dauphine, France.

[228] E.W. Montroll and M.F. Schlesinger (1989): On the ubiquity of $1/f$ Noise. *Int. J. Modern Phys.*, **3**, 795–819.

[229] H. Moscovici and A. Verona (1978): Coherent states and square-integrable representations. *Ann. Inst. Henri Poincaré* **29**, 139–156.

[230] C.I. Mullen, R.C. Micaletti, and A.S. Cakmak (1995): A simple method for estimating the maximum softening damage index. In *Soil Dynamics and Earthquake Engineering*

[231] R. Murenzi (1990): *Ondelettes Multidimensionnelles et à l'analyse d'images.* PhD Thesis, Louvain la Neuve, Belgium (in french).

[232] M.A. Muschietti and B. Torrésani (1995): Pyramidal algorithms for Littlewood-Paley decompositions. *SIAM J. Math. Anal.* **26** 925–943.

[233] E.A. Nadaraya (1989): Nonparametric estimation of probability densities and regression curves. in *Mathematics and its Applications.* Kluwer Dordrecht.

[234] G.P. Nason and B. Silverman (1995): The stationary wavelet transform and some statistical applications. In *Wavelets and Statistics, Lecture Notes in Statistics.* **103**, A. Antoniadis and G. Oppenheim, Eds., Springer Verlag, New York, 281–300.

[235] H.J. Newton (1988): *Timeslab: A Time Series Analysis Laboratory*, Wadsworth and Brooks/Cole.

[236] A.V. Oppenheim and R.W. Schafer (1989): *Discrete Time Signal Processing.* Prentice Hall.

[237] Z. Öri, G. Monir, J. Weiss, X. Sayhouni, and D.H. Singer (1992): Heart rate variability: frequency domain analysis. *Ambul. Electrocardiography* **10**, 499–537.

[238] T. Paul (1986): *Ondelettes et Mécanique Quantique*, Ph.D. Thesis, CPT Marseille (in french).

[239] D.B. Percival and A.T. Walden (1993): *Spectral Analysis for Physical Applications: Multitaper and Conventional Univariate Techniques.* Cambridge Univ. Press.

[240] V. Genon-Catalot, C. Laredo, and D. Picard (1990): Estimation non-paramétrique d'une diffusion par méthode d'ondelettes. *C.R. Acad. Sci. Paris, série 1*, **311**, 379–382.

[241] G. Kerkyacharian, D. Picard, and K. Tribouley (1995): L^p adaptive density estimation. *Bernoulli*, **2**, 229–247.

[242] B. Picinbono and W. Martin (1983): Représentation des signaux par amplitude et phase instantanées. *Annales des Télécommunications* **38**, 179–190.

[243] B. Picinbono (1995): On Instantaneous Amplitude and Phase of Signals. *IEEE Trans. Inf. Th.* (to appear).

[244] H.V. Poor (1988): *An Introduction to Signal Detection and Estimation.* Springer Verlag, New York.

[245] M.R. Portnoff (1980): Time-frequency representation of digital signals and systems based on short-time Fourier analysis. *IEEE Trans. ASSP* **28**, 55–69.

[246] M.R. Portnoff (1981): Short-time Fourier analysis of sampled speech. *IEEE Trans. ASSP* **29**, 374–390.

[247] W.H. Press, B.P. Flannery, S.A. Teukolsky, and W.T Wetterling (1986): *Numerical Recipes.* Cambridge Univ. Press, Cambridge, UK.

[248] M.B. Priestley (1965): Evolutionary Spectra and Non-Stationary Processes. *J. Roy. Stat. Soc.* **B27**, 204–237.

[249] M.B. Priestley (1965): Design relations for Non-Stationary Processes. *J. Roy. Stat. Soc.* **B28**, 228–240.

[250] M.B. Priestley (1981): *Spectral Analysis and Time Series.* 2 vols., Academic Press, London.

[251] M.B. Priestley (1988): *Nonlinear and Nonstationary Time Series Analysis.* Academic Press, London.

[252] T.F. Quatieri and R.J. McAulay (1989): Phase coherence in speech reconstruction for enhancement and coding applications. *Proc. IEEE Int. Conf. Audio, Speech and Sig. Proc.*, Glasgow, 207–209.

[253] J. Ramanathan and O. Zeitouni (1991): On the wavelet transform of fractional Brownian motion. *IEEE Trans. Info. Theory* **IT-37**, 1156–1158.

[254] M. Reed and B. Simon (1972): *Methods of Modern Mathematical Physics I: Functional Analysis.* Academic Press, New York.

[255] M. Reed and B. Simon (1976): *Methods of Modern Mathematical Physics II: Fourier Analysis and Self Adjointness.* Academic Press, New York.

[256] M. Rieffel (1981): Von Neumann algebras associated with pairs of lattices in Lie groups, *Math. Ann.* **257** 403–413.

[257] F. Riesz and B. S. Nagy (1955): *Functional Analysis.* Ungar, New York.

[258] O. Rioul and P. Duhamel (1992): Fast algorithms for discrete and continuous wavelet transforms. *IEEE Trans. Inf. Th.* **38**, 569–586.

[259] O. Rioul and P. Flandrin (1992): Time-scale energy distributions: a general class extending wavelet transforms. *IEEE Trans. Signal Proc* **40**, 1746–1757.

[260] N. Saito and G. Beylkin (1993): Multiresolution representations using the autocorrelation functions of compactly supported wavelets. *IEEE Trans. Signal Processing* **41**, 3584–3590.

[261] W. Schempp (1986): *Harmonic analysis on the Heisenberg Nilpotent Lie group.* Pitman series **147**, Wiley.

[262] E. Schrödinger (1926): Der stetige Übergang von der Mikro- zur Makromechanik. *Naturalwiss.* **14**, 664–666.

[263] M. Shensa (1993): Wedding the "à trous" and Mallat algorithms. *IEEE Trans. Inf. Th.* **40**, 2464–2482.

[264] B. Silverman (1957): Locally stationary random processes. *IRE Trans. Info. Th.* **3**, 182–187.

[265] B. Simon (1979) *Functional Integration.* Collection in Pure and Applied Mathematics, Academic Press.

[266] M.J.T. Smith and T. P. Barnwell (1986): Exact reconstruction techniques for tree-structured subband coders. *IEEE ASSP* **34**, 434–441.

[267] V. Solo (1992): Intrinsic random functions and the paradox of $1/f$ noise. *SIAM J. Appl. Math.* **52**, 270–291.

[268] G. Strang (1989): Wavelets and dilation equations: a brief introduction. *SIAM Review* **31**, 614–627.

[269] G. Strang (1994): Wavelets. *American Scientist* **31**, 250–255.

[270] J. O. Stromberg (1983): A modified Franklin system and higher order spline systems on $\mathbb{R}^n$ as unconditional bases for Hardy spaces. In *Conference on harmonic analysis in honour of A. Zygmund II*, Beckner et al., Eds., Belmont, CA; Wadsworth, 475–494.

[271] E. C. G. Sudarshan (1963): Equivalence of semiclassical and quantum mechanical description of statistical light beams. *Phys. Rev. Lett.* **10**, 277–279.

[272] Ph. Tchamitchian and B. Torrésani (1991): Ridge and Skeleton extraction from wavelet transform, In *Wavelets and their Applications*, M.B. Ruskai et al., Eds., Jones and Bartlett, Boston.

[273] B.A. Telfer, H.H. Szu, and G.J Dobeck (1994): Adaptive wavelet classification of acoustic backscatter. *SPIE* **2242**, 661–668.

[274] B.A. Telfer, H.H. Szu, and G.J Dobeck (1995): Adaptive time-frequency classification of acoustic backscatter. *SPIE* **2491**, 451–460.

[275] A.H. Tewfik and M. Kim (1992): Correlation structure of the discrete wavelet transform of fractional Brownian motion. *IEEE Trans. Info. Theory* **IT-38**, 904–909.

[276] J.R. Thompson and R.A. Tapia (1990): *Nonparametric Function Estimation, Modeling and Simulation.* SIAM, Philadelphia.

[277] B. Torrésani (1991): Wavelets associated with representations of the Weyl-Heisenberg group. *J. Math. Phys.* **32**, 1273–1279.

[278] B. Torrésani (1992): Time-frequency distributions: wavelet packets and optimal decompositions. *Annales de l'Institut Henri Poincaré* **56#2**, 215–234.

[279] B. Torrésani (1992): Some geometrical aspects of wavelet decompositions. In *Wavelets: Theory, Algorithms and Applications*, Wavelet Analysis and Its Applications **5**, Academic Press, San Diego.

[280] M. Unser and A. Aldroubi (1994): Fast algorithms for running wavelet analysis. In *Wavelet Applications in Signal and Image Processing*, **2242** SPIE, 308–319.

[281] M. Unser, A. Aldroubi, and M. Eden (1993): B-spline signal processing I: Theory. *IEEE Trans. Signal Processing* **41**, 821–833.

[282] M. Unser, A. Aldroubi, and M. Eden (1993): B-spline signal processing II: efficient design and applications. *IEEE Trans. Signal Processing* **41**, 834–848.

[283] M. Unser, A. Aldroubi, and S.J. Schiff (1994): Fast implementation of continuous wavelet transform with integer scales. *IEEE Trans. Signal Processing* **42**, 3519–3523.

[284] I. Vajda (1889): *Theory of Statistical Inference and Information.* Kluwer Academic.

[285] R.J. Vanderbei (1997): *Linear Programming: Foundations and Extensions.* Kluwer.

[286] M. Vetterli (1986): Filter banks allowing perfect reconstruction. *Signal Processing* **10**, 219–244.

[287] M. Vetterli and C. Herley (1992): Wavelets and filter banks: theory and design. *IEEE Trans. Signal Processing* **40**, 2207–2232.

[288] J. Ville (1948): Théorie et applications, de la notion de signal analytique. *Cables et Transmissions* **2**, 61–74. Translated into English by I. Selin, RAND Corp. Report T-92, Santa Monica, CA (August 1958).

[289] G. Wahba (1988): *Spline Models for Observational Data.* CBMS-NSF Reg. Conf. Ser. in Applied Math. # **59**. SIAM.

[290] P.D. Welch (1967): The use of fast Fourier transform for the estimation of power spectra: a method based on time averaging over short modified periodograms. *IEEE Trans. Audio,* **AU-15**, 70–73.

[291] B.J. West and M.F. Schlesinger (1982): On $1/f$ Noise and other Distributions with Long Tails. *Proc. Nat. Acad. Sci.,* **79**, 3380–3383.

[292] R. von Sachs and K. Schneider (1996): Wavelet smoothying of evolutionary spectra by non-linear thresholding. *Appl. and Comp. Harmonic Anal.* **3**, 268–282.

[293] E.P. Wigner (1932): On the quantum corrections for the thermodynamic equilibrium. *Phys. Rev.***40**, 749–759

[294] C.H. Wilcox (1991): The synthesis problem for radar ambiguity functions. in [50], 229–260.

[295] W. Willinger, M.S. Taqqu, R. Sherman, and D.V. Wilson (1997): Self-Similarity Through High-Variability: Statistical Analysis of Ethernet LAN Traffic at the Source Level. *IEEE/ACM Trans. on Networking* **5**, 71–86.

[296] W. Willinger, V. Paxson, and M.S. Taqqu (1997): Self-Similarity and Heavy Tails: Structural Modeling of Network Traffic. in *A Practical Guide to Heavy Tails: Statistical Techniques for Analyzing Heavy Tails Distributions*, R. Adler, R. Feldman and M.S. Taqqu eds.

[297] G.W. Wornell (1990): A Karhunen-Loève-like expansion for $1/f$ processes via wavelets. *IEEE Trans. Info. Theory* **IT-36**, 859–861.

[298] G.W. Wornell (1993): Wavelet based representation for the $1/f$ family of fractal processes. *Proc. IEEE* **81**, 1428–1450.

[299] G.W. Wornell and A.V. Oppenheim (1992): Estimation of fractal signals from noisy measurements using wavelets. *IEEE Trans. Sign. Proc.* **SP-40**, 611–623.

[300] V.K Yeragani, K. Srinivasan, R. Pohl, R. Berger, R. Balon, and C. Ramesh (1994): Effects of nortriptyline on heart rate variability in panic disorder patients: a preliminary study using power spectral analysis of heart rate. *Biological Psychiatry* **29**, 1–7.

[301] A.L. Yiulle and T.A. Poggio (1996): Scaling theorems for zero crossings. *IEEE Trans. on Pattern Anal. and Mach. Intell.* **8**, 15–25.

[302] A. Zygmund (1959): *Trigonometric Series*, Cambridge University Press.

Wavelet Books

[Aldroubi96] A. Aldroubi and M. Unser (1994): *Wavelets in Medicine and Biology*, CRC Press, Boca Raton.

[Antoniadis94] A. Antoniadis and G. Oppenheim, Eds. (1994): Proceedings of the Conference *Wavelets and Statistics*, Villard de Lans, France, *Lecture Notes in Statistics.* **103**, Springer Verlag, New York.

[Arneodo95] A. Arneodo, F. Argoul, E. Bacry, J. Elezgaray, and J.F. Muzy (1995): *Ondelettes, Multifractales et Turbulence*, Diderot éditeur des Arts et Sciences, Paris.

[Benedetto94] J.J. Benedetto and M.W. Frazier (1994): *Wavelets, Mathematics and Applications*, CRC Press.

[Chui92a] C.K. Chui (1992): *An Introduction to Wavelets.* Vol. 1, Wavelet Analysis and Its Applications, Academic Press, San Diego.

[Chui92b] C.K. Chui Ed. (1992): *Wavelets: A Tutorial in Theory and Applications*, Vol. 2, Wavelet Analysis and Its Applications, Academic Press, San Diego.

[Chui95] C.K. Chui, L. Montefusco, and L. Puccio Eds. (1994): *Wavelets: Theory, Algorithms and Applications*, Vol. 5, Wavelet Analysis and Its Applications, Academic Press, San Diego.

[A.Cohen95] A. Cohen and R.D. Ryan (1995): *Wavelets and Multiscale Signal Processing.* Vol. 11, Applied Mathematics and Mathematical Computation, Chapman and Hall, London.

[L.Cohen95] L. Cohen (1995): *Time-frequency Analysis.* Prentice Hall, Englewood Cliffs (NJ).

[Combes89] J.M. Combes, A. Grossmann, and Ph. Tchamitchian, Eds. (1989): *Wavelets, Time-Frequency Methods and Phase Space*, IPTI Springer Verlag.

[Daubechies92a] I. Daubechies (1992): *Ten Lectures on Wavelets.* Vol. 61, CBMS-NFS Regional Series in Applied Mathematics.

[Daubechies92b] I. Daubechies, S. Mallat, and A.S. Willsky, Eds. (1992): *IEEE Transactions on Information Theory* **38**, Special Issue on Wavelet Transforms and Multiresolution Signal Analysis.

[Daubechies93] I. Daubechies Ed. (1993): *Different Perspectives on Wavelets.* Proc. Symposia in Applied Math. #**47**, Amer. Math. Soc., Providence.

[Duhamel93] P. Duhamel, P. Flandrin, T. Nishitani, and M. Vetterli Eds. (1993): *IEEE Transactions on Signal Processing* **41**, (12). Special Issue on Wavelets and Signal Processing.

[Erlebacher96] G. Erlebacher, M.Y. Hussaini, and L.M. Jameson (1996): *Wavelets Theory and Applications.* ICASE/LaRC Series in Computational Science and Engineering, Oxford University Press.

[Flandrin93] P. Flandrin (1993): Temps-Fréquence. *Traité des Nouvelles Technologies, série Traitement du Signal, Hermes.*

[Hernández97] E. Hernández and G. Weiss (1997): *A First Course on Wavelets*, CRC Press.

[Holschneider95] M. Holschneider (1995): *Wavelets: An Analysis Tool.* Oxford Mathematical Monographs, Clarendon Press, Oxford.

[Hubbard96] B.B. Hubbard (1996): *The World According to Wavelets*, A. K. Peters, Wellesley.

[Jetter93] K. Jetter and F.I. Utreras (1993): Vol. 3, *Multivariate Approximation: From CAGD to Wavelets.* Series in Approximations and Decompositions, World Scientific, Singapore.

[Joseph94] B. Joseph and R.L. Motard (1994): *Wavelet Applications in Chemical Engineering*, Kluwer Academic Publishers, Boston.

[Kahane96] J.P. Kahane and P.G. Lemarié (1996): *Fourier Series and Wavelets*, Gordon and Breach.

[Kaiser94] G. Kaiser (1994): *A Friendly Guide to Wavelets*, Birkhäuser, Boston.

[Klauder85] J.R. Klauder and B.-S. Skagerstam (1985): *Coherent States*, World Scientific, Singapore.

[Koornwinder93] T. Koornwinder Ed. (1993): *Wavelets: An Elementary Treatment of Theory and Applications*, Vol. 1, Series in Approximations and Decompositions, World Scientific.

[Laine93] A.F. Laine Ed. (1993): *SPIE Proceedings* #**2034**: Mathematical Imaging: Wavelet Applications in Signal and Image Processing.

[Laine94] A.F. Laine and M.A. Unser Eds. (1994): *SPIE Proceedings* #**2303**: Wavelet Applications in Signal and Image Processing II.

[Lemarié90] P.G. Lemarié Ed. (1990): *Les Ondelettes en 1989.* Lecture Notes in Mathematics, **1438**.

[Laurent94] P.J. Laurent, A. Le Méhauté, and L.L Schumaker (1994): *Wavelets, Images, and Surface Fitting*, A.K. Peters, Wellesley, Massachusetts.

[Louis94] A.K. Louis, P. Maaß, and A. Rieder (1994): *Wavelets: Theorie und Anwendungen*, Teubner Studienbücher Mathematik, B.G. Teubner, Stuttgart.

[Mallat97] S. Mallat (1998): *A Wavelet Tour of Signal Processing.* Academic Press, New York, N.Y.

[Massopust93] P.R. Massopust (1993): *Fractal Functions, Fractal Surfaces and Wavelets.* Academic Press, 1993.

[Meyer89a] Y. Meyer (1989): *Ondelettes et opérateurs*,I: Ondelettes; II: Opérateurs de Calderón-Zygmund; III: (with R. Coifman) Opérateurs multilinéaires, Hermann. (English translation of first volume is published by Cambridge University Press.)

[Meyer91] Y. Meyer Ed.(1991): *Wavelets and Applications*, Proceedings of the Second Wavelet Conference, Marseille, Research Notes in Applied Mathematics **20** Springer-Verlag/Masson.

[Meyer93a] Y. Meyer and S. Roques Eds. (1993): *Progress in Wavelet Analysis and Applications.* Editions Frontières.

[Meyer93b] Y. Meyer (1993): *Wavelets, Algorithms and Applications.* Translated and revised by R.D. Ryan, SIAM.

[Ogden96] T. Ogden (1996): *Essential Wavelets for Statistical Applications and Data Analysis.* Birkhäuser, Boston.

[Qian96] S. Qian and D. Chen (1996): *Joint Time-Frequency Analysis: Method and Application.* Prentice Hall, Englewood Cliffs, NJ.

[Ruskai91] M.B. Ruskai, G. Beylkin, R. Coifman, I. Daubechies, S. Mallat, Y. Meyer, and L. Raphael, Eds. (1991): *Wavelets and Their Applications.* Jones and Bartlett, Boston (1991).

[Schumaker93] L.L. Schumaker and G. Webb, Eds. (1993): *Recent Advances in Wavelet Analysis*, Academic Press.

[Strang96] G. Strang and T. Nguyen (1996): *Wavelets and Filter Banks.* Wellesley–Cambridge Press.

[Torrésani96] B. Torrésani (1996): *Analyse Continue par Ondelettes.* Inter Editions/Editions du CNRS (in French). English translation to appear at SIAM (translated and revised by R.D. Ryan).

[Vaidyanathan92] P. P. Vaidyanathan (1992): *Multirate Systems and Filter Banks.* Prentice Hall, Englewood Cliffs, NJ.

[Vetterli96] M. Vetterli and J. Kovacevic (1996): *Wavelets and SubBand Coding*, Prentice Hall, Englewood Cliffs, NJ.

[Walter 94] G.G. Walter (1994): *Wavelets and other Orthogonal Systems with Applications.* CRC Press, Boca Raton.

[Wickerhauser94] M.V. Wickerhauser (1994): *Adapted Wavelet Analysis, from Theory to Software.* A.K. Peters Publ.

[Wornell95] G. Wornell (1995): *Signal Processing with Fractals: A Wavelet–Based Approach.* Signal Processing Series, Prentice Hall, Englewood Cliffs, NJ.

[Young92] R.K. Young, (1992): *Wavelet Theory and its Applications.* Kluwer International Series in Engineering and Computer Science, **SECS 189**, Kluwer Academic Publishers.

Splus Books

[Bruce94] A. Bruce and H. Gao (1994): *S+WAVELETS User's Manual.* StatSci Division, MathSoft Inc.

[Becker88] R.A. Becker, J.M. Chambers, and A.R. Wilks (1988): *The S Langage.* Wadsworth and Brooks Cole.

[Chambers90] J.M. Chambers and T.J. Hastie, Eds. (1990). *Statistical Models in S.* Wadsworth and Brooks Cole.

[Spector94] P. Spector (1994): *An Introduction to S and S-Plus.* Duxbury Press.

[Venables95] W.N Venables and B.D. Ripley (1995): *Modern Applied Statistics with S-plus.* Springer Verlag, New York.

[Hardle95] W.Härdle (1995): *Smoothing Techniques with Implementation in S.* Springer Verlag, New York.

Indexes

Notation Index

Author Index

S Functions and Utilities

Subject Index

WAVELET ANALYSIS AND ITS APPLICATIONS

CHARLES K. CHUI, SERIES EDITOR

1. Charles K. Chui, *An Introduction to Wavelets*
2. Charles K. Chui, ed., *Wavelets: A Tutorial in Theory and Applications*
3. Larry L. Schumaker and Glenn Webb, eds., *Recent Advances in Wavelet Analysis*
4. Efi Foufoula-Georgiou and Praveen Kumar, eds., *Wavelets in Geophysics*
5. Charles K. Chui, Laura Montefusco, and Luigia Puccio, eds., *Wavelets: Theory, Algorithms, and Applications*
6. Wolfgang Dahmen, Andrew J. Kurdila, and Peter Oswald, eds., *Multirate Wavelet Methods for PDEs*
7. Yehoshua Y. Zeevi and Ronald R. Coifman, eds., *Signal and Image Representation in Combined Spaces*
8. Bruce W. Suter, *Multirate and Wavelet Signal Processing*
9. René Carmona, Wen-Liang Hwang, and Bruno Torrésani, *Practical Time-Frequency Analysis: Gabor and Wavelet Transforms, with an Implementation in S*